KT-563-828

Nuclear and Particle Physics

A: Background and Symmetries

Hans Frauenfelder
University of Illinois

Ernest M. Henley
University of Washington

1975
W. A. Benjamin, Inc.
ADVANCED BOOK PROGRAM
Reading, Massachusetts
London · Amsterdam · Don Mills, Ontario · Sydney · Tokyo

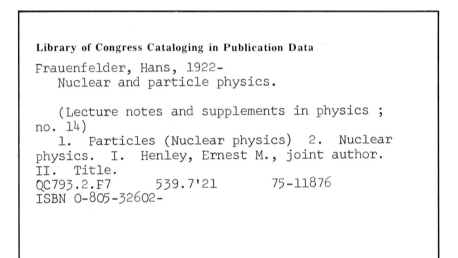

Library of Congress Cataloging in Publication Data

Frauenfelder, Hans, 1922-
    Nuclear and particle physics.

    (Lecture notes and supplements in physics ;
no. 14)
    1. Particles (Nuclear physics) 2. Nuclear
physics. I. Henley, Ernest M., joint author.
II. Title.
QC793.2.F7   539.7'21   75-11876
ISBN 0-805-32602-

Copyright © 1975 by W. A. Benjamin, Inc.
Published simultaneously in Canada.

All rights reserved. No part of this publication may be reproduced, stored in
a retrieval system, or transmitted, in any form or by any means, electronic,
mechanical, photocopying, recording, or otherwise, without the prior written
permission of the publisher, W. A. Benjamin, Inc., Advanced Book Program,
Reading, Massachusetts 01867, U.S.A.

Manufactured in the United States of America.

Nuclear
and
Particle Physics

A: Background and Symmetries

LECTURE NOTES AND SUPPLEMENTS IN PHYSICS
John David Jackson and David Pines, *Editors* (Nos. 1–13)

Volumes of the Series published from 1962–1974 are not officially numbered. The parenthetical numbers shown are designed to aid librarians and bibliographers to check the completeness of their holdings.

(1)	John David Jackson	Mathematics for Quantum Mechanics: An Introductory Survey of Operators, Eigenvalues, and Linear Vector Spaces, 1962
(2)	Willem Brouwer	Matrix Methods in Optical Instrument Design, 1964
(3)	R. Hagedorn	Relativistic Kinematics: A Guide to the Kinematic Problems of High-Energy Physics, 1964 (3rd printing, with corrections, 1973)
(4)	Robert S. Knox and Albert Gold	Symmetry in the Solid State, 1964
(5)	David Pines	Elementary Excitations in Solids: Lectures on Phonons, Electrons, and Plasmons, 1964
(6)	Gabriel Barton	Introduction to Dispersion Techniques in Field Theory, 1965
(7)	David Bohm	The Special Theory of Relativity, 1965
(8)	David Park	Introduction to Strong Interactions: A Lecture-Note Volume, 1966
(9)	Hans A. Bethe and Roman W. Jackiw	Intermediate Quantum Mechanics, 1968 (2nd printing, with corrections, 1973)
(10)	Gordon Baym	Lectures on Quantum Mechanics, 1969 (3rd printing, with corrections, 1974)
(11)	K. Nishijima	Fields and Particles: Field Theory and Dispersion Relations, 1969 (3rd printing, with corrections, 1974)
(12)	B. H. Bransden	Atomic Collision Theory, 1970
(13)	R. D. Sard	Relativistic Mechanics: Special Relativity and Classical Particle Dynamics, 1970

LECTURE NOTES AND SUPPLEMENTS IN PHYSICS
David Pines, *Editor*

Volumes published from 1975 onward are being numbered as an integral part of the bibliography:

Other volumes in preparation

Nuclear and Particle Physics

Hans Frauenfelder and Ernest M. Henley

A: Background and Symmetries, 1975

*B: Electromagnetic and Weak
Interactions, in preparation*

This book is dedicated to our children, Anne, Bradford, Karen, Kätterli, and Uli, four of whom have grown to adulthood while we were writing the many drafts that led to this final version.

CONTENTS

form the building blocks of atoms and nuclei.
Until 1932, only two particles were known, the
proton and the electron. A nucleus of charge Ze
and mass number A was assumed to consist of A
protons and (A-Z) electrons. The existence of
electrons in the nucleus, however, became harder
and harder to reconcile with quantum mechanics.[9]
A second puzzle involving electrons was posed by
the continuous energy spectrum of the electrons
emitted in beta decay. Pauli's neutrino hypothesis in 1930[10] and Chadwick's discovery of the
neutron in 1932[11] opened the way for further
progress in the understanding of nuclear structure. Before the end of 1932, Heisenberg put
forward the revolutionary proposal that protons
and neutrons, and not protons and electrons, are
the nuclear constituents.[12] The emissions of an
electron-neutrino pair, however, remained a puzzle
until 1933 when Fermi produced a beta-decay theory
as an exercise in the application of creation and

9. G. Gamow, Constitution of Atomic Nuclei and Radioactivity (Oxford University Press, New York, 1931).
10. Pauli suggested the existence of light un-charged particles in December 1930 in an open letter to friends taking part in a meeting. He called these particles "neutrons". When Fermi was later asked about the difference between Pauli's and Chadwick's "neutron", he said, "Pauli's is a little one, a neutrino."
11. J. Chadwick, Nature 129, 312 (1932); Proc. Roy. Soc. 137A, 229 (1932).
12. W. Heisenberg, Z. Physik 77, 1 (1932).

destruction operators and second quantization.[13)]
He assumed that a neutron in the decaying nucleus
transforms into a proton by emitting an electron-
neutrino pair. For the interaction energy density
$\mathcal{H}$ between the nucleons and the electron-neutrino
field Fermi assumed an expression of the form

$$\mathcal{H}_w(\vec{x}) = G_w Q \; \bar{e}(\vec{x}) \quad \nu(\vec{x}), \tag{1.4}$$

where G_w is a coupling constant, Q an operator
that changes a neutron into a proton, and $\bar{e}(\vec{x})$ and
$\nu(\vec{x})$ are operators describing the creation of an
electron and a neutrino, respectively. (The particle created in such a transformation is actually
an antineutrino. We discuss the distinction between neutrinos and antineutrinos later.) Fermi's
beta-decay interaction is vastly different from
the gravitational or the electromagnetic interaction. In particular, it is a contact (or zero-
range) force that acts only if all particles are
at the same point in space.

Fermi's theory introduced a third interaction into physics. While it led to an understanding of beta decay, it turned out to be too
weak to account for bound nuclei. To explain

---

13.   E. Fermi, Ric. Scient. 4, 491 (1933); Z.
      Physik 88, 161 (1934). Reprinted in The
      Collected Papers of Enrico Fermi, Vol. 1
      (University of Chicago Press, Chicago,
      1962).

FOREWORD

Everyone concerned with the teaching of
physics at the advanced undergraduate or graduate
level is aware of the continuing need for a modern-
ization and reorganization of the basic course
material. Despite the existence today of many
good textbooks in these areas, there is always an
appreciable time-lag in the incorporation of new
viewpoints and techniques which result from the
most recent developments in physics research.
Typically these changes in concepts and material
take place first in the personal lecture notes of
some of those who teach graduate courses. Eventually, printed notes may appear, and some fraction
of such notes evolve into textbooks or monographs.
But much of this fresh material remains available
only to a very limited audience, to the detriment
of all. Our series aims to fill this gap in the
literature of physics by presenting occasional
volumes with a contemporary approach to the classical topics of physics at the advanced undergraduate and graduate level. Clarity and sound-
ness of treatment will, we hope, mark these
volumes, as well as the freshness of the approach.

Another area in which the series hopes to
make a contribution is by presenting useful sup-
plementing material of well-defined scope. This
may take the form of a survey of relevant mathematical principles, or a collection of reprints of
basic papers in a field. Here the aim is to provide the instructor with added flexibility through
the use of supplements at relatively low cost.

The scope of both the lecture notes and supplements is somewhat different from the FRONTIERS IN PHYSICS Series. In spite of wide variations from institution to institution as to what comprises the basic graduate course program, there is a widely accepted group of "bread and butter" courses that deal with the classic topics in physics. These include: mathematical methods of physics, electromagnetic theory, advanced dynamics, quantum mechanics, statistical mechanics, and frequently nuclear physics and/or solid-state physics. It is chiefly these areas that will be covered by the present series. The listing is perhaps best described as including all advanced undergraduate and graduate courses which are at a level below seminar courses dealing entirely with current research topics.

The above words were written in 1962 in collaboration with David Jackson who served as co-editor of this Series during its first decade. They serve equally well as a Foreword for the present volume, which offers to graduate students and to research physicists alike an unusually lucid and unified view of basic concepts common to nuclear and particle physics.

Hans Frauenfelder and Ernest Henley have made many important contributions to both nuclear and particle physics; gifted pedagogues as well as researchers, they are thus especially well qualified to write a text which describes in depth those features which are important to both of these major fields of physics. It is a pleasure to welcome them as contributors to LECTURE NOTES AND SUPPLEMENTS IN PHYSICS.

David Pines

PREFACE

NUCLEAR AND PARTICLE PHYSICS have grown so much in the last few decades that it is impossible to give a complete in-depth treatment in one or even a few volumes. Any text thus must compromise between breadth and depth, favoring one or the other. In an earlier book, SUBATOMIC PHYSICS[1], we have surveyed the entire field, sacrificing depth. In the present volume we lean towards the other extreme by selecting some topics and filling the gap between an introductory text and research reviews. We stress features that are common to nuclear and particle physics and emphasize concepts that we expect to remain important. The book is intended for a graduate post-quantum mechanics course and for students who wish to learn the

1. H. Frauenfelder and E. M. Henley, Subatomic Physics (Prentice-Hall, Inc., Englewood Cliffs, New Jersey, 1974).

subject independently.

　　We should like to thank the many colleagues
and students who have helped with criticisms and
valuable suggestions. We thank Diane Florian for
the careful typing of the final version. We are
grateful to the Universities of Illinois and
Washington, the Aspen Center for Physics, CERN, and
the Los Alamos Scientific Laboratory for making
joint work possible. Finally we acknowledge sup-
port from the John Simon Guggenheim Memorial
Foundation.

HANS FRAUENFELDER

ERNEST M. HENLEY

CHAPTER 1. HISTORICAL SURVEY AND INTRODUCTION

1.1 RADIOACTIVITY AND NUCLEAR INTERACTIONS

The end of the nineteenth century produced three successive vintage years in physics: Roentgen discovered X rays in 1895, Becquerel detected radioactivity in 1896,[1] and J. J. Thomson, Wiechert, and Kaufmann showed the existence of electrons in 1897.[2] The dream, or nightmare, that physics was an almost finished field of study was shattered. Before Becquerel's discovery only two types of interactions had been known, gravitation and electromagnetism. While vastly different in strength, their spatial dependence is the same. Newton's law of gravitation and Coulomb's law of electrostatics (in rationalized

---

1. H. Becquerel, Compt. Rend. 122, 501 (1896). This and many other classical papers on radioactivity are translated in ref. 4.
2. None of these years produced superb wines, however!

c.g.s. units) lead to the following expressions
for the energy of interaction:

$$H_{grav}(r) = -G\, m_1 m_2/r, \tag{1.1}$$

$$H_{em}(r) = q_1 q_2/4\pi r. \tag{1.2}$$

Here, q_1 and q_2 are the two interacting charges,
m_1 and m_2 the two interacting masses, r their
separation, and G the universal gravitational
constant. Newton's and Coulomb's laws did not
explain the new discoveries; radioactivity and
the existence of electrons could not be fitted
into electromagnetism and mechanics. Many ingenious experiments and considerable theoretical
work were required before the significance of
the three discoveries became clear and the existence of additional interactions was realized. A
complete understanding of particles and of the
strong (hadronic) and the weak interactions is
still lacking today.

At first the search for an understanding of
the new phenomena was dominated by investigations
involving X rays and radioactivity was ignored.[3]
Interest in radioactivity was revived in 1898
through the discovery of polonium and radium by
Marie and Pierre Curie.[4] In 1903 Rutherford and

---

3. L. Badash, Am. J. Phys. 33, 128 (1965).
4. A. Romer, The Discovery of Radioactivity and
    Transmutation (Dover Publications, Inc., New
    York, 1964).

Soddy[5] put forward the transformation theory according to which atoms of the radioactive elements, unlike ordinary atoms, can undergo spontaneous transformations with the emission of alpha or beta radiation. They also showed that the energies involved in these transformations are vastly larger than those characteristic of chemical reactions.

The next essential step occurred in 1911, when data on the scattering of alpha particles by atoms led Rutherford to postulate that the mass of the atom is predominantly concentrated in the center, the <u>nucleus</u>.[6] In 1919, Rutherford systematically studied the scattering of alpha particles from RaC (^{214}Bi) by light elements and found the first example of a nuclear disintegration.[7] The process was later recognized to be the reaction

$$^{4}_{2}\text{He} + ^{14}_{7}\text{N} \rightarrow ^{1}_{1}\text{H} + ^{17}_{8}\text{O}, \tag{1.3}$$

or written concisely

$$\alpha \; ^{14}\text{N} \rightarrow p \; ^{17}\text{O},$$

or

5. E. Rutherford and F. Soddy, Phil. Mag. <u>5</u>, 445, 576 (1903). Reprinted in 4.
6. E. Rutherford, Phil. Mag. <u>21</u>, 669 (1911).
7. E. Rutherford, Phil. Mag. <u>37</u>, 581 (1919).

$$^{14}\mathrm{N}(\alpha,\mathrm{p})\,^{17}\mathrm{O}.$$

The discovery of the disintegration of nitrogen
can be considered the beginning of nuclear physics.
However, even 15 years later nuclei remained
mysterious and two central questions were unanswered: Are the nuclear forces of electromagnetic
origin? Can quantum mechanics be used to describe
nuclear processes? The second question was answered, at least to a certain extent, by Gamow,
and Condon and Gurney:[8] Geiger and Nuttall had
earlier found an empirical relation between the
lifetime of alpha emitters and the energy of the
emitted alpha particles. No explanation for the
relation could be discovered until Gamow, and
Condon and Gurney, independently showed that the
Geiger-Nuttall relation could be derived by
treating alpha emission as a quantum mechanical
tunneling effect.

The question of what interactions are responsible for nuclear phenomena turned out to be
complex. Alpha particle scattering from light
nuclei indicated that non-Coulombic forces were
involved at distances less than about 10^{-12} cm,
but detailed information was lacking. To understand nuclear interactions, it was clearly important to identify the "fundamental" particles that

8.   G. Gamow, Z. Physik 51, 204 (1928); R. W.
     Gurney and E. U. Condon, Nature 122, 439
     (1928); Phys. Rev. 33, 127 (1929).

binding, Heisenberg[12] had assumed the existence
of "exchange" forces between the nucleons in a
nucleus, giving rise to an interaction energy H_h.
The form of H_h and the mechanism responsible for
it remained uncertain. In 1934, Yukawa combined
some features of the theories of Heisenberg and
Fermi[14]: He suggested that nucleons interact
strongly through the exchange of mesons, particles
with a mass m between that of an electron and a
nucleon. The interaction energy that results from
the exchange of such mesons is of the form

$$H_h(r) = -g_h^2 \, e^{-kr}/4\pi r, \quad k = mc/\hbar. \qquad (1.5)$$

Its strength is given by the coupling constant g_h
and its range, 1/k, is determined by the meson
mass m. The force is of very short, but non-zero,
range. Although the actual nuclear forces cannot
be explained by the exchange of only one type of
meson, Yukawa's theory represented an enormous
advance in understanding the strong interaction
and forms the basis of most of the later work.
    In the present section, we have followed the
history of radioactivity and nuclear physics to
the point where two new interactions, often
characterized by the names Fermi and Yukawa, have

---

14. H. Yukawa, Proc. Phys. Mat. Soc. (Japan) 17,
     48 (1935); Am. J. Phys. 18, 154 (1950); Rev.
     Mod. Phys. 29, 213 (1957).

emerged. In the next section, we start anew from
a different point of view and discuss nuclei and
particles.

1.2 NUCLEI AND PARTICLES

In the early 1930's, detailed studies of
nuclear properties were very difficult. Only a
limited number of nuclear species were available
and only few excited states appeared in the decay
of naturally radioactive nuclei. The discovery of
artificial radioactivity,[15] the use of neutrons[16]
and of charged particles for nuclear transmuta-
tions, and the development of better radiation
detectors and analyzers changed the situation
drastically. Today, well over a thousand nuclear
species are known, the number of observed excited
states is very much larger, and a wealth of systematic data on nuclear properties has been amassed.
Studies of nuclear levels and of nuclear reactions are carried out with particles that vary
in energy over a wide range, from neutrons with
energies of a few eV to protons and electrons with
energies of a few GeV. In the "low energy" realm,
below about 100 MeV, a vast amount of data concerning nuclear energy levels has already been accum-

15. I. Curie and F. Joliot, Compt. Rend. 198,
        254 (1934).
16. E. Fermi, Nature 133, 898 (1934).

ulated (spins, parities, moments, level spacings
and distributions, transition matrix elements,
etc.), and this information has led to a rather
successful description of nuclei. Investigation
with "medium energy" projectiles, (100 MeV to 1
GeV), can give considerably finer details of the
nucleus because of the shorter wave lengths involved. Studies with electrons, for instance,
have provided descriptions of nuclear charge
distributions in terms of form factors, i.e., the
Fourier transforms of the square of the ground
state nuclear wave functions. We expect medium-
and high-energy (> 1 GeV) nuclear structure work
to increase greatly in the decade ahead.

The increase in the experimental knowledge
concerning fundamental particles has also been very
large during the past three or four decades. In
1928, it was thought that only two fundamental
particles existed in nature, the light negative
electron and the heavy positive proton. In 1932
and 1933 came the first additions, the neutrino,[10]
the neutron,[11] and the positron.[17] The positron
had already been predicted by Dirac as the antiparticle to the electron, with opposite charge, but
with the same mass and spin. The experimental
discovery of the positron raised a new question:
Does each particle have an antiparticle? The
discovery of the antiproton[18] suggested an affir-

17. C. D. Anderson, Phys. Rev. 43, 491 (1933).
18. O. Chamberlain, E. Segre, C. Wiegand, and T.
     Ypsilantis, Phys. Rev. 100, 947 (1955).

Table 1			
Particle	Spin	Charge (e)	Mass (MeV/c^2)
Photon γ	1	0	0
Neutrino ν_e	1/2	0	0
ν_μ	1/2	0	0
Electron e^-	1/2	-1	0.5110
Muon μ^-	1/2	-1	105.66
Pion π^+	0	+1	139.57
π^0	0	0	134.96
Kaon K^+	0	+1	493.71
K^0	0	0	497.70
K_S^0	0	0	
K_L^0	0	0	
Eta η	0	0	548.8
Proton p	1/2	+1	938.28
Neutron n	1/2	0	939.57
Lambda Λ	1/2	0	1115.6
Sigma Σ^+	1/2	+1	1189.4
Σ^0	1/2	0	1192.5
Σ^-	1/2	-1	1197.4
Cascade Ξ^0	1/2	0	1315
Ξ^-	1/2	-1	1321
Omega Ω^-	3/2	-1	1672

Row group labels (left margin): LEPTONS; HADRONS → MESONS; HADRONS → BARYONS → NUCLEONS (Proton, Neutron), HYPERONS (Lambda through Omega).

For each of the particles listed there exists an
antiparticle. The antiparticle is, with the exception

Particles stable against decay through strong interactions.	
Mean life (sec)	Main decay modes
stable	
stable	
stable	
stable	
2.20×10^{-6}	$e^- \bar{\nu}_e \nu_\mu$
2.60×10^{-8}	$\mu\nu$ (100%)
0.84×10^{-16}	$\gamma\gamma$ (99%), $\gamma e^+ e^-$ (1%)
1.24×10^{-8}	$\mu\nu$ (64%), $\pi^+\pi^o$ (21%)
$\approx 50\% K_S, \approx 50\% K_L$	
0.89×10^{-10}	$\pi^+\pi^-$ (69%), $\pi^o\pi^o$ (31%)
5.18×10^{-8}	$\pi^o\pi^o\pi^o$ (21%), $\pi^+\pi^-\pi^o$ (12%)
	$\pi\mu\nu$ (28%), $\pi e\nu$ (39%)
$\Gamma = 2.6$ keV	$\gamma\gamma$ (38%), $3\pi^o$ (30%)
	$\pi^+\pi^-\pi^o$ (24%)
stable	
0.92×10^3	$pe^-\nu$ (100%)
2.6×10^{-10}	$p\pi^-$ (64%), $n\pi^o$ (36%)
0.80×10^{-10}	$p\pi^o$ (52%), $n\pi^+$ (48%)
$<1.0 \times 10^{-14}$	$\Lambda\gamma$ (100%)
1.48×10^{-10}	$n\pi^-$ (100%)
2.96×10^{-10}	$\Lambda\pi^o$ (100%)
1.65×10^{-10}	$\Lambda\pi^-$ (100%)
1.3×10^{-10}	$\Xi\pi$ ($\approx 50\%$), $\Lambda\bar{K}$ ($\approx 50\%$)

of the photon, neutral pion, and η, different
from the particle.

mative answer. At present antiparticles to each known particle have been found. (Neutral particles can be identical with antiparticles.)

Yukawa's explanation of nuclear forces as being caused by the exchange of a meson with mass about 200 m_e started the search for such a particle. The first candidate found, the muon, turned out to be an impostor. It had about the right mass but not the required strong interaction with nucleons.[19,20] The muon was the first particle that showed up unexpectedly; it was not required to patch up some inconsistencies or lend credence to some hypothesis and it could not be fitted into the existing picture. Today we still cannot fathom the role of the muon in the structure of matter.

The true Yukawa particle, now called the pion, was discovered in 1947,[21] just after the situation appeared most desperate because the interaction of the muon with nuclei was at least a factor 10^{10} weaker than predicted.[20] The appearance of new particles did not stop with the pion, however. Unexpected new ones were found, first in cosmic rays and later also in accelerator experiments. With the increasing energies and improved techniques that have become available in the last

19. M. Conversi, E. Pancini, and O. Piccioni, Phys. Rev. 71, 314 (1947).
20. E. Fermi, E. Teller, and V. Weisskopf, Phys. Rev. 71, 314 (1947).
21. C. M. G. Lattes, H. Muirhead, G. P. S. Occhialini, and C. F. Powell, Nature 159, 694 (1947).

decade, the number of "fundamental" particles has
increased at an alarming rate. As a first introduction, we list in Table 1.1 the particles that
are stable or decay only through the weak or
electromagnetic interactions.

The particles in Table 1.1 are subdivided into
various classes, according to their interactions
and their spins and statistics. The photon, a
boson, is in a class by itself and subject only to
electromagnetic interactions. The leptons are
fermions without strong interactions. The strongly
interacting particles, called hadrons, interact
also electromagnetically and weakly; they are subdivided into mesons (bosons) and baryons (fermions).

In addition to the particles listed in Table
1.1, many more exist that can decay through strong
(hadronic) interactions. The question then arises
as to which are fundamental, which are composite
and what is meant by "fundamental". Why, for
instance, do we consider the neutron (that decays)
more fundamental than the deuteron (that is stable)?

A second set of questions related to the many
"fundamental" particles comes to mind: Is there a
scheme analogous to the periodic system of elements
according to which all these particles can be phy-
sicalltheory? Group theoretical ideas[22] bring

---

22. M. Gell-Mann, California Institute of Technology
     Report CTSL-20 (1961), unpublished. Y. Ne'eman,
     Nucl. Phys. 26, 222 (1961). Both papers are re-
     printed in M. Gell-Mann and Y. Ne'eman, The
     Eightfold Way (W. A. Benjamin, New York, 1964).

order into chaos, but it is not yet understood why such a group theoretical approach is successful and what the underlying principles are.

1.3   INTERACTIONS, FIELDS, AND QUANTA

In Section 1.1, we mentioned the four interactions that are known at present, namely the gravitational, the electromagnetic, the weak, and the hadronic ones. It is not clear whether these exhaust all possibilities; speculations as to additional force fields (for instance, superweak) have been put forward a few times, but no unambiguous evidence has been found. At the present time, we are forced to treat the four interactions separately,since the ultimate goal, a single theory for all, is still far away. Recent work, in which the weak and electromagnetic forces are unified,[23)] looks promising and may be the beginning of a synthesis.

Our understanding of the four basic forces is not uniform. We know the electromagnetic interaction best. Its laws, as summarized by the Maxwell equations in classical physics and by the laws of quantum electrodynamics in quantum physics,

---

23.   S. Weinberg, Phys. Rev. Letters 19, 1264
      (1967) and Phys. Rev. Letters 27, 1688 (1971).
      A. Salam, Elementary Particle Physics, N.
      Svartholm, ed. (Almavist and Wiksell, Stockholm, 1968), p. 367.

are well established. Experiments agree with theoretical predictions to about 1 part in 10^5.[24] Furthermore we know the role of this interaction in the structure of atoms and solids; chemistry and very likely biology are ruled by it. A major unsolved problem is the limit of validity of quantum electrodynamics. Experiments indicate that it holds down to distances of about 10^{-15}cm,[24] but we have yet to learn if the laws break down at shorter distances. The <u>gravitational interaction</u> is understood in most large-scale manifestations, but there still exist many unsolved questions, particularly those associated with gravitational waves and with quantization. The electromagnetic and the gravitational interaction are long-range ones and their essential features can therefore be found by observations at macroscopic distances. Only the study of their behavior at short distances requires high energies. The <u>hadronic</u> and the <u>weak interactions</u>, however, act only over short distances; their study generally demands high energies.

An aspect that will occupy us frequently concerns the explanation of interactions in terms of fields and quanta. It is generally assumed today that there exist no interactions-at-a-distance. How then do, for instance, two electrons repel each other when not in contact? Classically, the electromagnetic field is introduced as an intermediary.

---

24. S. J. Brodsky and S. D. Drell, Ann. Rev. Nucl. Sci. <u>20</u>, 147 (1970); B. L. Beron, et al., Phys. Rev. Letters <u>33</u>, 663 (1974).

Quantum mechanically, the quanta of the electromagnetic field, the photons, are held responsible.
The interaction process is pictured as follows:
Electron A emits a photon that in turn is absorbed
by electron B, as indicated in Fig. 1.1. Emission
by electron B and absorption by electron A is
equally likely, as demanded by symmetry.

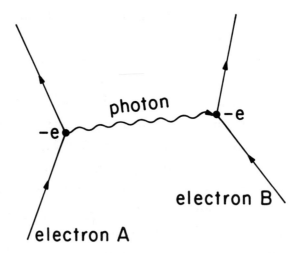

Fig. 1.1   Virtual photon exchange. The "strength"
of the force between the two electrons
is proportional to e^2. We assign a
coupling constant -e to each vertex to
characterize this proportionality. A
vertex is a point at which an interaction occurs.

The emission and absorption of these field quanta
occurs in such a way that they cannot be observed
directly. The photon shown in Fig. 1.1 is not a
real one, but is said to be "virtual". Virtual
processes are allowed by the Heisenberg uncertainty relation: An observation to measure the

energy of a system to within ΔE requires at least
a time $\Delta T \gtrsim \hbar/\Delta E$, where $2\pi\hbar$ is Planck's constant.
Hence if a particle with total energy E lives only
for a time T, we will <u>not</u> be able to observe it if
T and E satisfy the relation

$$T\ E \lesssim \hbar. \tag{1.6}$$

We have reversed the sign of the inequality to
express the fact that the time T is less than that
required to observe the particle. According to
an argument by Wick,[25] Eq. (1.6) can be used to
estimate the range R of an interaction produced
by field quanta of mass m: The quanta propagate
at best with the velocity c of light; during a
time T they travel at most a distance R = cT.
With Eq. (1.6), the distance becomes $R = \hbar c/\Delta E$.
In order to produce a particle of mass m, at least
an energy mc^2 is required: $E \gtrsim mc^2$. The range R
then is given by

$$R \approx \hbar/mc, \tag{1.7}$$

the Compton wavelength of the field quantum. The
expression (1.7) gives an upper limit on R because
a particle of mass m and total energy $\approx mc^2$ will
travel with a velocity that is smaller than c.
The question now arises: Are all interactions
mediated by quanta? Quantum field theory in its

---

25. G. C. Wick, Nature <u>142</u>, 993 (1938).

present form asserts that the answer is yes. Experimentally, the situation is reasonably clear
only for the electromagnetic interaction, where
the photon is the field quantum (Fig. 1.1). For
the other forces, difficulties of one kind or
another exist. The gravitational force is so weak
that the postulated "gravitons", will be hard to
detect individually.. The weak force poses a
different problem. As mentioned in Section 1.1
and expressed by Eq. (1.4), the weak interaction
in the form assumed by Fermi is of the contact
type. Such a "four-fermion" interaction can be
represented by the graph shown in Fig. 1.2. Although
it is known that the weak interaction cannot be
correctly described by Fig. 1.2, it is not estab-

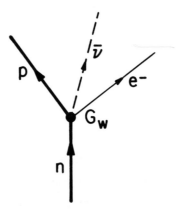

Figure 1.2 Four-fermion interaction. The strength
            of the weak force is taken to be pro-
            portional to the coupling constant G_w,
            assigned to the single vertex.

lished that it is of the type shown in Fig. 1.1.
The hypothetical quantum involved in the latter
case is called "intermediate boson" or "W particle."
If indeed such a quantum exists, then the form of
the interaction is not given by Eq. (1.4), but by
an expression similar to the Yukawa interaction,
Eq. (1.5); the coupling constant then appears
squared in a weak interaction.

The explanation of the hadronic interactions
in terms of an exchange of field quanta is also
not yet satisfactory. We mentioned in Section 1.1
that Yukawa postulated the pion to explain the
interaction between nucleons. One corresponding
diagram is shown in Fig. 1.3a. However, despite
valiant efforts, theoretical physicists never
succeeded in explaining quantitatively the observed
features of the nuclear forces with a diagram of
the type Fig. 1.3a. We know one reason for this
failure: the pion is not the only meson responsible for the hadronic forces; others such as the
ρ and the ω mesons also play a role. Computations
involving all of these together with possible
excited states of the nucleon are extremely difficult; in addition, the form of the interaction
between these mesons and nucleons is not sufficiently well known. The baryons also can be
considered as quanta of the hadronic forces.
Consider Fig. 1.3b: One contribution to the interaction between pions and nucleons comes from a
process that can be represented as the exchange
of a nucleon.

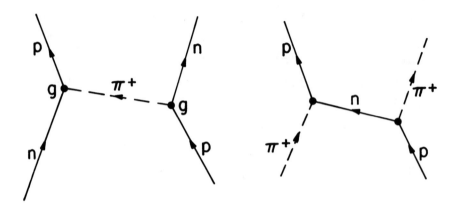

Fig. 1.3 a) Exchange of virtual pions between
            nucleons. b) Nucleons can also be
            considered as field quanta, contributing
            to the pion-nucleon interaction.

In Table 1.2, we summarize some of the information that we have discussed so far.

For many arguments, the relative strengths
of the four interactions are important. A straightforward comparison is possible only for the electromagnetic and the gravitational force because
only they have the same spatial dependence. The
ratio of interaction energies that follows from
Eq. (1.1) and (1.2) still depends on the masses
and charges of the particle involved. For protons
we find for instance $|H_{grav}/H_{em}| = 8 \times 10^{-37}$.
Similar relations hold for other particles, and
they all indicate that the gravitational interaction is vastly weaker than the electromagnetic
one.

Table 1.2   Interactions and Quanta

The list of the possible quanta of the hadronic field is not complete; in particular, no baryons are listed.   (X indicates that the property cannot be assigned a unique value.)

Inter-action	Quantum	Approx. Mass (MeV/c^2)	Compton Wave-length $(10^{-13}cm)$	Spin	Par-ity	Iso-Spin
Hadronic	Pion	137	1.44	0	-	1
	Kaon	496	0.40	0	-	1/2
	ρ meson	770	0.26	1	-	1
	ω meson	783	0.25	1	-	0
Electro-magnetic	Photon	0	∞	1	-	X
Weak	W Par-ticle (?)	$\gtrsim 5000$	<0.1	1	X	X
Gravi-tational	Graviton (P)	0	∞	2 (?)	X	X

The comparison between any other pair of forces is not as straightforward as that between electromagnetic and gravitational ones.   Range and strength must be taken into account and any comparison of two forces of different range depends on the phenomenon that is being studied.   Consider on the one hand the lifetimes of processes characteristic for the three forces that interest us:

The charged pion decays weakly in about 10^{-8} sec, the neutral pion decays electromagnetically in about 10^{-16} sec, and the first excited state of ^{8}Be breaks up strongly into two alpha particles in about 10^{-22} sec. On the other hand, no system bound by weak forces alone exists; the electromagnetic two-body system, the hydrogen atom, has an infinite number of bound states while the nuclear two-body system, the deuteron, has only one bound state. In the following we discuss possible ways to compare interaction strengths.[26]

In the usual analysis, the strengths are defined according to the squares of the coupling constants regardless of the range of the forces.[27] This approach is meaningful in field theory and particle physics where processes are analyzed by considering basic vertices characterized by the coupling constants that appear in the interaction Hamiltonian. In Fig. 1.1, for instance, the two vertices are characterized by the coupling constant e, in Fig. 1.3a by g.

---

26. Note that transition rates are proportional to the square of interaction energies.
27. Some justification for the use of coupling constants as sole strength criterion comes from the rules for Feynman graphs. For four-momentum transfers q such that $q^2 \gg (mc)^2$, the propagators are almost independent of the mass m of the field quanta and the transition probability is predominantly determined by the coupling constants.

To simplify the comparison of forces, we make the coupling constants dimensionless. The constants e^2 and g_h^2 have the dimension of $\hbar c$ (see Eqs. (1.2) and (1.5)) and we have for the electromagnetic force constant

$$\frac{e^2}{4\pi\hbar c} \approx 1/137. \qquad (1.8)$$

The corresponding quantity for the hadronic interaction depends on the theory; one finds

$$\frac{g_h^2}{4\pi\hbar c} \approx 1 - 10. \qquad (1.9)$$

The higher number is characteristic of the pion-baryon couplings and the lower one may be appropriate for pion-boson couplings. The ratio of squared coupling constants becomes

$$g_h^2/e^2 \approx 10^2 - 10^3. \qquad (1.10)$$

The weak interaction introduces new complications. The hadronic and the electromagnetic forces are ascribed to the exchange of virtual particles and the coupling constants appear squared, once from each vertex. The weak force, as represented by Fig. 1.2, differs in that the coupling constant appears only once. Thus we must compare G_w and not its square, with e^2 and g_h^2.

We next note that G_w has the dimension of $\hbar^3/m^2c$
or $(\hbar c)^3/(mc^2)^2$. To make it dimensionless, we
must divide not only by $(\hbar c)^3$, but also multiply
with the square of an energy, mc^2. At the present
time we do not know what energy to select. To get
an order of magnitude, one usually takes mc^2 to be
the rest energy of the proton. Experiment shows
that

$$G_w \approx 10^{-49} \text{ erg cm}^3, \tag{1.11}$$

and the dimensionless coupling constant becomes

$$G_w/(\hbar^3/m_p^2c) \approx 10^{-5}. \tag{1.12}$$

Alternatively we may assume that the weak force
is of the type shown in Fig. 1.1 with the exchange
of a W particle, with mass m_W. The corresponding
coupling constant g_w, which now appears once at
each vertex, is connected to G_w by

$$\frac{g_w^2}{4\pi \, \hbar c} = \frac{m_W^2}{m_p^2} \frac{G_w \, m_p^2 c}{\hbar^3} \approx \frac{m_W^2}{m_p^2} 10^{-5}. \tag{1.13}$$

For a mass m_W of about 30 GeV/c^2, the dimensionless coupling constant $g_w^2/4\pi \, \hbar c$ would be of the
order of the electromagnetic one; for a mass of
about 300 GeV/c^2 the coupling constant becomes of
order one. The weakness of the weak interaction
would then only be a low-energy phenomenon. At

momentum transfers q^2 roughly equal to $m_W^2 c^2$ the
weak interaction would become strong.

The four forces and the approximate spatial
dependences and relative strengths of their
interaction energies are shown in Table 1.3. The
relative strengths are taken to be determined by
the coupling constants; the interaction energy
(or the interaction potential) is the one that
appears in the Schroedinger (or Klein-Gordon or
Dirac) equation Hu = Eu.

Table 1.3 The Four Basic Interactions.		
Interaction	Approximate Spatial Dependence	Approximate Relative Strength of the Interaction
Hadronic	e^{-kr}/r, $k = mc/\hbar$	1 - 10
Electromagnetic	1/r	10^{-2}
Weak (Fermi)	contact (?)	10^{-5}
Gravitational (for proton)	1/r	10^{-38}

In order to take the range of the forces into
account, we consider as an example a nonrelativistic system under the influence of a spherically
symmetric local short-range force, H_{int}. The

number of bound states and the low-energy scattering cross section is determined by the dimensionless quantity

$$\frac{M^{\frac{1}{2}}}{\hbar} \int dr \left[H_{int}(r) \right]^{\frac{1}{2}}, \tag{1.14}$$

where M is the reduced mass of the system. The
strength of the force is therefore no longer determined by the coupling constant alone, the
spatial extension of $H_{int}(r)$ comes into play. More
generally various moment of the interaction energy
can be compared. For a zero-range force, only
the second moment (volume integral) is meaningful
and it is given by

$$\int dr \ r^2 \ H_{int}(r) = \int d^3x \ \delta^3(\vec{x}) H'(r) = H'(0).$$

All other moments vanish, or are infinite. For a
long-range force, such as the Coulomb one, all
positive moments may be infinite and cut-offs
are required. Such cut-offs may be introduced
naturally in a problem by the relevant quantum-
mechanical wavefunctions Ψ. For a finite-range
force, the wavefunctions may determine what spatial
part of the force is sampled.

Actually, even the arguments given so far do
not describe all difficulties that arise in
comparing forces. We have assumed that the force
laws and coupling constants are independent of
the energies of the particles. But for instance
the coupling constants may be functions of the

interacting particle energies (or, more correctly, of the invariant four-momenta). At very high energies, unforeseen problems could thus arise! At present, very little is known about these aspects.

Finally, without going into details, we can ask for the raison d'être for each interaction. Without the gravitational force, there could be no large scale universe. Without the electro-magnetic force, there would be no atoms, no solids, no plasmas, and no light. Without the nuclear force, there would be no matter. The weak interaction plays an important role in the crea-tion of elements.[28] The processes that very likely formed the elements would have stopped with-out beta decay. Similarly, the carbon and hydrogen burning cycles responsible for a large fraction of the energy production in the sun and the stars could not proceed without beta decay.[29] One more aspect of the weak interaction has become clear only within the last few years; with increasing energy, the weak interaction becomes stronger and stronger. As indicated after Eq. (1.13), it may

---

28. E. M. Burbidge, G. R. Burbidge, W. A. Fowler, and F. Hoyle, Rev. Mod. Phys. 29, 547 (1957); W. A. Fowler and F. Hoyle, Nucleosynthesis in Massive Stars and Supernovae (University of Chicago Press, 1965), and J. W. Truran, "Theories of Nucleosynthesis" in Symposium on Cosmochemistry (Cambridge, Mass., 1972).
29. E. E. Salpeter, Ann. Rev. Nucl. Sci. 2, 41 (1953), and D. D. Clayton, Principles of Stellar Evolution in Nucleosynthesis (McGraw-Hill, New York, 1968).

even become dominant at very high energies. ("And
God has chosen the weak things of the world to
confound the things that are mighty.")[30] The weak
interactions may be crucial in understanding the
strong ones.

1.4 THE EXPERIMENTAL APPROACH

Three different types of experiments are
involved in the study of elementary particles and
of nuclei: the investigation of collision processes, the study of decays, and the determination
of static properties.

Collision experiments (scattering, production
of new particles and nuclear reactions) have yielded by far the largest body of information and it
is nearly impossible to overestimate the importance
of these processes for the development of nuclear
and elementary particle physics. A direct line
leads from Rutherford's first scattering experiment
to today's investigations at very high energies.
In order to discuss the information that can be
gained from collision experiments, we sketch a
typical process in Fig. 1.4, where the target
represents a single particle. From the experimental point of view, such a collision process allows
many different and often very specific questions
to be asked by selecting the proper conditions.

---

30. 1 Corinthians 1:27.

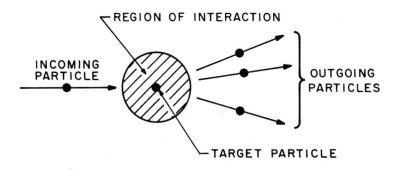

Fig. 1.4 Collision process - highly magnified.

The type, energy, and polarization of the incoming
particles and the type and polarization of the
target nuclei can be chosen within wide limits.
For a given choice of these parameters, the experi-
menter tries to observe the nature of the outgoing
particles and their energy, angular, and polarization distributions. At high energies, where many
particles may be produced in a reaction, one often
detects and measures properties of only one particle. Such a process is called an inclusive
reaction. An example is π + p → p + X, where X
is any undetected particle or set of particles
and only an outgoing proton is detected.

From the theoretical standpoint, the significance of collision processes lies in the dynamical
information gained therefrom. The kinematics of
scattering is determined by energy and momentum
conservation laws; the dynamics is governed by
the interactions among the collision partners and
by their structure. Hence the study of the dynamic

features of scattering can lead to information on
the structure of the particles involved and on
the forces between them. If the forces are known,
as for instance in electrodynamics, then the
scattering process is especially useful for studying the structure. If the force laws are unknown,
then it is usually a formidable task to unravel
the results in order to distinguish structure
effects and force characteristics. It is in great
part this obstacle that plagues strong interaction
physics. To illustrate these ideas, consider the
collision of electrons, photons, protons, and
neutrinos with protons:

(a) $e^- + p \rightarrow$
(b) $\gamma + p \rightarrow$
(c) $p + p \rightarrow$
(d) $\nu + p \rightarrow$

The first two collisions (a) and (b) involve the
well-known electromagnetic interaction and thus
yield information on the structure of the proton.
(Although weak forces may be present, they are too
weak to make their effects felt.) It is usually
assumed that the electron has no structure and
all structure effects are ascribed to the proton.
All present scattering data are consistent with
such an assumption. If the electron has any
structure, it must be smaller than about
10^{-15} cm.[24] At low energies, where the De Broglie
wavelength is large, structure effects cannot be

observed directly, but it is possible to gain
some information on the nature of the strong
forces from reaction (c). High energies are
needed, however, to learn about the short-range
properties of these forces. Under such circum-
stances it is difficult or even impossible to
disentangle structure effect and force properties
unless the structure is already known, for instance from experiments of the type (a) or (b).
The last reaction is governed by the Fermi interaction. It is experimentally difficult to observe,
but yields information on weak forces that cannot
be obtained in any other way.

The discovery of <u>radioactive decay</u>[1] was the
origin of nuclear physics. The properties of decay processes (lifetimes, branching ratios, spectrum shapes, angular corrections, angular distributions, and polarizations) yield information
about weak, electromagnetic, and strong interactions.

The investigations of the <u>static properties</u>
of nuclei and elementary particles have also added
much to our knowledge. Information on nuclear and
particle sizes, deformations, masses, spins,
charges, magnetic and electric moments is essential
in predicting and understanding reactions that the
particles may initiate or undergo.

The rapid increase in the knowledge of nuclear
and elementary particle physics is in great part
due to the enormous improvements of the <u>experi-</u>
<u>mental tools</u>. We will not give a systematic

survey of these tools, but will discuss the major
ones in context of important experiments. Furthermore, many problems will require the study of
experimental equipment.

1.5 THEORIES AND MODELS

The ultimate goal of studies in nuclear and
particle physics is to obtain a theory that explains the properties of the weak and strong
interactions and the structure of nuclei and
elementary particles. We are a long way from this
goal. On the path to it there is a strong interplay between experiment and theory.[31] A successful interplay requires that the experimental
physicist has more than a casual knowledge of the
theory and that the theoretical physicist knows
the experimental data, their reliability, and their
limitations. In the construction of a new theory,
certain guidelines and boundary conditions are
present; we summarize some salient features here.

Invariance Principles and Conservation Laws

Invariance principles and conservation laws

---

31. M. Born, Experiment and Theory in Physics
     (Dover Publications, Inc., New York, 1956.)

are closely related (Noether's theorem[32]): Whenever a law is invariant under a certain symmetry there exists a corresponding conservation law. Energy conservation follows from the invariance under time translation, linear momentum conservation from the invariance under spatial translations. The original derivation of Noether's theorem is based on classical mechanics, but it is also valid in quantum mechanics. The values of conserved quantities can often be used as quantum numbers to label states of a quantum mechanical system, such as a nucleus or an elementary particle. Some of these labels can take on continuous values (e.g., the energy); others can assume only discrete values and, in transitions, do not change or change by finite amounts determined by selection rules. There exists thus a chain: symmetry principle - conservation law - selection rule. The experimental discovery of a selection rule will induce us to search for the corresponding conserved quantity and symmetry principle. One of the important developments during the past decade has been the discovery that symmetries are not always universally valid. Some conservation laws depend on the nature of the forces. The electromagnetic interaction, for instance, violates isospin invariance and the weak interaction does not conserve parity

32. Emmy Noether, Nachrichten Akad. Wiss. Goettingen, Math. Physik. Kl. IIa, Math. Phys. Chem. Abt. 1918, 235 (1918).

and charge conjugation.

Relativity and Causality

While some problems may exist,[33] there is
at the present time no experimental evidence to
indicate the failure of the special theory of
relativity in the realm of short distances. It
is therefore generally assumed that all physical
laws must be invariant under proper Lorentz
transformations. This condition is weak, but it
places certain restrictions on any theory. One
of these is that no signal can travel faster than
the velocity of light. (It is amusing to note
that particle physics has led to a direct experimental verification of a crucial postulate of
special relativity - the independence of the velocity of light on the velocity of the source. A
CERN group[34] used high energy neutral pions
with $v/c = 0.99975$ as source of gamma rays and
determined the laboratory velocity c_γ of these
gamma rays directly. They found $c_\gamma = 299\ 770\ {}^+_-$
40 km/sec, in good agreement with the laser value
(299 792 456 $^+_-$ 1 m/sec.).

33. D. I. Blokhintsev, Sov. Phys.-Usp. 9, 405
     (1966).
34. T. Alväger, F. J. M. Farley, J. Kjellman,
     and I. Wallin, Phys. Letters 12, 206 (1964).
     T. Alväger, J. M. Bailey, F. J. M. Farley,
     J. Kjellman, and I. Wallin, Arkiv Fysik 31,
     145 (1966).

The survey given in this chapter should make it clear that many problems remain to be solved in particle and nuclear physics. Indeed, we are still far from a unified theory.

SUGGESTED READING

The book, <u>Nature of Matter</u>, written by thirty leading theoretical physicists and edited by L. C. L. Yuan, provides a most fascinating survey of many aspects of high energy physics. (Clearinghouse for Federal Scientific and Technical Information, National Bureau of Standards, Springfield, Virginia, 1965).

<u>The World of the Atom</u> (Basic Books, Inc., New York, 1966), in the words of the editors, H. A. Boorse and L. Motz "presents the actual texts of the landmark documents in the history of atomic physics". The editors have written brief introductions to each document and the two volumes make fascinating reading, starting with Lucretius and winding up with Weisskopf.

The journal, "Comments on Nuclear and Particle Physics" (Gordon and Breach, New York) provides bimonthly columns, written by distinguished physicists, "noted for their lucidity of exposition."

E. Fermi's book, <u>Elementary Particles</u>, (Yale University Press, New Haven, 1951) remains a masterpiece of clear writing; the physical ideas in it are explained with a minimum of mathematical complexity.

A number of <u>serial publications</u> provide up-to-date information on theoretical and experimental aspects of nuclear and particle physics:

Annual Review of Nuclear Science, E. Segrè, ed. (Ann. Revs. Inc., Palo Alto, California).

Progress in Nuclear Physics, O. R. Frisch, ed. (Pergamon Press, London).

Progress in Elementary Particle and Cosmic Ray Physics, J. G. Wilson and S. A. Wouthuysen, eds. (North Holland Publishing Co., Amsterdam).

Progress in Nuclear Techniques and Instruments, F. J. M. Farley, ed., (North Holland Publishing Co., New York).

Advances in Nuclear Physics, M. Baranger and E. Vogt, eds. (Plenum Publishing Co., New York).

High Energy Physics, E. H. S. Burhop, ed. (Academic Press, New York).

Advances in Particle Physics, R. L. Cool and R. E. Marshak (J. Wiley and Sons, Inc., New York).

PROBLEMS

1.1   Sketch arguments against the existence of electrons in nuclei.

1.2   Assume that electrons are used to study the structure of a) atoms, b) nuclei as a whole, c) nuclear features of linear dimensions comparable to the distance between the centers of two nucleons, d) nucleons. Calculate the laboratory kinetic energies at which the center-of-mass DeBroglie wavelengths of the electrons is approximately equal to the linear dimensions of the system under study.

1.3   Repeat 1.2 for muons.

1.4   Discuss the carbon cycle. Why would it not work if there were no weak interactions?

1.5 It has been conjectured that a particle with
     mass > 5 GeV/c^2 (W particle) is responsible
     for weak interactions.

     a) What limit does such a particle set on
         the range of weak forces?
     b) By means of simple arguments indicate
         the approximate effect that the finite
         range of weak forces would have in the
         decays n → p + e$^-$ + $\bar{\nu}$ and K → e + ν.

1.6 Use dimensional analysis to show that Eq.
     (1.12) gives the dimensionless value for the
     weak coupling constant.

1.7 One number that is sometimes used to char-
     acterize the strength of a force is the
     ratio of the potential to the kinetic energy.

     a) Use the ground states of the hydrogen
         atom and of the nucleus of deuterium to
         compare the ratio

$$\int \psi^* H_{int} \psi \, d^3x \bigg/ \int \psi^* T\psi \, d^3x \quad ,$$

         where H$_{int}$ and T are the potential and
         kinetic energy operators, respectively.
     b) By contrast, compare the binding energies
         of the two systems.

1.8 Compare the zeroth, first, and second moments
     of the electromagnetic and the strong forces;
     introduce and justify a reasonable cut-off
     for the former interaction. Explain the
     meaning of your results.

1.9 Assume a three-dimensional square well of
     radius R = 10^{-13} cm. How deep must the well
     be to just

     a) have one bound state,
     b) have six bound states?

1.10   Discuss techniques that can be used to
        measure lifetimes of the order of
        a) 10^6 years, b) years, c) nsec, d) 10^{-16}
        sec, and e) 10^{-22} sec.

1.11   Use the dimensionless strength parameters
        and simple arguments to obtain rough orders
        of magnitude for the lifetimes τ (in sec)
        and the widths Γ (in MeV) for

    a)   a resonance that decays strongly.
          Assume that the lifetime is roughly the
          time required to cross the interaction
          region.
    b)   a particle that decays electromagnetic-
          ally into two photons, e.g., the neutral
          pion.
    c)   a particle decaying by weak interactions
          (e.g., the Σ^+). Compare your answers
          with typical experimental data.

1.12   Use the uncertainty principle $\Delta p \Delta x \geq \hbar$,
        together with the definitions $(\Delta p)^2 = <p^2>$
        $- <p>^2$ and $(\Delta x)^2 = <x^2> - <x>^2$ to derive
        Eq. (1.7).

CHAPTER 2   CLASSICAL CONCEPTS AND CONSERVATION
          LAWS

2.1   UNITS

Units are an eternal source of confusion,
particularly if fate forces one to change from
one system to another.   In nuclear and particle
physics, most calculations are performed in
"natural" units, where $\hbar = c = 1$.   These units
are not very convenient when the length of a
magnet or the resolving time of a coincidence
circuit is measured; the experimental physicist
hence uses the metric system in the laboratory.

A.   The "Experimental" Units

Lengths are measured in decimal fractions
or multiples of m (meters), times in decimal
fractions of sec, energies in decimal multiples
of eV, momenta in decimal multiples of eV/c, and
masses in decimal multiples of eV/c^2.   The pre-
fixes used are $c = 10^{-2}$, $m = 10^{-3}$, $\mu = 10^{-6}$,
$n = 10^{-9}$, $p = 10^{-12}$, $f = 10^{-15}$, $k = 10^{3}$, $M = 10^{6}$,

$G = 10^9$, $T = 10^{12}$. In particular, fm can be read
as "Fermi" or as femtometer. We realize that we
violate the rule requiring capitalization if fm
is to be read as "Fermi", but prefer to use the
international nomenclature. We also use GeV instead of BeV.

The "experimental" system has two advantages.
The accelerator energy is usually given in powers
of eV. (In electrostatic accelerators, this
energy corresponds numerically to the applied
voltage for particles of charge ±e.) Magnets
select particles of a given momentum. The energy
of these particles then follows directly from
the relativistic equations connecting energy E,
momentum $|\vec{p}|$, and mass m,

$$E^2 = (\vec{p}c)^2 + (mc^2)^2, \qquad\qquad (2.1)$$

provided the rest mass m is known.

In Table 2.1, we collect some useful numbers.
The electric charge is given in <u>rationalized</u> units.
This choice introduces fewer factors of 4π. Also,
rationalized units are easier to recognize: whenever the charge appears in the combination $e/\sqrt{4\pi}$,
rationalized units have been used. The sign of
e often leads to confusion. We denote the
(positive) elementary quantum by e and the charge
of a particle under discussion by q. In particular, we have q(electron) = -e and q(positron) = e.

Table 2.1.   Useful Numbers. We quote the numerical value here
             only to the accuracy to which we need them in most
             discussions.

Quantity	Symbol	Numerical Value
Electric charge/$\sqrt{4\pi}$	$e/\sqrt{4\pi}$	4.803×10^{-10} (erg cm)$^{1/2}$
Planck's constant/2π	$\hbar$	6.5822×10^{-22} MeV sec
		1.0546×10^{-27} erg sec
Velocity of light	c	2.9979×10^{10} cm/sec
		2.9979×10^{23} fm/sec
Boltzmann constant	k	8.6174×10^{-11} MeV/K
	$\hbar c$	197.33 MeV fm
	$e^2/4\pi$	1.4399 MeV fm
Fine structure constant	$e^2/4\pi \hbar c$	$1/137.036$
Bohr radius	$4\pi\hbar^2/m_e e^2$	5.2918×10^4 fm
Classical electron radius r_e	$e^2/4\pi\, m_e c^2$	2.8179 fm
Compton wavelength of electron	$\hbar/m_e c$	386.16 fm
charged pion	$\hbar/m_\pi c$	1.4138 fm ($\approx \sqrt{2}$ fm)
proton	$\hbar/m_p c$	0.210 fm
Thomson cross section	$(8/3)\pi\, r_e^2$	0.6652×10^{-24} cm^2
		$= 0.6652$ barn
Nuclear magneton	$e\hbar/\sqrt{4\pi}\; 2m_p c$	3.1524×10^{-18} MeV/gauss
Avogadro's number	N	6.0220×10^{23} molecules/mole
Neutron-proton mass difference	$m_n - m_p$	1.2934 MeV

B. Natural Units

"Keeping track of $\hbar$ and c is a complete
waste of time." FEYNMAN

For theoretical calculations, it is convenient to use rationalized natural units, where
$\hbar$ = c = 1, and $e^2/4\pi \approx 1/137$. The use of these
units almost always encounters resistance at
first and two arguments are brought into play to
defend the conservative attitude. It is claimed
that dimensional checks are lost and that the
quantum properties of an expression are no longer
recognizable. The first argument is a fallacy.
An $\hbar$-c check to see whether an equation has the
right dimensions tells us only whether the right
number of $\hbar$ and c are present. The same end is
achieved by omitting them. The second argument
is valid, but not very important. After a final
result has been found in natural units, it is
easy to return to conventional units by a
dimensional analysis.

To justify the introduction of natural units,
we note that cgs or SI units are out of place
for describing nuclear phenomena. It is more
convenient to select the mass m_o of a "standard
particle" as mass unit and express the masses
of all other particles in terms of m_o. An
appropriate unit of length is the Compton wavelength $\hbar/m_o c$ of the standard particle; a proper
unit of time is the time required by light to

travel the unit of length. Hence we have the
three equations defining the units

$$m_o = 1$$

$$\hbar/m_o c = 1$$

$$\hbar/m_o c^2 = 1 \quad ,$$

whence we get $\hbar = c = m_o = 1$. These units are
sometimes used in the literature, but we prefer
to keep the symbols for all masses. We shall
therefore set $\hbar = c = 1$, but not $m_o = 1$. (If
the pion is used as "standard" particle, the unit
of length becomes 1.41 fm, the unit of time
4.7×10^{-24} sec.)

To convert from conventional units to natural
units, every $\hbar$ and c is set equal to one. For
instance the relativistic energy-momentum relation
Eq. (21) becomes in natural units:

$$E^2 = \vec{p}^2 + m^2. \tag{2.2}$$

To return from natural units to conventional
ones, we multiply by the appropriate powers of $\hbar$
and c. (If one has forgotten the appropriate
expressions, he can easily produce the correct
powers of $\hbar$ and c by a dimensional analysis.)
For instance, in units of energy Eq. (2.2) is

$$E^2 = (\vec{p}c)^2 + (mc^2)^2 ,$$

in units of momenta it is

$$(E/c)^2 = \vec{p}^2 + (mc)^2,$$

and finally in mass units it is

$$(E/c^2)^2 = (\vec{p}/c)^2 + m^2 .$$

The dimension of some quantities are given in Table 2.2 in natural and in conventional units. Numerical values of the natural units are also given, assuming that either the electron, the charged pion, or the proton, respectively, is taken to be the "standard" particle.

Results of experiments or calculations are often expressed in natural units. The numerical value of a result then depends on the choice of the one remaining unit. Momentum, for instance, has the dimension of m in natural units. Table 2.2 then shows that it can be expressed in terms of GeV, fm^{-1}, or sec^{-1}, with 1 GeV = 5.07 fm^{-1}, 1 fm = 0.1973 GeV^{-1}, and 1 GeV = 1.53 x $10^{24} sec^{-1}$.

2.2 METRIC AND NOTATION[1]

We denote ordinary three-vectors by an arrow

$$\vec{x} = (x, y, z), \qquad\qquad (2.3)$$

and four-vectors by x or by x^{μ}. The magnitude of

Table 2.2   Dimensions and values of some obser-
            vables in conventional and natural units.

Observable	Natural units	Conventional units	Numerical values of the units for		
			e	$\pi^{\pm}$	p
mass	m	m	0.511	139.6	938.3 MeV/c^2
momentum	m	mc	0.511	139.6	938.3 MeV/c
energy	m	mc^2	0.511	139.6	938.3 MeV
length	m^{-1}	$\hbar/mc$	386.12	1.414	0.211 fm
time	m^{-1}	$\hbar/mc^2$	1.28 $\times 10^{-21}$	4.71 $\times 10^{-24}$	2.62 $\times 10^{-25}$ sec

1.   Electrodynamics is not the only field where
     different units cause confusion. Field theory
     and elementary particle physics are plagued by
     the fact that three different metrics are widely
     used. Each choice has its advantages and is
     fiercely defended by its adherents. We have se-
     lected the system described in the present sub-
     section for the following reasons: 1.All coor-
     dinates are real and the reality properties of
     operators are easy to see; 2.The invariant four-
     momentum squared for a real particle is equal to
     the square of the particle rest mass; 3.Many of
     the current textbooks of relativistic quantum
     mechanics or field theory do use it.

a three-vector $\vec{p}$ is denoted by $|\vec{p}|$. We often write $|\vec{x}|$ as r. Any four-vector can be written either in contravariant or in covariant form.[2,3] Consider the transformation from one coordinate system

$$x = x^\mu = (x^0, x^1, x^2, x^3) = (t,\vec{x}) \qquad (2.4)$$

to another, $x' = x'^\mu = (x'^0, x'^1, x'^2, x'^3) = (t', \vec{x}')$. When we transform the coordinates, their differentials transform according to

$$dx^\mu = \sum_{\nu=0}^{3} \frac{\partial x^\mu}{\partial x'^\nu} dx'^\nu. \qquad (2.5)$$

Every quantity A^μ ($\mu = 0,1,2,3$) that transforms like the coordinate differentials is called a <u>contravariant</u> four-vector,[4] and we denote it by

---

W. Pauli, "<u>Relativitätstheorie</u>." Enzyklopädie der Mathematischen Wissenschaften, Band V, Art. 19, 1921. Reprinted 1963 by Boringhieri, Torino, 1963. (Translation: Pergamon Press, London, 1958).

3. L. D. Landau and E. M. Lifshitz, <u>The Classical Theory of Fields</u> (Addison-Wesley Publishing Co., Inc., Reading, Massachusetts, 1962), §83.

4. As Pauli[2] pointed out, it would be more logical to call it a <u>covariant</u> four-vector, since it transforms like the <u>coordinates</u>, but usage is otherwise.

a superscript. Now consider a scalar Φ. Its
gradient transforms according to

$$\frac{\partial \Phi}{\partial x^\mu} = \sum_{\nu=0}^{3} \frac{\partial \Phi}{\partial x'^\nu} \frac{\partial x'^\nu}{\partial x^\mu} \qquad . \qquad (2.6)$$

Any quantity A_μ ($\mu = 0,1,2,3$) that transforms like
the derivative of a scalar is called a covariant
four-vector, and denoted by a subscript. For the
proper-time interval ds we write

$$ds^2 = dt^2 - |d\vec{x}|^2 = g_{\mu\nu} \, dx^\mu \, dx^\nu , \qquad (2.7)$$

where $g_{\mu\nu}$ is the metric tensor

$$g^{\mu\nu} = g_{\mu\nu} = \begin{pmatrix} 1 & & & \\ & -1 & & \\ & & -1 & \\ & & & -1 \end{pmatrix} , \qquad (2.8)$$

$$g_{\mu\nu} g^{\mu\nu} = I, \qquad (2.9)$$

and where we use the Einstein summation convention:
Repeated indices are summed, but only if one index
appears as a superscript, the other as a subscript.
Exceptions are indicated. In general, Greek indices are used for any of the four components
0,1,2,3, and Latin ones for 1,2,3. The metric
tensor changes contravariant to covariant vectors

and vice versa:

$$x_\mu = g_{\mu\nu} x^\nu, \quad x^\mu = g^{\mu\nu} x_\nu \quad . \qquad (2.10)$$

Similarly for tensors

$$A_{\mu\nu} = g_{\mu\rho} g_{\nu\sigma} A^{\rho\sigma} \quad . \qquad (2.11)$$

The contravariant and covariant forms of the most often occurring vectors are given in Table 2.3. Note particularly the sign of the term

$$\vec{\nabla} = \left(\frac{\partial}{\partial x^1}, \frac{\partial}{\partial x^2}, \frac{\partial}{\partial x^3} \right) \quad \text{in the gradient.}$$

Table 2.3 Contravariant and covariant vectors.

Vector	Contravariant Form	Covariant Form
Coordinate x	$x^\mu = (t, \vec{x})$	$x_\mu = (t, -\vec{x})$
Momentum p	$p^\mu = (E, \vec{p})$	$p_\mu = (E, -\vec{p})$
Gradient ∇	$\nabla^\mu = \frac{\partial}{\partial x_\mu} = (\frac{\partial}{\partial t}, -\vec{\nabla})$	$\nabla_\mu = \frac{\partial}{\partial x^\mu} = (\frac{\partial}{\partial t}, \vec{\nabla})$

The <u>invariant (scalar) product</u> of two four-vectors is given by

$$p \cdot q = g_{\mu\nu}p^{\mu}q^{\nu} = g^{\mu\nu}p_{\mu}q_{\nu}$$

$$= p^{\mu}q_{\mu} = p_{\mu}q^{\mu} \quad . \tag{2.12}$$

In particular we have

$$p^2 = E^2 - \vec{p}^2 \quad , \tag{2.13}$$

$$p \cdot x = Et - \vec{p} \cdot \vec{x} \quad . \tag{2.14}$$

For a real free particle, the square of the four-momentum, Eq. (2.13), becomes with Eq. (2.2)

$$p^2 = m^2 \quad . \tag{2.15}$$

A four-vector is called time-like if its square is positive and space-like if its square is negative. The conditions for space-like and time-like four-vectors are summarized in Table 2.4 for x and p.

Table 2.4   Space-like and time-like four vectors.

Four-vector	"Space-like"	"Time-like"
Space-time x	$t^2 < |\vec{x}|^2$	$t^2 > |\vec{x}|^2$
Energy-momentum p	$E^2 < |\vec{p}|^2$	$E^2 > |\vec{p}|^2$

The momentum operator in coordinate representation is given by

$$p = i \nabla, \tag{2.16}$$

in particular

$$p^\mu = i \frac{\partial}{\partial x_\mu} = i \nabla^\mu, \quad p_\mu = i \frac{\partial}{\partial x^\mu} = i \nabla_\mu, \tag{2.17}$$

and

$$p \cdot p = p^\mu p_\mu = - \nabla^\mu \nabla_\mu = - \square. \tag{2.18}$$

If A represents a matrix with elements a_{ik}, we call A^*, with elements a^*_{ik}, its complex conjugate matrix, $\tilde{A}$ with elements a_{ki} its transposed matrix, $A^\dagger$, elements a^*_{ki}, its Hermitian conjugate (H.c.) matrix. We denote the unit matrix of any order by I. Some particular matrices are given in Table 2.5. We denote the complex conjugate of an expression with c.c. and the Hermitian conjugate with H.c. Note that $(AB)^\dagger = B^\dagger A^\dagger$. In the case of a matrix, c.c. means complex conjugation of all elements, without interchange of rows and columns.

Table 2.5 Special Matrices.		
Matrix	Special Property	
Symmetric	$A = \widetilde{A}$	$a_{ik} = a_{ki}$
Hermitian	$A = A^{\dagger}$	$a_{ik} = a_{ki}^{*}$
Unitary	$A^{\dagger} A = A A^{\dagger} = I$	
Orthogonal	$A \widetilde{A} = \widetilde{A} A = I$	

2.3   CONSERVATION OF ENERGY, MOMENTUM, AND ANGULAR MOMENTUM

Invariance principles and conservation laws play an increasingly important role in physics. Indeed, as Wigner points out:[5]   "Laws of Nature could not exist without principles of invariance... If the correlations between events changed from day to day, and would be different for different points of space, it would be impossible to discover them. Thus the invariance of the laws of nature with respect to displacements in space and time are almost necessary prerequisites for our ability to discover, or even catalogue, the correlations between events which are the laws of nature."

5.   E. P. Wigner, in _Proceedings of the International School of Physics Enrico Fermi_, Course 29, 1964 (Academic Press, New York, 1964), pp. IX-XVI.

Many of the symmetries and their conservation laws, such as that of invariance under displacement and the consequent principle of conservation of linear momentum, are well known to all physicists. They are classical, exact, and universal. In fact, they can be taken over without change to the quantum mechanical framework. Other examples are the invariance under rotations (angular momentum conservation) and Lorentz invariance (relativity). In addition to these conservation laws, there exist invariances in nature that are specific to certain interactions, but are not universally valid. Our primary concern will be with these invariances, since they serve as guides in the investigations of phenomena at small distances.

The power of invariance arguments lies in their generality. Usually we do not have to know the specific form of the interaction involved. The existence of certain symmetries alone permits us to draw conclusions, to find conserved quantum numbers, and to establish selection rules. Many aspects of the dynamics of elementary particles and nuclei can be deduced in this manner without more detailed knowledge of the dynamics.

In the present section, we briefly introduce the connection between invariance properties and conservation laws in classical physics.[6]

---

6.   E. L. Hill, Rev. Mod. Phys. 23, 253 (1951).

We first consider a classical conservative system
that can be described by a Lagrangian L. If q
are the generalized coordinates, V(q) the potential energy, T(q,$\dot{q}$) the kinetic energy, then
the Lagrangian is given by

$$L(q, \dot{q}) = T(q, \dot{q}) - V(q) \quad . \qquad (2.19)$$

The generalized momentum p is

$$p = \frac{\partial L}{\partial \dot{q}} \qquad (2.20)$$

and the Lagrangian equation of motion reads

$$\frac{d}{dt} \left(\frac{\partial L}{\partial \dot{q}} \right) - \frac{\partial L}{\partial q} = 0 \quad . \qquad (2.21)$$

If we now assume that the Lagrangian is independent of the coordinate q,

$$\frac{\partial L}{\partial q} = 0 \quad , \qquad (2.22)$$

then we find

$$\frac{d}{dt} \left(\frac{\partial L}{\partial \dot{q}} \right) = 0 \qquad (2.23)$$

and hence p is constant. Thus if the Lagrangian
is independent of the generalized coordinate q,

the generalized momentum p is a constant of the
motion. This connection between an invariance
principle and a conservation law can be written
differently by using the Poisson brackets. To
do so, we first introduce the Hamiltonian

$$H = T + V = H(p, q), \qquad (2.24)$$

which must be expressed in terms of p and q.
The equations of motion then become

$$\dot{p} = - \frac{\partial H}{\partial q} , \quad \dot{q} = \frac{\partial H}{\partial p} . \qquad (2.25)$$

If F is a dynamical variable of the system and
does not explicitly depend on time, we find for
its rate of change from Eq. (2.25)

$$\dot{F} = \{F,H\} \qquad (2.26)$$

where the Poisson bracket is defined by

$$\{F,H\} = \frac{\partial F}{\partial q} \frac{\partial H}{\partial p} - \frac{\partial F}{\partial p} \frac{\partial H}{\partial q}. \qquad (2.27)$$

Equation (2.26) shows that F is a constant of
motion if the Poisson bracket of F with the
Hamiltonian of the system vanishes. Using Eq.
(2.26), the connections shown in Table 2.6 can
be established. Experimentally, no evidence
for violation of either of the three conservation
laws listed in Table 2.6 has been found. We

therefore assume that the three invariance laws
listed hold.

Table 2.6 Connection between invariance proper-
           ties and conservation laws in classical
           physics.

Invariance under	Conserved quantity
spatial translation time translation rotation	linear momentum $\vec{p}$ energy E angular momentum $\vec{L}$

2.4 KINEMATICS

Two arguments make kinematical considerations
important in the treatment of collision problems.
Energy-momentum conservation restricts the possible configurations between ingoing and outgoing
particles in a collision. Moreover, experiments
are performed in the laboratory system in which
the target particle is often, but not always, at
rest. Calculations, on the other hand, are
mostly performed in the center-of-momentum (c.m.)

system, since the dynamics depends only on the relative motion of the colliding particles. To compare theoretical computations and experimental results, we must transform from one frame to the other.

A. Cross Sections

The flux F of a beam of particles is defined as the number of particles crossing unit area perpendicular to the beam per unit time. If the beam is uniform and contains n_i particles per unit volume that move with velocity v with respect to a stationary target, the flux is given by

$$F = n_i v. \tag{2.28}$$

If the target is not at rest, but moves towards the beam, the number of beam particles intersecting unit target area per unit time is

$$F = n_i (v_1 + v_2), \tag{2.29}$$

where v_1 and v_2 are particle and target velocities, respectively. Both velocities are measured in the laboratory and no relativistic factor appears in Eq. (2.29).

If the beam impinges on one scattering center, then $d\mathcal{N}$, the number of particles observed

per unit time in a detector located in angles
(θ,ϕ) relative to the incident beam and subtending
a solid angle $d\Omega$ (Fig. 2.1) is given by

$$d\mathcal{N} = F \, \sigma(\theta,\phi) d\Omega = F \, d\sigma(\theta,\phi).$$

This equation <u>defines</u> the differential scattering
cross section $d\sigma(\theta,\phi) = \sigma(\theta,\phi)d\Omega$. The total
cross section is given by

$$\sigma_{tot} = \int \sigma(\theta,\phi) d\Omega. \qquad\qquad (2.30)$$

If neither beam nor target are polarized,
and if no external fields are applied, the only
physical direction is that of the incident beam.
If this direction is taken as z axis, then no
azimuthal dependence of $\sigma(\theta,\phi)$ is expected and the
angle ϕ can be omitted, $\sigma(\theta,\phi) = \sigma(\theta)$.

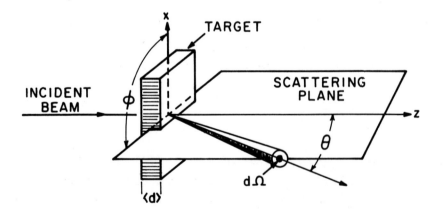

Figure 2.1   Elastic scattering.   The polar angle
            θ and the azimuth ϕ are indicated in
            the figure.

Actual targets contain many scattering centers. If N_s is the number of independent scattering centers intercepted by the beam, then the number of particles detected per unit time becomes

$$d\mathcal{N} = F\, N_s \sigma(\theta, \phi)\, d\Omega. \tag{2.31}$$

The total number of particles scattered per unit time is given by

$$\mathcal{N} = F\, N_s\, \sigma_{tot}. \tag{2.32}$$

If A is the area of the beam (assumed to be smaller than the target), n the number of scattering centers per unit volume, and d the target thickness, then

$$N_s = A\, n\, d. \tag{2.33}$$

Usually the product nd is measured in scattering centers/cm^2 and the cross section, which has the dimension of an area, is given in cm^2, in barns (1 barn = 10^{-24} cm^2), mb(10^{-27} cm^2), or μb(10^{-30} cm^2). (Note that $1/m_\pi^2 \approx 20$ mb).

Equation (2.32) is valid only if each incident particle is at most scattered once. This condition limits the thickness of the target

$$\sigma_{tot}\, nd \ll 1. \tag{2.34}$$

We also note that N_s, the number of scattering centers, is equal to the number of intercepted target particles only if each of these particles acts as an independent scatterer. This condition requires that the wavelength, $\bar{\lambda}$, of the incident beam particles be small compared to the inter-particle spacing in the target. Otherwise several target particles can scatter coherently and the coherent parts must be treated as units in the definition of the cross section. Coherent effects occur, for example, in the scattering of X rays, of low-energy electrons, or low-energy neutrons from crystals. The appearance of diffraction patterns is a clear indication of the existence of coherence effects.

   In theoretical calculations, it is not the cross section that is usually computed first, but the transition probability per unit time, i.e., the transition rate w. From the definition of the cross section it is clear that transition rate and cross section coincide if one particle per second falls on the target. In general, transition rate and cross section are connected by

$$w = d\sigma\ F\ ,\ \ \ \ \ \ \ \ \ \ \ \ \ \ \ \ \ \ \ (2.35)$$

where F is the flux of incident particles.

B. Laboratory, Center-of-Momentum, and Breit Frames

Consider the following two-body reaction

$$a + b \rightarrow c + d. \qquad (2.36)$$

What can we say about energies and momenta of the particles involved if we do not know the dynamics, i.e., the laws of interaction? In studying the kinematics, two different coordinate systems are used predominantly, the laboratory and the center-of-momentum system. If the target particle is free, it is at rest in the laboratory system (superscript L), which can then be defined by the condition

$$\vec{p}_b^{\,L} = 0, \qquad E_b^{\,L} = m_b. \qquad (2.37)$$

Most observations are performed in the laboratory system, with particle b being the target. Colliding beam experiments and high-energy collisions on nucleons bound in nuclei are exceptions. In the second case, the motion of the nucleons in the nucleus must be taken into account. The center-of-momentum system (no superscript, or superscript c.m.) is defined by

$$\vec{p}_a + \vec{p}_b = 0. \qquad (2.38)$$

Calculations are usually performed in the c.m.

system. Since the dynamical forces do not affect
the center-of-momentum, it moves freely and the
separation of the c.m. motion does not result
in any loss of information.

In some instances, particularly for the
interpretation of form factors, the Breit (or
brick wall) frame system is introduced. If a
is the particle of interest, the corresponding
Breit system is defined by

$$\vec{p}_a^B + \vec{p}_c^B = 0 \quad .$$ (2.39)

If b is to be studied in detail, the proper Breit
system is defined by

$$\vec{p}_b^B + \vec{p}_d^B = 0 \quad .$$ (2.39')

To link experiments and computations, it is
necessary to find the connections between energies, momenta, angles, and cross sections in the
two systems, both for relativistic and nonrelativistic problems, before and after collision. To
establish these connections, we use the Lorentz
transformation and conservation of energy and
momentum.

C. Nonrelativistic Transformation

Since nucleons have a rest mass of about 1
GeV, collisions involving nucleons can be treated

nonrelativistically at energies below about 100 MeV. In the nonrelativistic case, the transformation from one coordinate system to the other is given by a Galilean transformation: If $\vec{v}_o$ is the velocity of the c.m. system as seen in the laboratory system, we have for the velocity of any particle, as seen in the two systems:

$$\vec{v}^L = \vec{v} + \vec{v}_o. \tag{2.40}$$

In addition, the time is the same in both systems,

$$t^L = t. \tag{2.41}$$

Consider the collision of a particle a with momentum $\vec{p}_a^L$ and mass m_a with a target particle, mass m_b, at rest. The initial momentum of the entire system is $\vec{p}_a^L$; momentum conservation demands that the total momentum of the system be always the same, hence $\vec{p}_o = \vec{p}_a^L$. Using the Eqs. (2.37), (2.38), and (2.40), we first get the velocity of the c.m. system,

$$\vec{v}_o = \frac{1}{m_a + m_b} \vec{p}_a^L. \tag{2.42}$$

The c.m. momenta of the two particles then are

$$\vec{p}_a = -\vec{p}_b = \frac{m_b}{m_a + m_b} \vec{p}_a^L. \tag{2.43}$$

Kinetic energy T and momentum $\vec{p}$ are related by

$T = \vec{p}^2/2m$; we find for the total kinetic energy in the c.m. system

$$T = \frac{m_b}{m_a + m_b} T^L. \tag{2.44}$$

In an <u>elastic collision</u>, we set a = c, b = d. Energy and momentum in the c.m. system then show that

$$|\vec{p}_a| = |\vec{p}_c|, \quad |\vec{p}_b| = |\vec{p}_d|. \tag{2.45}$$

As a result of an elastic collision, the magnitudes of the momenta do not change; the momenta simply rotate through the c.m. scattering angle θ^{cm}. (Fig. 2.2). With Eq. (2.40) one obtains

$$\tan \theta^L = \frac{\sin \theta}{\frac{m_a}{m_b} + \cos \theta}. \tag{2.46}$$

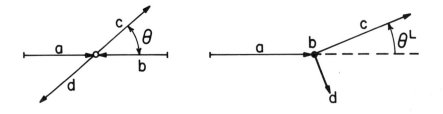

Figure 2.2   Elastic nonrelativistic collision.

The __total cross section__ for the collision
is represented by an area and must be the same
in both systems.   The differential cross section
is then obtained from the condition that the
__number__ $d\mathcal{N}$ of particles scattered into the same
solid angle must be the same (see Eq. (2.31)):

$$\sigma^L \, d\Omega^L = \sigma \, d\Omega$$

$$\sigma^L(\theta^L,\phi^L) \, \sin \theta^L \, d\theta^L \, d\phi^L = \sigma(\theta,\phi) \, \sin \theta \, d\theta \, d\phi.$$

Using Eq. (2.46), we compute (with $\phi^L = \phi$)

$$\sigma^L(\theta^L,\phi^L) = \frac{\left[1 + \left(\dfrac{m_a}{m_b}\right)^2 + 2 \dfrac{m_a}{m_b} \cos\theta\right]^{3/2}}{\left|1 + \dfrac{m_a}{m_b} \cos\theta\right|} \sigma(\theta,\phi).$$

$$(2.47)$$

D.   Relativistic transformation

Most important scattering experiments today
are performed at energies where relativistic
effects come into play; the Lorentz transformation
is then needed.   Since all four-vectors transform
the same way we write down the transformation
equation for the energy-momentum four-vector
$(E,\vec{p})$:

$$\vec{p} = \vec{p}^L + \vec{v}_o \gamma \left[\frac{\gamma}{1 + \gamma} \vec{v}_o \cdot \vec{p}^L - E^L \right],$$

$$E = \gamma \left[E^L - \vec{v}_o \cdot \vec{p}^L \right]$$

$$\vec{p}^L = \vec{p} + \vec{v}_o \gamma \left[\frac{\gamma}{1 + \gamma} \vec{v}_o \cdot \vec{p} + E \right],$$

$$E^L = \gamma \left[E + \vec{v}_o \cdot \vec{p} \right] \qquad (2.48)$$

$$\gamma = \frac{1}{\sqrt{1 - v_o^2}} .$$

Here $\vec{v}_o$ is the velocity of the c.m. in the laboratory system. Equation (2.48) applies to any four-vector. To calculate $\vec{v}_o$ and γ, we compute the total four-momentum of the system of particles in the laboratory, $P^L = (E^L, \vec{p}^L)$ and set

$$P^2 = M^2. \qquad (2.49)$$

Equation (2.49) defines the total mass of the system. For $\vec{v}_o$ and γ we then get

$$\vec{v}_o = \vec{p}^L / E^L, \quad \gamma = E^L / M. \qquad (2.50)$$

For a system of two particles, we find with $\vec{p}_b = 0$

$$p^2 = (E_a^L + m_b)^2 - (\vec{p}_a^L)^2 \qquad (2.51)$$

and

$$\vec{v}_o = \frac{\vec{p}_a^L}{E_a^L + m_b}, \qquad \gamma = \frac{E_a^L + m_b}{\sqrt{(E_a^L + m_b)^2 - (\vec{p}_a^L)^2}}.$$

$$(2.52)$$

E. Transformation of Cross Sections

The **total** cross section is an area perpendicular to the momentum of the incoming particle.
The change from the laboratory to the c.m. system
is effected by a Lorentz transformation along
this direction; such a transformation does not
change areas that are perpendicular. The total
cross section therefore is invariant under
transformations between c.m. and laboratory system.

The differential cross section can also be
expressed in terms of invariants. We will discuss
the conventional choice of these in Subsection G.
Often, however, the differential cross section is
not expressed in invariant form and must be transformed. We sketch some cases here and give the
relevant transformation equations.

The definition (2.31) of the differential cross section $d\sigma = \sigma(\theta,\phi)d\Omega$ is sufficient for a two-particle collision where the energies and momenta of the two particles in the final state are uniquely determined by kinematics. In a reaction, where additional particles are produced and more than two particles are present in the final state, the definition must be generalized. Consider for example the reaction $\pi^- p \to \pi^- \pi^+ n$. The number of positive pions going per unit time into the solid angle $d\Omega$ and having a momentum between $|\vec{p}|$ and $|\vec{p}| + |d\vec{p}|$, is given by

$$d\mathcal{N} = \frac{d^2\sigma}{d|\vec{p}|d\Omega} \, d|\vec{p}| \, d\Omega \; F \; N_s \quad .$$

Similarly, the number of positive pions going per unit time into the same solid angle, but having energies between E and E + dE is

$$d\mathcal{N}^1 = \frac{d^2\sigma}{dEd\Omega} \, dE \; d\Omega \; F \; N_s \quad ,$$

with

$$\frac{d^2\sigma}{dEd\Omega} = \frac{d^2\sigma}{d|\vec{p}|d\Omega} \, \frac{E}{|\vec{p}|} \quad . \tag{2.53}$$

Relations between cross sections in different systems are derived most easily by noting that the ratio d^3p/E is an <u>invariant</u>,

$$\frac{d^3p^L}{E^L} = \frac{d^3p}{E} = \text{invariant.} \qquad (2.54)$$

The proof of this relation is left as a problem. Since σ is also an invariant under Lorentz transformations along the direction of the incoming particle, we get with $d^3p = \vec{p}^2\, d|\vec{p}|\, d\Omega$

$$(2.55)$$

$$\frac{E}{\vec{p}^2} \frac{d^2\sigma}{d|\vec{p}|\, d\Omega} = \frac{E}{\vec{p}^2} \left(\frac{d^2\sigma}{d|\vec{p}|\, d\Omega}\right)^L = \text{invariant.}$$

With Eq. (2.53) it then follows that

$$\frac{1}{|\vec{p}|} \frac{d^2\sigma}{dE\, d\Omega} = \frac{1}{|\vec{p}|}\left(\frac{d^2\sigma}{dE\, d\Omega}\right)^L = \text{invariant.} \quad (2.56)$$

F. <u>Relativistic Invariants</u>[7]

The Lorentz transformation, Eq. (2.48), can always be performed. Often, however, its use is a waste of time and energy: we introduce quantities, such as γ, that disappear in the final expression. If we calculate the value of an invariant, the use of the Lorentz transformation

---

7.   G. F. Chew and F. Low, Phys. Rev. <u>113</u>, 1640 (1959).

is unnecessary. Relativistic invariants have the
same value in any system and one hence expresses
them in that system where they are simplest. Consider two particles a and b, with four-momenta
(E_a, p_a) and (E_b, p_b), respectively. From these
two four-momenta, we can form the invariants
(which are not all independent):

$$\left.\begin{array}{l} p_a^2 = m_a^2, \quad p_b^2 = m_b^2 \quad, \\[2ex] p_a \cdot p_b \quad, \\[2ex] (p_a + p_b)^2, \quad (p_a - p_b)^2. \end{array}\right\} \text{invariants} \qquad (2.57)$$

In addition, we write for any particle
$j(j = a, b, c, d)$

$$\vec{p}_j = m_j \vec{v}_j \gamma_j \quad, \tag{2.58}$$

$$E_j = m_j \gamma_j \quad . \tag{2.59}$$

With the invariants (2.57) and equations (2.58)
and (2.59), many transformation problems can be
solved without ever resorting to the explicit
Lorentz transformation. We give two examples:

Invariance of $(p_a + p_b)^2$. Assume that we
know energy and momentum of the two colliding
particles a and b in the laboratory system. What
is the total energy W in the cm system? We first
write in the two systems (see Eqs. (2.37) and
(2.38))

$$p = (p_a + p_b) = (E_a + E_b, \vec{0}) \equiv (W, \vec{0}) \quad ,$$

$$p^L = (p_a + p_b)^L = (E_a^L + m_b, \vec{p}_a^L) \quad .$$

Then, since p^2 is an invariant, we have $p^2 = (p^L)^2$, or with Eq. (2.49)

$$M^2 = p^2 = W^2 = (E_a^L + m_b)^2 - (\vec{p}_a^L)^2 \quad \text{invariant.}$$

$$(2.60)$$

Note that the "total mass" M is equal to the sum of the masses of the two particles a and b only if these particles are at rest with respect to each other. The generalization of the right-hand-side of Eq. (2.60) to the case where both particles are moving in the laboratory system is clear.

Invariance of $p_a \cdot p_b$. We have

$$(p_a \cdot p_b)^L = E_a^L m_b, \quad (p_a \cdot p_b) = E_a E_b + \vec{p}^2$$

where $|\vec{p}|$ is the magnitude of the c.m. three-momentum of one particle. Invariance of the scalar product gives

$$E_a^L m_b = E_a E_b + \vec{p}^2 \quad . \tag{2.61}$$

The right-hand side of Eq. (2.61) can be expressed entirely in terms of $\vec{p}^2$

$$E_a^L \, m_b = \sqrt{\vec{p}^2 + m_a^2} \; \sqrt{\vec{p}^2 + m_b^2} \; + \vec{p}^2 \; . \; (2.62)$$

We find for the c.m. momentum

$$|\vec{p}| = \frac{m_b}{\sqrt{m_a^2 + m_b^2 + 2E_a^L \, m_b}} \; |\vec{p}_a^L| \quad , \quad (2.63)$$

and we also get for the c.m. energies of the two particles

$$E_a = \frac{m_b E_a^L + m_a^2}{\sqrt{m_a^2 + m_b^2 + 2E_a^L \, m_b}} \quad , \quad (2.64)$$

$$E_b = \frac{E_a^L + m_b}{\sqrt{m_a^2 + m_b^2 + 2E_a^L \, m_b}} \; m_b \quad . \quad (2.65)$$

G. Kinematical Invariants[8-10)]

In a two-body collision, we can ask as to the best variables and to the number of such variables

8.   C. Møller, Ann. Physik 14, 531 (1932).
9.   S. Mandelstam, Phys. Rev. 112, 1344 (1958).
10. P. Nyborg, Am. J. Phys. 34, 932 (1966).

that are needed to characterize the process completely. To answer these two questions, we assume
a scattering process as indicated in Fig. 2.3.
The four-momenta of the particles are denoted by
p_i, the masses by m_i. The scattering

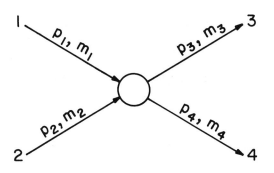

Figure 2.3   Two-body scattering.

can be described by an invariant scattering amplitude $\mathcal{M}$; the transition rate from a given initial
to a given final state is then proportional to
$|\mathcal{M}|^2$. If we disregard spin and isospin, the
invariant amplitude can only depend on the four
four-vectors shown in Fig. 2.3:

$$\mathcal{M} = \mathcal{M}(p_1, p_2, p_3, p_4). \tag{2.66}$$

Moreover, since $\mathcal{M}$ is a relativistic scalar, it
cannot depend on all sixteen components of the
four four-vectors, but only on the ten invariants

that can be formed from them (compare Eq. (2.57)):

$$p_i \cdot p_k, \quad i, k = 1, 2, 3, 4.$$

These ten invariants are not all useful for describing the scattering process; the restrictions

$$p_i^2 = m_i^2, \quad i = 1 - 4 \quad ,$$

remove four invariants. Four-momentum conservation,

$$p_1 + p_2 = p_3 + p_4 \quad , \tag{2.67}$$

yields four more conditions and we are left with
<u>two</u> independent scattering variables. The question
then reduces to the choice of the two most convenient ones. In low-energy scattering it is
customary to take the energy E_1^L of the incoming
particle and the scattering angle θ^L as variables,
as shown in Fig. 2.4. If we know E_1^L, θ^L, and the
masses, we can calculate all other parameters from
energy and momentum conservation.

    In computations of high-energy collisions the
use of E_1 and θ is inconvenient; these two vari-
bles are not Lorentz invariant. To describe the
scattering we want to find the two most useful
Lorentz <u>scalars</u>. One could be similar to the energy
of the incoming particle, the other to the scattering angle. The first is easy to find: Eq.

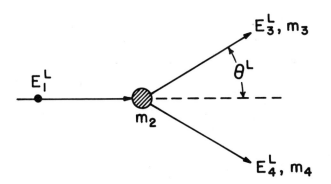

Figure 2.4 Scattering variables in low-energy processes.

(2.60) gives the square of the total energy W in the c.m. system. W^2 is an invariant and it is related to the energy E_1^L of the incoming particle in the lab system. To find the second variable, we notice that the square of the three-momentum transfer

$$\Delta \vec{p} = \vec{p}_4 - \vec{p}_2 = \vec{p}_1 - \vec{p}_3$$

is related to the scattering angle:

$$(\Delta \vec{p})^2 = (\vec{p}_1 - \vec{p}_3)^2 = \vec{p}_1^2 + \vec{p}_3^2 - 2|\vec{p}_1| \, |\vec{p}_3|\cos\theta .$$

The square of the four-momentum transfer, $(p_1 - p_3)^2$, is a logical generalization of $(\Delta p)^2$; it is an invariant and it is related to the scattering angle.

The two scalars, $W^2 = (p_1 + p_2)^2$ and

$(p_1 - p_3)^2$ are sufficient to describe scattering
kinematics. However, it is useful to introduce
a third scalar, the exchange momentum transfer
$(p_1-p_4)^2$; we will see its meaning in subsection H.
Since these three kinematical invariants are used
frequently, standard symbols are introduced:

$$s = (p_1 + p_2)^2 = (p_3 + p_4)^2,$$
$$t = (p_1 - p_3)^2 = (p_4 - p_2)^2, \qquad (2.68)$$
$$u = (p_1 - p_4)^2 = (p_3 - p_2)^2.$$

The relation

$$s + t + u = m_1^2 + m_2^2 + m_3^2 + m_4^2, \qquad (2.69)$$

which follows from Eq. (2.68), shows that only
two of the three invariants are independent.

Experimentally, E_1^L, m_1, and m_2 are fixed; for
given masses m_3 and m_4 and for a particular scattering angle θ^L, the energies E_3^L and E_4^L can be
computed from four-momentum conservation, Eq.
(2.67). The kinematical invariants then are given
by

$$s = W^2 = (E_1^{cm} + E_2^{cm})^2 = (E_3^{cm} + E_4^{cm})^2$$

$$= m_1^2 + m_2^2 + 2m_2 E_1^L \quad ,$$

$$t = m_1^2 + m_3^2 + 2(\vec{p}_1 \cdot \vec{p}_3 - E_1 E_3)$$

$$= m_2^2 + m_4^2 - 2m_2 E_4^L \quad , \tag{2.70}$$

$$u = m_1^2 + m_4^2 + 2(\vec{p}_1 \cdot \vec{p}_4 - E_1 E_4)$$

$$= m_2^2 + m_3^2 - 2m_2 E_3^L .$$

For given values of s, t, and u, momenta and energies can be calculated in a straightforward way. Inserting $E_i^{cm} = ((\vec{p}_i^{cm})^2 + m_i^2)^{1/2}$ into the expression for s yields for the magnitude of the c.m. momenta

$$|\vec{p}_1^{cm}| = |\vec{p}_2^{cm}| = \frac{1}{2s^{1/2}} \lambda(s, m_1^2, m_2^2) \quad ,$$

$$\tag{2.71}$$

$$|\vec{p}_3^{cm}| = |\vec{p}_4^{cm}| = \frac{1}{2s^{1/2}} \lambda(s, m_3^2, m_4^2) \quad ,$$

where

$$\lambda(x,y,z) = (x^2 + y^2 + z^2 - 2xy - 2xz - 2yz)^{1/2}.$$

$$\tag{2.72}$$

With Eqs. (2.70) and (2.71), we further get

$$E_1^{cm} = (s + m_1^2 - m_2^2)/2s^{1/2},$$

$$E_1^L = (s - m_1^2 - m_2^2)/2m_2,$$

$$E_2^{cm} = (s - m_1^2 + m_2^2)/2s^{1/2}, \quad E_2^L = m_2,$$

$$E_3^{cm} = (s + m_3^2 - m_4^2)/2s^{1/2},$$

$$E_3^L = (m_2^2 + m_3^2 - u)/2m_2, \qquad (2.73)$$

$$E_4^{cm} = (s - m_3^2 + m_4^2)/2s^{1/2},$$

$$E_4^L = (m_2^2 + m_4^2 - t)/2m_2,$$

$$|\vec{p}_1^L| = \lambda(s,m_1^2,m_2^2)/2m_2,$$

$$|\vec{p}_3^L| = \lambda(u,m_2^2,m_3^2)/2m_2, \qquad (2.74)$$

$$|\vec{p}_4^L| = \lambda(t,m_2^2,m_4^2)/2m_2.$$

With Eqs. (2.70) to (2.74) we express the scattering angles in the c.m. and the laboratory systems in terms of the kinematical invariants:

$$\cos\theta^{cm} = \frac{s^2 + s(2t - m_1^2 - m_2^2 - m_3^2 - m_4^2) + (m_1^2 - m_2^2)(m_3^2 - m_4^2)}{\lambda(s,m_1^2,m_2^2)\ \lambda(s,m_3^2,m_4^2)},$$

$$\cos\theta^L = \frac{(s - m_1^2 - m_2^2)(m_2^2 + m_3^2 - u) + 2m_2^2(m_2^2 + m_4^2 - s - u)}{\lambda(s,m_1^2,m_2^2)\ \lambda(u,m_2^2,m_3^2)}.$$

$$(2.75)$$

Finally, we consider cross sections. In Subsection E, we pointed out that the differential cross section $d\sigma/d\Omega$ is not an invariant. To obtain an invariant cross section, we express σ in terms of s and t instead of E and θ, and then calculate $d\sigma(s,t)/dt$. For a two-body reaction, $d\sigma/dt$ and $d\sigma/d\Omega$ are connected by

$$\frac{d\sigma}{dt} = 2\pi\frac{d(\cos\theta)}{dt}\frac{d\sigma}{d\Omega} \quad . \tag{2.76}$$

Equations (2.75) and (2.76) immediately give

$$\left(\frac{d\sigma}{d\Omega}\right)^{cm} = \frac{1}{4\pi s}\ \lambda(s,m_1^2,m_2^2)\ \lambda(s,m_3^2,m_4^2)\ \frac{d\sigma}{dt} \quad . \tag{2.77}$$

The connection between $d\sigma/dt$ and $(d\sigma/d\Omega)^{L}$, obtained by computing $d(\cos\theta^{L})/dt = -d(\cos\theta^{L}/du)$ for fixed s, is complicated if expressed in terms of s and t. With the help of Eqs. (2.73) and (2.74), we introduce laboratory quantities and get

$$\frac{d\sigma}{dt} = \frac{\pi}{|\vec{P}_1^{L}||\vec{P}_3^{L}|}\left\{1 + \frac{E_1^{L}}{m_2} - \cos\theta^{L}\frac{E_3^{L}}{m_2}\frac{|\vec{P}_1^{L}|}{|\vec{P}_3^{L}|}\right\}\left(\frac{d\sigma}{d\Omega}\right)^{L} \quad . \tag{2.78}$$

Relations similar to Eqs. (2.70) to (2.78) can also be established for variables measured in the Breit frame.

H.   The Physical Regions for the Kinematical
     Variables[9-11]

   The variables s, t, and u are not only helpful
because they are relativistic invariants, but also
because they can be used to describe the analytic
properties of scattering amplitudes. In such investigations, the definition of s, t, and u is
slightly different from the one given in the
previous subsection: All four particles shown in
Fig. 2.3 are assumed to be <u>incoming</u>. Momentum
conservation then reads $\sum p_i = 0$, $i = a,b,c,d$.
(For convenience we call the four particles here
a,b,c, and d.) The four-momenta of physically
outgoing particles are denoted by primes:

$$p'(\text{outgoing}) = -p(\text{incoming}). \qquad (2.79)$$

The invariants t and u then are defined as
$t = (p_a + p_c)^2$ and $u = (p_a + p_d)^2$. Since all four
momenta are "ingoing", we can take any two as
those corresponding to the incident and the target
particles. We then say that the scattering "occurs
in the s channel" if a and b (or c and d) are
projectile and target, respectively, "in the t
channel" if a and c (or b and d) are the incident

---

11.   T. W. B. Kibble, Phys. Rev. <u>117</u>, 1159 (1960).

particles, and "in the u channel" if a and d (or
b and c) are the incident particles.   In each
case, the other two particles then correspond to
two physically outgoing <u>antiparticles</u>.   (We discuss
antiparticles in Chapter 3.)   As an example,
consider pion-proton scattering, as shown in Fig.
2.5.   This diagram describes six different processes, depending on how the time axis is chosen.

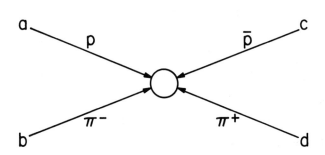

Fig. 2.5.   Pion-proton scattering. $\bar{p}$ denotes an
antiproton, the π^{+} is the antiparticle
of the π^{-}.   When the momentum arrow is
reversed, an incoming particle changes
into an outgoing antiparticle and vice-
versa; the four-momentum is reversed.

These six processes are summarixed in Table 2.7.

Table 2.7. Pion-proton scattering. Labels refer to
            Fig. 2.5.

Channel	Incident Particles	Reactions
s	ab or cd	$\pi^- p \to \pi^- p$ and $\pi^+ \bar{p} \to \pi^+ \bar{p}$
t	ac or bd	$p\bar{p} \to \pi^+ \pi^-$ and $\pi^+ \pi^- \to p\bar{p}$
u	ad or bc	$\pi^+ p \to \pi^+ p$ and $\pi^- \bar{p} \to \pi^- \bar{p}$

We now ask: What is the physical significance
of s,t, and u for each of the three cases shown in
Table 2.7? How do we know in which "channel" we
are? The first question is easy to answer by
looking at Eqs. (2.68) and Table 2.7: For each
channel, one kinematic invariant gives the square
of the total c.m. energy, a second the square of
the four-momentum transfer, and the third has no
direct physical meaning in this channel. The proper
choices are shown in Table 2.8.

Table 2.8. Kinematic invariants that give the
            square of the total c.m. energy and the
            square of the four-momentum transfer for
            the three channels of Fig. 2.5.

Channel	Incident Particles	$(\text{c.m. Energy})^2$	$(\text{Four-momentum Transfer})^2$
s	ab	s	t
t	ac	t	s
u	ad	u	t

Each reaction corresponds to a range of values of s, t, and u. The values of the kinematic invariants for which the particular reaction can actually take place is the <u>physical region</u> for that reaction. To plot the physical regions, we use the property of an equilateral triangle that the sum of the perpendicular distances of any point from the three sides is a constant equal to the height Q of the triangle:

$$d_1 + d_2 + d_3 = Q. \tag{2.80}$$

This theorem is also valid for points outside the triangle, if the sign of the distance d_i is taken to be negative. Since s, t, and u for elastic pion-nucleon scattering by Eq. (2.69) satisfy Eq. (2.80) with $Q = 2m_\pi^2 + 2m_p^2$, each possible scattering event is indicated in the two-dimensional plot Fig. 2.6. To find the boundaries of the physical region for the s channel, we write for t with Eqs. (2.70) and (2.71)

$$t = - \frac{[s-(m_\pi+m_p)^2] \, [s-(m_\pi-m_p)^2]}{2s} (1-\cos\theta). \tag{2.81}$$

Since θ is a physical angle, it can vary only from 0 to 180°. The two extreme values, $\cos\theta = 1$ and $\cos\theta = -1$, give the two boundaries for the s channel shown in Fig. 2.6. The other boundaries can be found similarly.

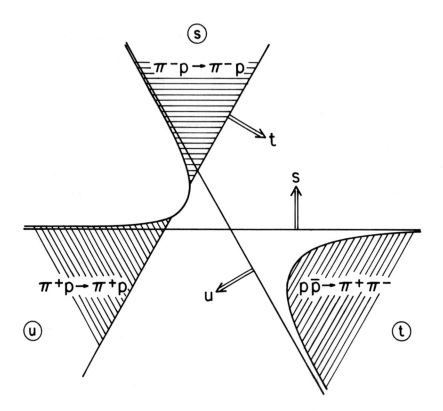

Fig. 2.6.   Physical regions for the processes
            shown in Fig. 2.5. The shaded
            areas give the physical regions for
            the three channels (s) (t) and (u).

2.5   ACCELERATORS AND BEAM OPTICS

    Most experiments in particle and nuclear
physics involve particle beams produced by accelerators. The basic set-up for such experiments

is shown in Fig. 2.7.   Particles are accelerated
and eventually hit a target.   The desired

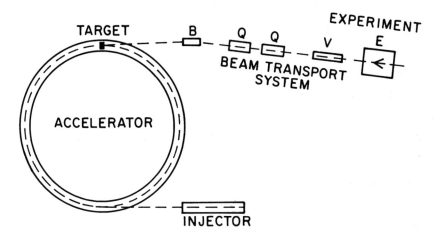

Fig. 2.7.   Basic set-up for experiments in nuclear
           and particle physics.   The accelerator
           can be circular or linear; B: bending
           magnet, Q: quadrupole magnet; V: veloc-
           ity selecting device; E: experiment,
           e.g., scattering chamber or counter
           array.

particles, either the primary ones, or secondary
ones produced in the target or through decay, are
selected according to momentum, mass, and lifetime,
and brought to the "experiment" by a beam transport system.
     Most nuclear and particle physicists have
little direct responsibility for the planning and
the construction of accelerators and the components

of the beam system. One physicist said: "The
beautiful thing about a Van de Graaff is that you
write a purchase order. Some time later, a truck
arrives, they unload the gadget, and finally tell
you which knob to turn in order to change intensity
and energy." Reality is not so simple and the
experimental physicist must know at least the
characteristics of his accelerator (e.g., energy,
energy spread, intensity, duty cycle, beam structure and emittance) and of the beam transport
system (e.g., momentum, momentum spread, angular
divergence, focusing properties, particle composi-
tion). In most cases, he will be actively involved
in planning and setting up the beam transport
system.

In the present section, we mention some
properties of accelerators and of the main components of beam transport systems. Some detection
devices will be treated later, in the context of
experiments.

A. Accelerators[12]

Since the first accelerator was built, in
1928, particle energies attained have increased

---

12. M. S. Livingston and J. P. Blewett, Particle
     Accelerators (McGraw-Hill Publishing Co.,
     New York, 1962). E. Persico, E. Ferrari, and
     S. E. Segrè, Principles of Particle Acceler-
     ators (W. A. Benjamin, Inc., Reading, Mass.,
     1968).

by one order of magnitude about every six years.
Whenever it appeared that unsurmountable limits
were reached, a new idea was born and accelerator
development continued. R. R. Wilson compares the
obsession of building bigger and better accelerators with the magnificent construction of the
Gothic cathedrals. Without this drive towards
higher and higher energies, our knowledge of the
fundamental laws governing the microcosmos would
soon stagnate.

Accelerators and beam transport systems are
based on two equations; the Lorentz equation

$$\frac{d\vec{p}}{dt} = q(\vec{E} + \vec{v} \times \vec{B}) \qquad (2.82)$$

governs the motion of a particle with charge q
and momentum $\vec{p}$ in the presence of the electric
field $\vec{E}$ and the magnetic field $\vec{B}$, and momentum and
velocity are connected through the relation (2.58),

$$\vec{p} = \frac{1}{\sqrt{1 - \vec{v}^2}} \vec{v} \quad .$$

In all accelerators, the particles gain energy by
falling through an electric field. At low energies,
d.c. fields can be used, but at higher ones, linear
and circular machines work with radio-frequency
fields. Focusing and guidance of the beam is
achieved by electric and/or magnetic fields. The
strong focusing principle, which we will sketch in

Subsection B, has made the construction of huge
accelerators economically feasible.[13]

In accelerators with a final energy below a
few hundred MeV, electrons, protons, deuterons,
^{3}H, ^{3}He, alpha particles, and heavy ions are
frequently used. At higher energies, electrons
and protons predominate; positrons (pair produced
and then accelerated) are also available. Although
most beams are unpolarized, considerable effort
has gone into producing polarized beams, and such
beams are now becoming available at high energies.

Secondary beams of muons, pions, kaons,
neutrons, and antiprotons are obtained from reactions on internal or external targets. Photons
are produced either by bremsstrahlung or through
the decay of neutral pions, "neutrino beams"
through the decay of pions and kaons.

Although accelerator energies have increased
one order of magnitude every few years, the useful
energy has not grown at the same rate: The energy
available for a reaction, for instance the production of a new particle, is not the laboratory
energy of the projectile, but the c.m. energy of
the colliding particles. Consider for instance

---

13.   E. D. Courant, M. S. Livingston, and H. S.
      Snyder, Phys. Rev. 88, 1190 (1952). The
      strong focusing principle had been discovered
      two years earlier by N. Christofilos in
      Greece, but the work was not known to Courant
      et al.

electrons or protons of energy E^L reacting with a
proton at rest. At energies where E^L is much
larger than the proton mass m_p, we get from Eq.
(2.60) for the total energy in the c.m.

$$W = \sqrt{2\ E^L\ m_p} \qquad .$$
(2.83)

The total energy in the c.m. increases only as the
square root of the accelerator energy! This difficulty led Kerst et al.[14] and O'Neill[15] to
suggest the attainment of very high energies by
using colliding beams. Various groups improved
the early designs and at the present time, such
systems are in operation.

Figure 2.8 shows the planned storage ring for
electrons and positrons at the Stanford Linear
Accelerator Center. This storage ring will have
circulating beams of electrons and positrons of
15 GeV, so that a c.m. energy of 30 GeV is
achieved. To obtain this energy through collisions
of positrons with stationary electrons would require about 900,000 GeV in the laboratory! The
figure also shows that the presently existing
storage ring, SPEAR, with a circulating beam of
4.2 GeV, is dwarfed by the proposed facility. The

---

14. D. W. Kerst, F. T. Cole, H. R. Crane, L. W.
     Jones, L. J. Laslett, T. Ohkawa, A. M. Sessler,
     K. R. Symon, K. M. Terwilliger, and N. Vogt
     Nilsen, Phys. Rev. 102, 590 (1956).
15. G. K. O'Neill, Phys. Rev. 102, 1418 (1956).

design of the latter permits the installation in
the same tunnel of an additional superconducting
proton storage ring, for 200 GeV protons. This
explains the acronym PEP for positrons, electrons,
protons.

The electrons and positrons are injected into
the storage ring via two beam transport paths after
being accelerated in the present two mile long
linear accelerator. The two beams circulate in
opposite directions and intersect in six regions
where collisional studies can be carried out. It
is planned to store electrons and positrons corresponding to currents of about 100 mA; it requires
approximately 10-15 minutes to reach this goal.
The proposed facility will allow quantum electrodynamics to be probed to extremely short distances,
$\lesssim 10^{-16}$cm. In addition to studying the production
of lepton pairs and hadrons, it should be possible
to detect the weak interactions and investigate
their relation to the electromagnetic forces.

B.   Beam Transport

In studying reactions, collisions, or decays,
it is desirable to have an intense monoenergetic
beam containing only the particles in question.
If these particles happen to be charged and stable,
they can be accelerated directly in a machine,
and can then be extracted and transported to the
experimental set-up. However, if they are produced
in a reaction, they must be separated from the

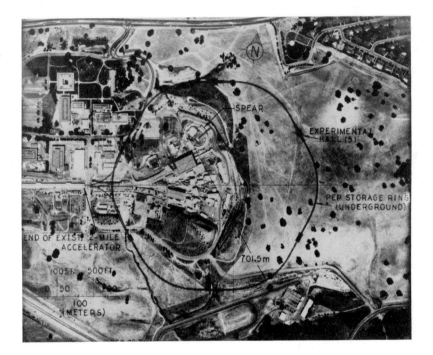

Fig. 2.8.   Proposed storage ring (PEP) for
electrons and positrons superimposed
on a photograph of SLAC. The present
storage ring (SPEAR) is also shown.
(Courtesy of the Stanford Linear
Accelerator Center).

primary beam and from other reaction products.
(Such a separation can be very difficult or even
impossible if the particles are very short-lived
or neutral, or both.) A beam transport system
hence should select particles according to momentum
and mass and then guide them to the experimental
set-up. In the present subsection, we sketch
the basic aspects of beam transport systems;
additional details can be found in the references
given at the end of Chapter 2.

Momentum selection of charged particles is
performed with bending magnets. The simplest
type is a rectangular magnet; the field $\vec{B}$ is
perpendicular to the horizontal plane and charged
particles are deflected as shown in Fig. 2.9.
For estimates it is assumed that the field region
is rectangular, with constant field B_{max} inside
and zero outside. The length of the rectangle
is given by

$$L_{eff} = \int B \ ds / B_{max} \ .$$
(2.84)

The effective length usually exceeds the physical
length of the iron pole face by about one gap
width. The trajectory becomes a circular arc,
and the radius of curvature ρ can easily be
calculated from Eq. (2.82). For numerical estimates, the relation

$$B \text{(in gauss)} \ \rho \text{(in cm)} = 3330 \ |\vec{p}| \text{(in MeV/c)}$$

(2.85)

is useful. From ρ and L_{eff}, the deflection angle θ can be computed.

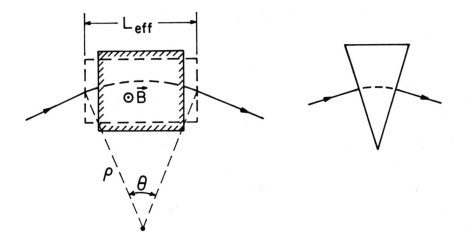

Fig. 2.9. Rectangular bending magnet. The magnetic field $\vec{B}$ points out of the plane, the deflection is appropriate for a positive particle. The terms are defined in the text. The optical analogue of a bending magnet is a prism (right).

A bending magnet selects particles of a given momentum, regardless of mass. If the beam

consists only of one type of particle, as in most
low-energy experiments, no problem arises. At
high energies, where particles of different masses
can be present in the same beam, a mass selection
must be performed. Equation (2.58) shows that if
the velocity is determined or selected, the mass
is known for a given momentum. The velocity can
be measured by a time-of-flight method or with a
Cerenkov counter. We will discuss these two
methods when we report the discovery of the antiproton in Chapter 3.

Beam focusing is needed because the distance
between accelerator and final detector can be
extremely long. Without focusing, the intensity
of beams at the experimental set-up would be
very much less than is desirable. Bending magnets
have some focusing qualities, but not enough to
produce well collimated beams. Fortunately,
strong focusing, introduced in 1952 by Courant,
Livingston, and Snyder[13], can not only be used
for accelerators, but also for beam transport.
The essential element is the quadrupole magnet,
sketched in Fig. 2.10. The field in a quadrupole
magnet vanishes along the center; a particle
passing through the center is not deflected. A
positive particle going into the magnet at the
point A, Fig. 2.10, experiences a Lorentz force
(Eq. (2.82)) that pushes the particle towards the
center. The particle at B, however, is pushed
away from the center. The quadrupole magnet of
Fig. 2.10 is focusing in the x direction, but

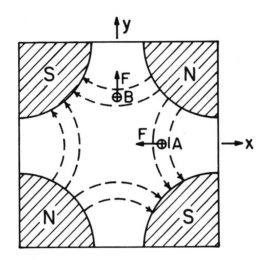

Fig. 2.10 Cross section through a quadrupole
magnet. Two positive particles, at
A and B, are going parallel to the
negative z axis. $\vec{F}$ indicates the
Lorentz force.

defocusing in the y direction.

The magnetic field strength in a quadrupole
field is given by

$$B_x(x,y) = gy, \quad B_y(x,y) = gx, \quad B_z(x,y) = 0 \quad ,$$

$$g = \frac{\partial B_y}{\partial x} = \frac{\partial B_x}{\partial y} = \text{const.} \qquad (2.86)$$

A beam transport system contains a number of quadrupole magnets of finite lengths, arranged in
doublets or triplets. The field then is no longer
completely described by Eq. (2.86) and the

behavior of a beam of charged particles is usually
studied by using a matrix formalism and by solving
the equation of motion with a computer. For crude
estimates, however, a thin-lens approximation is
sufficient. A quadrupole of length L is represented by a thin lens with focal length

$$f = 1/kL \qquad\qquad (2.87)$$

where k, the focusing strength for a particle of
momentum $|\vec{p}|$, is given by

$$k = eg/|\vec{p}| \quad . \qquad\qquad (2.88)$$

The thin-lens approximation applies if $Lk^{1/2} \ll 1$.
For numerical estimates we use[16]

$$k(\text{in } m^{-2}) = 3\frac{g(\text{in kG/cm})}{|\vec{p}| \text{ (in GeV/c)}} \quad . \qquad (2.89)$$

One quadrupole magnet alone is not very
useful because it focuses in only one direction.
However, it is known from classical optics that
the combination of a focusing and a defocusing
lens of equal focal length f, and separated by a
distance d, is focusing with an over-all focal
length f*. In the thin-lens approximation, f*

_____

16.   g is often expressed in T/m, where
      $1 \text{ T(Tesla)} = 10^4 \text{ G(gauss)}$.

is given by

$$f^* = f^2/d \quad .\qquad\qquad (2.90)$$

Two identical quadrupole magnets,[17] with one rotated by 90° about its central axis, will therefore act as a focusing system, independent of the order of the two magnets. The beam envelope for such a quadrupole doublet is shown in Fig. 2-11. The upper figure applies to motion in the yz plane, the lower to motion in the xz plane if the field in the first quadrupole magnet is chosen as in Fig. 2.10. The principal planes for the quadrupole doublet shown in Fig. 2.11 can be found in a straightforward way. It turns out that their positions depend on the order of the two magnets. In the thin-lens approximation, the distances of the principal planes from the center of the doublet are

$$h_{FD} = |f| - \frac{1}{2}d \qquad h_{DF} = |f| + \frac{1}{2}d. \quad (2.91)$$

Principal planes and focal points for the two cases are shown in Fig. 2.12. The two principal planes are separated by a distance $2|f|$; for most

_____

17. For the discussion here we have chosen two identical quadrupole magnets with equal focal strengths k. In actual beam transport systems, the two focal strengths are usually adjusted somewhat differently.

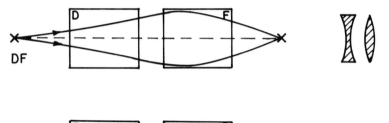

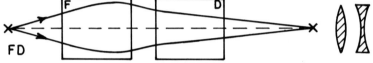

Fig. 2.11.   Quadrupole doublet, with optical
             analogue.   F indicates focusing,
             D defocusing.

doublets they lie on different sides of the
central plane.
     The wide separation of the principal planes
prevents the doublet from being represented by
a single thin lens of given position and a given
focal length.   In general, the focal points in
the FD and the DF planes do not coincide.   Moreover,

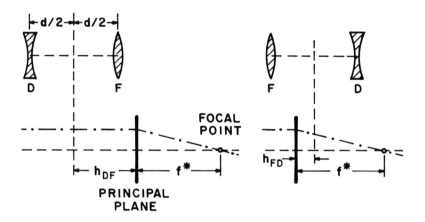

Fig. 2.12.   Principal planes and focal
            points for a quadrupole
            doublet.

Fig. 2.11 shows that the angular modification
in the plane DF is larger than in the plane FD.
Since the transverse magnification is the inverse
of the angular magnification, it is larger in the

plane FD. This property is often used in the
design of actual beam transport systems.

To set up a complete beam transport system,
the desired beam properties are first determined:
momentum, momentum spread, beam size, and beam
divergence. The system is then designed on paper,
using bending magnets and quadrupole doublets
(or sometimes triplets) as building blocks.
(Most large laboratories have standardized components and the properties of these are specificd
in "Users Handbooks.") The system is then set up
and the magnets are aligned to the desired
accuracy with survey instruments and lasers. (In
many experiments, the distances are so large that
walkie-talkies are used for communication during
the alignment.)

Once a system is set up, it is possible to
check many characteristics with the "floating
wire" technique: A current-carrying wire under
tension simulates the orbit of a charged particle
in a magnetic field provided tension and current
are adjusted properly. The current I (in A) and
the tension T (in grams of force) must satisfy

$$I(\text{in A})/T(\text{in g}) = 2.94\Big/|\vec{p}| \quad (\text{in MeV/c}), \quad (2.92)$$

where $|\vec{p}|$ is the particle momentum and where the
current I must flow in the direction opposite to
the beam current represented by the charged
particle.

Figure 2.13 shows a beam transport system

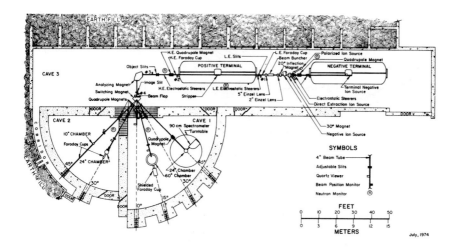

Fig. 2.13. Beam transport system of the tandem
Van de Graaff accelerator at the
University of Washington. (Courtesy
Nuclear Physics Laboratory, University
of Washington).

used at the University of Washington in connection
with a tandem Van de Graaff machine; Fig. 2.14
presents beam transport systems used at Brookhaven.

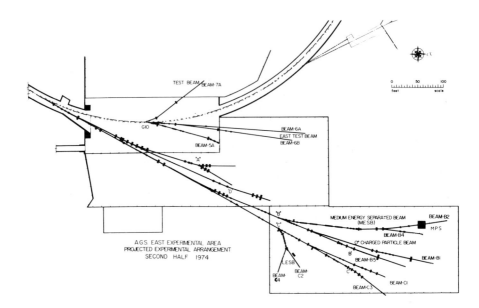

Fig. 2.14. Schematic of beam lines and target stations used at the AGS (Alternating Gradient Synchrotron) at Brookhaven in 1974. A small fraction of the circulating beam of typically 6×10^{12} protons interacts at the internal target G-10, producing particles used by beams 5, 6 and 7; the remainder of the protons are extracted from the machine and split into three beams which then strike targets A (or D), B and C. Secondary beams include: three electrostatically separated or enriched particle beams, giving high fluxes of $K^{\pm}$ mesons and anti-protons; two neutral beams of K^{O}_{S} and K^{O}_{L}; a hyperon beam; a μ-beam; and a 30 GeV/c unseparated charged particle beam. An alternative mode of operation is to extract the entire circulating proton beam and send it into the neutrino area, where neutrino, or antineutrino inter-actions are studied in bubble chamber, spark chamber-counter and calorimeter experimental setups. (Courtesy, R. Rau, Brookhaven National Laboratory)

PROBLEMS

2.1 In a beam, kaons of kinetic energy 6 GeV
are selected by a deflection magnet. What
is the kinetic energy and the momentum of
electrons, muons, pions, and protons that
can accompany the kaons?

2.2 Express in natural units $\hbar = c = 1$, with the
pion mass as unit mass, the following observables:

> The Bohr radius,
> the nuclear magneton,
> the electron mass,
> the decay half-life of the muon,
> the fine structure constant,
> the Thompson cross section.

2.3 Express the pion Compton wavelength m_π^{-1} in fm,

the cross sections m_e^{-2}, m_p^{-2}, and m_π^{-2}
in millibarns,

the lifetime $\Gamma^{-1} = 10$ MeV^{-1} in sec,
the momentum 100 MeV/c in fm^{-1}.

2.4 Establish the connections stated in Table 2.6.

2.5 Calculate the geometrical cross section of a
Cu nucleus. Assume a Cu target, 0.1 cm
thick. What fraction of the area is "covered"
by the Cu nuclei if no overlap is assumed?

2.6 The cross section for the reaction
$p^7Li \rightarrow n\ ^7Be$, for 2.5 MeV protons, is about
0.3 barns. What is the approximate maximum
thickness, in cm, of a "thin" Li target?

2.7 Justify Eq. (2.35).

2.8 Consider a non-relativistic two-body collision. Find a geometric construction to
transform the momenta from the laboratory
to the c.m. system.

2.9   Justify Eq. (2.43).

2.10   The scattering of low-energy neutrons from
       protons is isotropic in the c.m. system.

       a)   Show that the emerging particles must
            come off at 90° with respect to each
            other in the laboratory.
       b)   The differential cross section of neutrons
            of 5 MeV kinetic energy (c.m.) is about
            0.6 fm^2/sr. Calculate and plot the dif-
            ferential cross section in the laboratory.

2.11   Protons of 15 MeV kinetic energy in the
       laboratory are scattered by deuterium. The
       scattered protons are observed at a laboratory
       scattering angle of 20°. Calculate all rel-
       evant energies, angles, and momenta in the
       laboratory and the c.m. systems. Justify the
       non-relativistic approximation. Calculate
       the ratio of the differential cross sections
       at 20° in the laboratory and at the corre-
       sponding c.m. angle.

2.12   The Los Alamos Meson Factory will produce
       800 MeV protons, with a beam intensity of
       1 mA. Consider the reaction p ^{6}Li → p d X,
       where d is the deuteron.

       a)   What is the charge and atomic number of
            the nucleus X?
       b)   The differential cross section is expected
            to be of the order of 10 microbarns/sr^2
            MeV. The target contains 2 x 10^{20} Li^6
            nuclei/cm^2 and an energy interval of 3 MeV
            is selected. The protons and deuterons
            are observed in coincidence with two
            counters, each subtending an angle of
            10^{-4} sr. Compute the coincidence counting
            rate.

2.13   A kaon of 3.5 GeV kinetic energy hits a
       stationary proton. Calculate the total energy
       W in the c.m. system.

2.14 Justify Eq. (2.54).

2.15 Verify the following equations

a) (2.70), b) (2.71), c) (2.73),
d) (2.74), e) (2.75), f) (2.77) and
and (2.78).

2.16 a) If two particles emerge from a reaction,
what is the number of independent measurements that must be carried out to specify
the kinematics completely?
b) Repeat (a) for 3 particles.
c) Repeat (a) for n particles.

2.17 If a cross section for a reaction has a laboratory angular distribution of the form

$$\sin^4 \theta/2 \quad ,$$

and a counter 5 cm by 5 cm is placed 3 m away,
how important are solid angle corrections at
90°, at 45°, and at 1° in the laboratory?

2.18 Time intervals of 10^{-11} sec can be measured.
What kinetic energy is required to make a
lifetime determination of the π° feasible?
How far would the π° travel in one mean life?

2.19 a) Find the threshold for pion production in
the reaction pp → πd.
b) What are the c.m. and laboratory energies,
momenta, and velocities of the pion of
part (a) if the initial proton kinetic
energy in the laboratory is 100 MeV above
threshold?

c) If the laboratory cross section for (b)
is measured at 30°, what is the corresponding angle in the c.m.? Evaluate
the corresponding c.m. cross section in
terms of the laboratory cross section.

2.20 For elastic nonrelativistic p^3He scattering,
what is the maximum angle at which protons
will appear in the laboratory?

2.21 Verify, for elastic scattering, the following
      relations between the kinematical variables
      and the variables measured in the Breit frame:

$$|\vec{p}_2^{\ B}| = \frac{1}{2} \sqrt{-t} \quad .$$

$$E_2^{\ B} = E_4^{\ B} = \sqrt{m_2^{\ 2} - t/4} \quad ,$$

$$E_1^{\ B} = E_3^{\ B} = \frac{1}{\sqrt{4m_2^{\ 2} - t}} \{s + \frac{t}{2} - m_1^{\ 2} - m_2^{\ 2}\} .$$

$$\cos\theta^B = 1 + t/[2 \ (\vec{p}_1^{\ B})^2] \quad .$$

2.22 Pions of laboratory momentum 3 GeV/c are
      elastically scattered by protons. The
      scattered pions are observed at a laboratory
      angle of $5°$.

   a) Find s, t, u. Note signs. Is the pion-
       momentum space-like or time-like?
   b) What is the four-momentum transferred to
       the proton?
   c) What is t_{max}?
   d) What are the signs of s, t, and u for the
       $p\bar{p}$ annihilation to $\pi^+\pi^-$?
   e) Convert to the Breit frame for the proton
       and find E_π and E_p in this frame before
       and after scattering.

2.23 Draw a plot of the physical regions for s, t,
      and u channels, respectively, for elastic
      scattering of particles of equal mass.

2.24 Justify the boundaries of the physical regions
      shown in Fig. 2.6.

2.25 Protons incident on nucleons bound in a
      nucleus can produce antiprotons at kinetic
      energies of 2.9 GeV. Calculate the minimum
      internal momentum (in MeV/c) of the nucleons

that take part in such a process.

2.26 Assume that you teach physics at a small
      university and you would like to build a 1 MeV
      cyclotron with your students. Write the pro-
      posal that you would submit to a money-provi-
      ding agency. The proposal should contain:

    a) Your motivation for building the accel-
        erator, the advantages and disadvantages
        of this type of machine, and some possible
        experiments.
    b) A sketch of the proposed cyclotron, with
        a brief description of the underlying
        theory.
    c) A discussion of the various parts that are
        involved (magnets, power supplies, accel-
        erating chambers, vacuum system, ...).
    d) An estimate of the cost of building and
        running the accelerator.

2.27 Sketch the principle of the Van deGraaff
      accelerator.

2.28 For a 10 TeV proton accelerator

    a) Discuss the physical principles involved
        in the acceleration and the beam stability.
    b) Sketch the main components. Estimate
        size, cost.
    c) Discuss the various possible injectors
        and their advantages and disadvantages.
    d) What are the advantages of superconduc-
        ting magnets?

2.29 Discuss the focusing properties of a quadru-
      pole magnet.

2.30 Derive Eqs. (2.88), (2.90) and (2.91).

2.31 Verify Eq. (2.92).

2.32 Design a beam transport system to select
      (a) protons of energy 25 MeV and (b) pions
      of momentum 20 GeV/c, and to guide these
      particles over a distance of 60 m. Trace
      one central and one off-axis ray. Estimate

the magnetic field strengths necessary in the
bending magnets and estimate the design parameters for the quadrupole magnets.

2.33 Assume that two 30 GeV proton beams, each 20A
current, intersect over a distance of
20 cm. Beam diameter is 1 cm. How many
hadronic interactions do you expect per sec?
How many weak ones?

ADDITIONAL REFERENCES

Relativistic kinematics is treated in considerable detail and with many examples in:

R. Hagedorn, Relativistic Kinematics (W. A.
Benjamin, Inc., Reading, Mass., 1963).

E. Byckling and K. Kajantie, Particle Kinematics (J. Wiley and Sons, Inc., New York,
(1973).

Further information on non-relativistic and relativistic kinematics can be found in:

R. Herman and R. Hofstadter, High Energy Electron Scattering Tables (Stanford University
Press, Stanford, 1960).

A. M. Baldin, V. I. Goldanskii, and I. L.
Rozental, Kinematics of Nuclear Reactions
(Oxford University Press, Oxford, 1961).

W. S. C. Williams, Introduction to Elementary
Particles (Academic Press, Inc., New York,
1961), Appendix A.

J. D. Jackson, Classical Electrodynamics (John
Wiley and Sons, Inc., New York, 1962).

K. G. Dedrick, Rev. Mod. Phys. 34, 429 (1962);
35, 414 (1963). (Contains a list of kinematical tables.)

R. M. Sternheimer, "Nuclear Physics", in
Methods of Experimental Physics, Vol. 5B,
C. S. Wu and L. C. L. Yuan, eds. (Academic
Press, Inc., New York, 1963), Appendix 2.

A. Michalowicz, Cinématique des Réactions Nucleaires (Dunod, Paris, 1964).

Kinematics of Multiparticle Systems, M. Nikolic, ed. (Gordon and Breach, New York, 1968).

References concerning particle accelerators are listed and discussed in the Resource Letter PA-1 on particle accelerators, by J. P. Blewett, Am. J. Phys. 34, 9 (1966).

The motion of charged particles in electromagnetic fields and beam transport systems are treated in the following articles and books:

K. T. Bainbridge, in Experimental Nuclear Physics, Vol. I, E. Segrè, ed. (John Wiley and Sons, Inc., New York, 1953).

O. Chamberlain, Ann. Rev. Nucl. Sci. 10, 161 (1960).

D. Luckey, in Techniques of High Energy Physics, D. M. Ritson, ed. (Interscience Publishers, Inc., New York, 1961).

N. M. King, Progr. Nucl. Phys. 9, 73 (1963).

R. M. Sternheimer and B. Cork, "Nuclear Physics", in Methods of Experimental Physics, Vol. 5B, L. C. L. Yuan and C. S. Wu, eds. (Academic Press, Inc., New York, 1963).

K. G. Steffen, High Energy Beam Optics Interscience Publishers, Inc., New York, 1965).

A. P. Banford, The Transport of Charged Particle Beams (E. and F. N. Spon, Ltd., London, 1966).

Focusing of Charged Particles, A. Septier, ed. (Academic Press, New York, 1967), 2 Vols.

J. J. Livingood, The Optics of Dipole Magnets (Academic Press, New York, 1969).

CHAPTER 3. PARTICLES, ANTIPARTICLES AND FIELDS

How do we describe free particles? What distinguishes particles and antiparticles? In the present chapter, we discuss these questions. We begin with the photon, a massless boson with spin 1, then treat neutral and charged massive spinless bosons, and discuss nonrelativistic and relativistic spin 1/2 fermions. Finally we add some remarks concerning other particles. These examples do not cover all the particles that occur, but they should provide some insight into the mechanisms of describing free particles.

3.1 THE PHOTON

Are not the Rays of Light very Small Bodies emitted from Shining Substances?

NEWTON   Opticks

It is sometimes said that the photon was born in 1900, when Planck derived his radiation

law.[1)]   However, Planck assumed only that material
oscillators emit and absorb energy in discrete
quanta; he did not require the quantization of
light. Working along different and independent
lines, Einstein five years later explained the
experimental observations associated with black-
body radiation, fluorescence, and photoeffect by
postulating that light itself has discontinuous
properties, that a light ray consists of a finite
number of energy quanta that move without dividing
and can only be produced and absorbed as complete
units.[2)] It took many years until these postulates
were universally accepted. Even when Planck proposed Einstein for membership in the Prussian
Academy, he remarked: "That he may sometimes
have missed the target in his speculations, as
for example in his hypothesis of light quanta,
cannot really be held against him."

A.   Energy, Momentum, and Mass of the Photon

    A classical electromagnetic wave carries
energy and momentum and these two quantities are

---

1. M. Planck, Ann. Phys. 4, 561 (1901).
2. A. Einstein, Ann. Physik 17, 132 (1905).
    Reprinted in Dokumente der Naturwissenschaft,
    A. Hermann, ed. (E. Battenberg, Stuttgart,
    1965), Vol. 7. Translated in A. B. Arons
    and M. B. Peppard, Am. J. Phys. 33, 367 (1965).

related by

$$E = |\vec{p}| \quad . \tag{3.1}$$

If ω is the circular frequency of the beam and if there are n photons in the beam, the total energy is $E = n\omega$. For macroscopic observations, n is in general very large compared to one. In some experiments, however, the intensity can be made small enough so that observations can be ascribed to the interaction of individual photons for which we postulate the relations

$$E = |\vec{p}| = \omega \ , \qquad \vec{p} = \vec{k} \quad , \tag{3.2}$$

where $\vec{k}$ is the wave vector. These relations are confirmed by energy and momentum measurements using the Compton effect; from Eq. (2.2) it follows that

$$m_\gamma = 0 \quad . \tag{3.3}$$

A vanishing photon mass also is implied by Einstein's postulate that the speed of light in vacuum is identical in all inertial frames of reference connected by Lorentz transformations. The best experimental limits on the photon mass come from large-scale phenomena. Satellite measurements have established that the magnetic field of the earth decreases like r^{-3} out to several earth radii. The range of the field transmitted by the photon is hence larger than a few earth radii and we get from Eq. (1.7) as a crude upper limit $m_\gamma < 10^{-17} m_e$. A quantitative consideration of the influence of a finite photon mass on the radial decrease of the magnetic field yields a limit $m_\gamma \leq 10^{-15}$ ev.[3)]

B.   Maxwell's Equations

For an external charge density ρ and current density $\vec{j}$, Maxwell's equations for the electric field $\vec{E}$ and the magnetic field $\vec{B}$ read

$$\vec{\nabla} \times \vec{E} + \frac{\partial \vec{B}}{\partial t} = 0 \qquad \vec{\nabla} \cdot \vec{E} = \rho \quad ,$$

$$\text{(3.4)}$$

$$\vec{\nabla} \times \vec{B} - \frac{\partial \vec{E}}{\partial t} = \vec{j} \qquad \vec{\nabla} \cdot \vec{B} = 0 \quad .$$

(Since we deal primarily with light in free space, we do not distinguish between the electric field $\vec{E}$ and the displacement $\vec{D}$ or between the magnetic field $\vec{H}$ and the magnetic induction $\vec{B}$.) By virtue of the Eqs. (3.4), ρ and $\vec{j}$ obey the continuity equation

$$\vec{\nabla} \cdot \vec{j} + \frac{\partial \rho}{\partial t} = 0 \quad . \qquad\qquad \text{(3.5)}$$

3.   M. A. Gintsburg, Soviet Astronomy $\underline{7}$, 536 (1964). I. Yu Kobzarev and L. B. Okun, Soviet Phys. Uspekhi $\underline{11}$, 338 (1968). A. S. Goldhaber and M. M. Nieto, Rev. Mod. Phys. $\underline{43}$, 277 (1971).

A vector potential $\vec{A}$ and a scalar potential A^0 are introduced by the definitions

$$\vec{E} = - \frac{\partial \vec{A}}{\partial t} - \vec{\nabla} A^0 \quad , \quad \vec{B} = \vec{\nabla} \times \vec{A} \quad . \tag{3.6}$$

The total energy in the field is given by

$$W = \frac{1}{2} \int d^3 x \; (\vec{E}^2 + \vec{B}^2) \quad . \tag{3.7}$$

In the following discussions, we will often use the four-vectors $j^\mu = (\rho, \vec{j})$ and $A^\mu = (A^0, \vec{A})$. The continuity equation (3.5) then is

$$\nabla_\mu j^\mu = 0 \quad . \tag{3.5'}$$

The continuity equation in this form is <u>manifestly</u> <u>covariant:</u> both sides of Eq. (3.5') are scalars under Lorentz transformations and the continuity equation has the same form in every coordinate frame. In the future, we will call an equation "covariant" if it can be written in a form such that both sides have the same, well-defined, transformation property under Lorentz transformations. By a "manifestly covariant" equation we mean one that clearly shows its transformation properties. (This meaning of covariant should be distinguished from that used in Sec. 2.2.). The value of an <u>invariant</u> or <u>Lorentz scalar</u> is the same in any coordinate system.

In Sec. 1.5 we pointed out that physical laws

should have the same form in any Lorentz
frame (Einstein's first principle). The equations
describing physical laws should therefore be
covariant. The field equations of electromagnetism,
namely the Maxwell equations, are indeed covariant
and they can be written in a manifestly covariant
form. We will not need this form here.

The equations (3.6) define the vector potential A^μ incompletely; any <u>gauge transformation</u>

$$A^\mu \rightarrow A^\mu + \nabla^\mu \chi(x) \quad , \qquad (3.8)$$

where $\chi(x)$ is an arbitrary scalar function, leaves
the electric and magnetic field intensities unal-
tered. The invariance of the field equations
(3.4) under the gauge transformation (3.8) is
closely connected to the conservation of the electric charge, as we shall show in Chapter 4.

The freedom expressed by the gauge transformation (3.8) implies that we can impose additional
conditions on A^μ. The invariant <u>Lorentz subsidiary</u>
condition

$$\nabla_\mu A^\mu = 0 \qquad (3.9)$$

removes part of the arbitrariness. To eliminate
it completely, we can demand

$$\vec{\nabla} \cdot \vec{A} = 0 \quad . \qquad (3.10)$$

This <u>Coulomb</u> or <u>radiation gauge</u> is not covariant, but it is convenient to discuss the properties of the free radiation. For the free radiation field, where $\rho = 0$ and $\vec{j} = 0$, we can then set

$$A^0 = 0 \quad , \quad \left(\frac{\partial^2}{\partial t^2} - \nabla^2 \right) \vec{A} \equiv \square \vec{A} = 0 \quad . \quad (3.11)$$

The vector potential $\vec{A}$ becomes a solution of the homogeneous wave equation. We will use the Coulomb gauge most of the time.

C. Some Remarks Concerning Spin

Experience shows that the concept of photon spin often leads to confusion. We therefore discuss it in some detail and also make a few preliminary remarks here concerning the spin of particles. In 1916, Sommerfeld and Debye independently proposed that an angular momentum should only be able to assume certain orientations with respect to a given direction, for instance an external magnetic field.[4] In 1922, Stern and Gerlach beautifully demonstrated the existence of such an "azimuthal quantization".[5] In order to explain some mysterious hyperfine structure of spectral lines, Pauli suggested in 1924 that not only atoms, but also nuclei possess spin.[6]

---

4. A. Sommerfeld, Phys. Z. 17, 491 (1916); P. Debye, Phys. Z. 17, 507 (1916).
5. W. Gerlach and O. Stern, Z. Physik 8, 110 (1922).
6. W. Pauli, Naturwiss. 12, 741 (1924).

(Sommerfeld did not like this idea and tried
unsuccessfully to get Pauli to retract it.)   The
doubling of many spectral lines led to the introduction of the idea of the spinning electron.[7]
By now we know that each particle can be characterized by a unique value of its spin.

Spin is the intrinsic angular momentum of a
particle; its magnitude is constant.   Spin cannot
be expressed in terms of the classical position
and momentum coordinates and it has no analogue
in classical mechanics.   In the restframe of a
particle, any orbital contribution to the angular
momentum vanishes and the spin is hence the
angular momentum in the restframe.   If we denote
the restframe angular momentum operator by $\vec{s}$,
the commutation relations are given by

$$[s_x, s_y] = i\, s_z, + \text{cyclic permutations.} \quad (3.12)$$

From the commutation relations it follows that $\vec{s}^2$
and one component, usually taken to be s_z, can
be measured simultaneously; the corresponding
eigenvalue equations are

$$\vec{s}^2\, \chi_{s,m} = s(s+1)\, \chi_{s,m}$$

$$\quad (3.13)$$

$$s_z\, \chi_{s,m} = m\, \chi_{s,m} \quad ,$$

---

7.   G. E. Uhlenbeck and S. Goudsmit, Naturwiss.
    13, 953 (1925).

where s can take on the values 0, 1/2, 1, 3/2, ...
For a given particle, s is fixed and does not
change. For a given value of the spin s, the magnetic quantum number m can take on the 2s+1 values
s, s-1, ... -s. $\chi_{s,m}$ is the spin eigenfunction
belonging to the total spin s and the magnetic
quantum number m. To describe the orientation of
a particle with nonvanishing spin, we introduce
a <u>polarization</u> vector $\vec{P}$ by the definition

$$\vec{P} = \frac{1}{s} < \vec{s} > \; . \tag{3.14}$$

$\vec{P}$ is proportional to the expectation value of the
spin operator, it "points" in the direction of the
particle's spin.

Experimentally, the spin of a particle or a
state can be determined in various ways. The most
direct approach is to count the 2s+1 different m
states, i.e. the 2s+1 different orientations with
respect to a given direction. For atomic levels,
this counting can be done with the Zeeman effect,
for atomic beams with the Stern-Gerlach technique.
The Zeeman levels can be observed in most nuclear
ground states and, using the Mössbauer effect,
also in some excited nuclear states. However, for
most particles and excited nuclear states, the
Zeeman and the Stern-Gerlach techniques cannot be
used; the various m states cannot be separated
enough to permit counting. To determine the spin,
some spin-dependent feature, such as an angular
distribution of a reaction, must be observed.

The spin is then obtained by comparing calculated
and measured patterns.

D.   Statistics and Spin of the Photon

The statistics satisfied by photons is deduced
from the fact that a macroscopic electromagnetic
field exists; hence photons must obey Bose-Einstein
statistics. Particles obeying Fermi-Dirac statistics cannot give rise to a classically measurable
field of which they are the quanta.

We now turn to the spin of the photon. Since
the electromagnetic field can be deduced from a
four-vector, the vector potential A^μ, we call the
photon a vector particle. For any vector particle,
regardless of its mass, the subsidiary condition
(3.9), or an equivalent one, removes one of the
four degrees of freedom implied by the four components of A^μ.[8,9] The three remaining degrees of
freedom correspond to the three possible orientations of a spin 1 particle, for instance the ρ
meson. For such particles, the spin can be discussed following the lines of Subsection C. These
arguments can, however, not be applied to the
photon! The photon has mass zero and it cannot be
transformed to rest. Indeed, the vanishing mass
imposes an additional condition and the photon

---

8.   W. Pauli, Rev. Mod. Phys. 13, 203 (1941).
9.   R. Glauber, Prog. Theor. Phys. 9, 295 (1953).

has only two polarization states instead of the
three expected from Eq. (3.13). The restriction
to two polarization states applies not only to
the photon, but to any massless particle, regardless of its spin.[10]   It is not a property of the
spin, but a consequence of special relativity.[11]

The existence of two and only two polarization states for light is well known from classical
electrodynamics. The two states can be selected
in various ways. Consider a plane wave of momentum
$\vec{k}$ and arbitrary polarization. It can be expressed
as a superposition of two orthogonal linearly
polarized waves described by the vector potentials
$\vec{A}_1$ and $\vec{A}_2$,

$$\vec{A}_1 = \hat{\varepsilon}_1 \cos k \cdot x , \quad \vec{A}_2 = \hat{\varepsilon}_2 \cos k \cdot x \quad . (3.15)$$

Here $k \equiv (\omega, \vec{k})$ and $x \equiv (t, \vec{x})$ are four-vectors and
$\hat{\varepsilon}_i$ is a real unit vector pointing in the direction
of the vector potential $\vec{A}_i$ and hence by Eq. (3.6)
also in the direction of the electric vector $\vec{E}_i$.
It is called the electromagnetic polarization
vector and differs from the vector $\vec{P}$ defined by
Eq. (3.14). We will discuss the difference below.
The Coulomb gauge condition (3.10) gives

$$\hat{\varepsilon}_i \cdot \vec{k} = 0 \quad ; \qquad\qquad\qquad (3.16)$$

10. M. Fierz, Helv. Phys. Acta 13, 45 (1940).
11. E. P. Wigner, Ann. Math. 40, 149 (1939);
     Rev. Mod. Phys. 29, 255 (1957).

the vector potentials (3.15) describe transverse
waves, with $\hat{\epsilon}_1$, $\hat{\epsilon}_2$, and $\vec{k}$ forming a right-handed
orthogonal set:

$$\hat{\epsilon}_1 \times \hat{\epsilon}_2 = \hat{k} \quad . \tag{3.17}$$

The vector potentials (3.15) describe waves with
the electric vector oscillating along the direction $\hat{\epsilon}$. Such a description makes sense for a
macroscopic wave where the electric field can be
measured. For an individual photon, however, the
electric field cannot be determined and we would
like to find a more physical description. To do
so, we introduce two complex circular-polarization
vectors $\hat{\epsilon}_+$ and $\hat{\epsilon}_-$ as basic states:

$$\hat{\epsilon}_\pm = \mp \frac{1}{\sqrt{2}} (\hat{\epsilon}_1 \pm i \hat{\epsilon}_2) \ , \ \hat{\epsilon}_\pm^* = -\hat{\epsilon}_\mp \quad . \tag{3.18}$$

The minus sign for $\hat{\epsilon}$ is introduced so that the
phase is the standard one for an angular momentum
(See Eq. (4.83)). The unit vectors $\hat{\epsilon}_+$ and $\hat{\epsilon}_-$
satisfy the identities ($\lambda = \pm 1$)

$$\hat{\epsilon}_\lambda^* \cdot \hat{\epsilon}_{\lambda'} = \delta_{\lambda \lambda'} \ , \tag{3.19}$$

$$\hat{\epsilon}_\lambda^* \times \hat{\epsilon}_{\lambda'} = i \lambda \hat{k} \delta_{\lambda \lambda'} \quad . \tag{3.20}$$

Consider now a wave described by a vector potential

$$\vec{A}_{k\lambda}(t,\vec{x}) = \sqrt{\frac{n}{2\omega V}} \left\{ \hat{\epsilon}_\lambda \, e^{-ik \cdot x} + \text{c.c.} \right\} , \quad (3.21)$$

where we have added the complex conjugate term because $\vec{A}$ must be real. If we assume that the wave is restricted to the volume V and compute its total energy with Eqs. (3.6) and (3.7), we find W = nω. We interpret this result by saying that the vector potential (3.21) describes n photons, each with energy ω. To complete the discussion of the properties of these photons, we compute the angular momentum associated with the vector potential (3.21). The linear momentum density of an electromagnetic field is given by $\vec{E} \times \vec{B}$ and hence the total angular momentum is

$$\vec{J} = \int d^3x \; \vec{x} \times (\vec{E} \times \vec{B}) \quad . \qquad (3.22)$$

With $\vec{\nabla} \cdot \vec{E} = 0$, $\vec{J}$ can be rewritten as

$$\vec{J} = \vec{L} + \vec{S} \quad , \qquad (3.23)$$

where

$$\vec{L} = \sum_i \int d^3x \; E_i \; \vec{x} \times \vec{\nabla}A_i \quad , \qquad (3.24)$$

$$\vec{S} = \int d^3x \; \vec{E} \times \vec{A}. \qquad (3.25)$$

We have separated the total angular momentum into
a part $\vec{L}$ that depends on $\vec{x}$ and a part $\vec{S}$ that is
independent of $\vec{x}$. According to the definition of
spin in Subsection C, $\vec{S}$ can be taken to be the
photon spin and $\vec{L}$ the orbital angular momentum.
For a free particle, represented by a plane wave,
the orbital part $\vec{L}$ vanishes and we are left with
the spin angular momentum $\vec{S}$. With Eqs. (3.6) and
(3.20), we compute $\vec{S}$ for the vector potential
(3.21) and get

$$\vec{S} = n \; \lambda \; i \; \hat{\varepsilon}_\lambda^* \times \hat{\varepsilon}_\lambda = n \; \lambda \; \hat{k} \quad . \tag{3.26}$$

The photons described by Eq. (3.21) carry n units
of angular momentum along the direction of motion;
$\vec{A}_+$ hence is characteristic of right-handed and
$\vec{A}_-$ of left-handed photons. If we define the
helicity λ as the projection of the total angular
momentum along the direction of propagation,
we can say that $\vec{A}_+$ describes n photons, each
with helicity $\lambda = +1$ and $\vec{A}_-$ describes n photons,
each with helicity $\lambda = -1$. Because of Eq. (3.16),
$\vec{E}$ and $\vec{A}$ for a free electromagnetic wave are
transverse to $\hat{k}$ for any linear combination
of $\hat{\varepsilon}_1$ and $\hat{\varepsilon}_2$, or $\hat{\varepsilon}_+$ and $\hat{\varepsilon}_-$. No states with
$\vec{S}$ transverse can be formed and photons with
helicity 0 do not exist. The helicity of a photon
is either +1 or -1, but never 0. (Linear combinations with vanishing angular momentum can be
formed, but these states do not correspond to

single photons as we will see in Subsection 3.3 E.)

One final remark concerning the term "polarization vector" may be helpful. Equations (3.14), (3.16), and (3.26) give for $\hat{\epsilon}$ and $\vec{P}$:

$$\hat{\epsilon} \cdot \hat{k} = 0 \quad \text{and} \quad \vec{P} = \hat{k} \quad .$$

The electromagnetic polarization vector $\hat{\epsilon}$ is transverse, the vector $\vec{P}$ describing the spin lies along the momentum! The confusion arises from two different usages; the word "polarization" describes the direction of the electric vector and the direction of the angular momentum.

The angular momentum carried by circularly polarized light can be measured by the torque that is transferred if it is absorbed. One such experiment has been performed by Beth, who observed the torque exerted by a circularly polarized light beam on a doubly refracting quartz plate.[12] Visible light is, however, not the most convenient object. The ratio of absorbed angular momentum to absorbed energy is $1/\omega$. To get a maximum torque, the frequency of the electromagnetic wave should be as low as possible. Using microwaves, Allen has constructed a "radiation torque motor", sketched in Fig. 3.1.[13] A dipole is suspended in a circular waveguide and a circularly polarized microwave causes the dipole to rotate

---

12. R. A. Beth, Phys. Rev. 50, 115 (1936).
13. P. J. Allen, Am. J. Phys. 34, 1185 (1966).

continuously. The existence of the two helicity
states with angular momentum components $\lambda = \pm 1$
and the absence of the component $\lambda = 0$ has also
been seen directly by using the Mössbauer
effect.[14)]

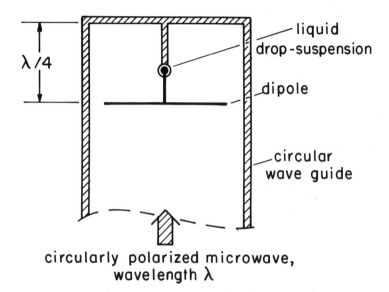

circularly polarized microwave,
wavelength λ

Fig. 3.1.   Allen's "radiation torque motor"
            transforms the angular momentum
            carried by a circularly polarized
            electromagnetic wave into a contin-
            uous rotation of the dipole that is
            suspended from a liquid drop.[13)]

14.   H. Frauenfelder, D. E. Nagle, R. D. Taylor,
      D. R. F. Cochran, and W. M. Visscher, Phys.
      Rev. 126, 1065 (1962).

3.2   DIRAC NOTATION

The state of a physical system can often be
described by a wave function $\psi(t,\vec{x})$. A more
general and more flexible approach is to represent
the state by a vector in Hilbert space.   In Dirac's
notation, such a state vector is written as a
ket $|>$ or a bra $<|$. If a bra and a ket describe
the same state, then one is the complex conjugate
of the other.   The quantum numbers characterizing
the state can be used to label bras and kets.
Using Dirac's notation, we can for instance rewrite
Eq. (3.13) as

$$\vec{s}^2 \quad |s,m> = s(s+1) \quad |s,m> \quad .$$

We often denote a complete set of quantum numbers
by one letter, for instance α. The scalar product
of a bra and a ket is denoted by $<\alpha|\alpha>$; in nonrel-
ativistic theory, the discrete state vectors are
usually chosen to be orthonormal,

$$<\alpha'|\alpha> = \delta_{\alpha',\alpha} \quad . \tag{3.27}$$

Most of the time we will work directly with
the state vectors, i.e., the bras and kets, just
as with vectors in normal vector algebra.   Sometimes, however, it is convenient to select a
particular coordinate system in Hilbert space and
to describe the state vector by its projection onto
all coordinate axes.   As basis of such a coordinate

system, any complete set of eigenfunctions can be used. Assume that the states $|m\rangle$ form such a complete orthonormal set and denote the projection onto a particular state $|m\rangle$ by $\langle m|\alpha\rangle$. The ket $|\alpha\rangle$ and the bra $\langle\alpha|$ are then given by the expansions

$$|\alpha\rangle = \sum_m |m\rangle \langle m|\alpha\rangle$$

$$\langle\alpha| = \sum_{m'} \langle\alpha|m'\rangle \langle m'| \tag{3.28}$$

with

$$\langle m|\alpha\rangle = \langle\alpha|m\rangle^* \quad . \tag{3.29}$$

If we select the eigenstates $|\vec{x}\rangle$ of the position operator $\vec{x}$ as basis, the state vector $|\alpha\rangle$ is specified by its components along these basic kets $|\vec{x}\rangle$. These components are just the Schrödinger wavefunctions

$$\psi_\alpha(\vec{x}) = \langle\vec{x}|\alpha\rangle \quad , \quad \psi_\alpha^*(\vec{x}) = \langle\vec{x}|\alpha\rangle^* = \langle\alpha|\vec{x}\rangle \quad .$$

The orthonormality of the states $|\alpha\rangle$ and $|m\rangle$ together with the expansions (3.28) lead to the closure relation

$$\sum_{m} |m> <m| = 1 \quad .$$
(3.30)

If a linear operator $\mathcal{O}(\vec{x})$ produces a transition between states α and β, specified by wavefunctions ψ_{α} and ψ_{β}, the transition amplitude is proportional to the integral

$$\int d^3x \ \psi_{\beta}^{*}(\vec{x}) \ \mathcal{O}\psi_{\alpha}(\vec{x}) \quad .$$

This integral is called the <u>matrix element</u> of the operator $\mathcal{O}$ between the states α and β; in Dirac's notation it is given by

$$<\beta| \ \mathcal{O} \ |\alpha> \quad .$$

Matrix elements of <u>linear</u> operators satisfy the relations

$$<\beta| \ \mathcal{O} \ |\alpha> = <\alpha| \ \mathcal{O}^{\dagger} \ |\beta>^{*} \quad ,$$
(3.31)

where the star denotes complex conjugation and the dagger Hermitian conjugation. Matrix elements of Hermitian operators, $H = H^{\dagger}$, satisfy

$$<\beta|H|\alpha> = <\alpha|H|\beta>^{*} \quad ;$$

expectation values ($\alpha=\beta$) of Hermitian operators are real.

3.3   QUANTIZATION OF THE FREE ELECTROMAGNETIC FIELD

A.   Second Quantization

In ordinary quantum mechanics, the starting point is the existence of particles. In classical physics, the motion of particles follows well-known laws, expressed in terms of the coordinates q_i and the momenta p_i. Experimental evidence from microscopic systems forces revision of the classical laws. The transition to quantum mechanics ("first quantization") is performed by making q_i and p_i operators and describing the state of a particle by a state vector $|>$ or a wavefunction ψ. The commutation relations satisfied by the operators p and q are postulated, guided by the correspondence principle; they are

$$[p_i, q_j] = -i \, \delta_{ij} \quad .\tag{3.32}$$

The situation in which we find ourselves with the electromagnetic field is different. The classical description of electromagnetism is not in terms of particles, but in terms of fields. Quantization is introduced to take care of the corpuscular aspect. The classical fields are given by $\vec{E}$ and $\vec{B}$, or by the vector potential A^{μ}; in the quantized description, these fields become operators that satisfy commutation relations.

Field quantization can also be applied to material particles, for instance electrons. In ordinary quantum mechanics, the behavior of a

particle is described by a wavefunction $\psi(\vec{x})$,
obtained from classical mechanics by quantization.
Can we regard $\psi(\vec{x})$ as a field and quantize a
second time? We can look at this problem in the
following way: Assume that the historical
development had gone differently and that electrons
had been discovered first by their diffraction
effects.[15] It would then have been necessary to
describe the phenomena by a field amplitude $\psi(\vec{x})$.
Later, with additional experiments such as electron
scattering by atoms and nuclei revealing the corpuscular properties of electrons, it would be
convenient to quantize $\psi(\vec{x})$. But since history
did not happen that way (it could have[16]) and
since ψ is already the result of a quantization
procedure, the quantization of the wavefunction
ψ is sometimes called "second quantization". The
fields $\vec{E}$ and $\vec{B}$ and the amplitude ψ are thus both
subjected to quantization, but there is a major
difference between the photon and the electron
fields: Because the photon obeys Bose statistics,
the number of quanta in one particular mode can be
large enough to give rise to a classically measurable field, and $\vec{E}$ and $\vec{B}$ can be observed directly.
The electron, however, is a fermion and its field

15.   C. J. Davisson and L. H. Germer, Phys. Rev.
      30, 705 (1927). G. P. Thomson, Nature 120,
      802 (1927); Am. J. Phys. 29, 825 (1961).
16.   See the footnote on page 247 of M. Jammer,
      The Conceptual Development of Quantum Mech-
      anics (McGraw-Hill Publishing Co., New York,
      1966).

amplitude can never be measured directly. Why,
then, is second quantization useful? Ordinary
quantum mechanics is well suited to single
particle problems, but for many-particle systems
and for systems where particles are created and
destroyed, second quantization becomes indispens-
able. Field quantization is used, for instance,
to describe the creation and destruction of
photons and material particles; in nuclear physics,
it is applied to many-particle systems.

In the following subsections, we outline
elementary aspects of the quantization of the
electromagnetic field. Formally, it would be
simpler to consider the quantization of the neutral
scalar field first, but conceptually the quantization of the electromagnetic field is easier to
understand.

B. The Number Representation

The state vector $|\rangle$ in ordinary quantum
mechanics is usually normalized to a fixed number
of particles per volume V and is hence not well
suited to describe processes in which the number
of particles changes. There exists, however, a
different representation of states, called the
number representation, with which creation and
annihilation processes can be treated naturally.
To introduce it, we consider photons in a cubical
box $V = L^3$ and assume periodic boundary conditions
at the surfaces of the box. The photon wave
vectors then can only assume values

$$\vec{k} = \frac{2\pi}{L} (\nu_1, \nu_2, \nu_3) \quad , \qquad\qquad (3.33)$$

where ν_1, ν_2, and ν_3 are integers. Moreover, to characterize each photon completely, its polarization state must also be indicated, for instance by its helicity λ.

A photon is then described by k and λ, or by $(\nu_1, \nu_2, \nu_3; \lambda)$. We can now enumerate the allowed states in some defined order $(\nu_1, \nu_2, \nu_3; \lambda)$, for instance: state 1 = (1,0,0; 1), state 2 = (1,0,0; -1), state 3 = (0,1,0; 1), ... A many-photon system with n_1 photons in state 1, n_2 photons in state 2, and n_i photons in state i is then denoted by the state vector

$$|n_1, n_2, \ldots n_i \ldots> \qquad . \qquad\qquad (3.34)$$

The plane wave for which $\vec{A}$ is given by Eq. (3.21) can be characterized by the state vector

$$|0, 0, \ldots 0, n, 0, \ldots>$$

where the n occurs in the position corresponding to the particular wave vector $\vec{k}$ and the helicity λ. The order in which the various states are enumerated in Eq. (3.34) is useful for bookkeeping but has no physical consequences since the Bose-Einstein statistics allows interchange of any two particles. For particles obeying Fermi-Dirac statistics, the order is important and must be specified.

The state vectors (3.34) are taken to be orthonormal,

$$\langle n_1, n_2, \ldots n_i \ldots | n_1', n_2', \ldots n_i' \ldots \rangle$$

$$= \prod_{i=1}^{\infty} \delta_{n_i, n_i'} \tag{3.35}$$

where $\delta_{n_i, n_i'}$ is the Kronecker symbol.

C.   Creation and Annihilation Operators

We know from experiment that photons can be created and annihilated and obey Bose-Einstein statistics. We therefore need a formalism in which the creation and annihilation can be expressed and in which the number of photons in any given state is not restricted. The state (3.34) satisfies the latter criterion; to express the annihilation of a photon we <u>define</u> an annihilation operator $a_{k\lambda}$ by

$$a_{k\lambda} | n_1 \ldots n_i \ldots \rangle = \sqrt{n_i} \, | n_1 \ldots, n_i - 1, \ldots \rangle \; . \tag{3.36}$$

We will see in the following that this definition leads to a consistent theory for free fields. According to Eq. (3.36), $a_{k\lambda}$ destroys <u>one</u> photon with four-momentum $k = (\omega, \vec{k})$ and helicity λ. The factor $\sqrt{n_i}$ in the definition (3.36) yields simple commutation relations, Eq. (3.40) below. With the normalization (3.35) we find that all but one matrix elements of $a_{k\lambda}$ vanish; the non-vanishing element is given by

$$\langle n_1, \ldots, n_i-1, \ldots | a_{k_i \lambda_i} | n_1, \ldots, n_i, \ldots \rangle = \sqrt{n_i} \, .$$

$$(3.37)$$

To arrive at a creation operator, we use the relation (3.31). Since the matrix element (3.37) is real, we find

$$\langle n_1, \ldots, n_i, \ldots | a^{\dagger}_{k_i \lambda_i} | n_1 \ldots, n_i-1, \ldots \rangle = \sqrt{n_i} \, .$$

$$(3.38)$$

The operator $a^{\dagger}_{k_i \lambda_i}$ hence satisfies the relation

$$a^{\dagger}_{k_i \lambda_i} | n_1, \ldots n_i, \ldots \rangle$$

$$= \sqrt{n_i+1} \, | n_1 \ldots, n_i+1, \ldots \rangle \quad . \qquad (3.39)$$

The operator $a^{\dagger}_{k\lambda}$ increases the number of photons with four-momentum k and helicity λ by one; it creates one photon with these properties. Equations (3.37) and (3.38) show that $a_{k\lambda}$ is not Hermitian.

The commutation relations satisfied by a and

$a^\dagger$ are easily found from Equations (3.36) and (3.39); they are

$$[a_{k\lambda}, a_{k'\lambda'}^\dagger] = \delta_{k,k'} \, \delta_{\lambda,\lambda'} \quad . \tag{3.40}$$

To get the commutation relations between two creation operators, we note that the interchange of any two photons leaves the arbitrary state function $|>$ unchanged:

$$a_{k\lambda}^\dagger \, a_{k'\lambda'}^\dagger |> = a_{k'\lambda'}^\dagger \, a_{k\lambda}^\dagger |> \quad .$$

A similar equation holds for two annihilation operators and we thus get

$$[a_{k\lambda}, a_{k'\lambda'}] = [a_{k\lambda}^\dagger, a_{k'\lambda'}^\dagger] = 0 \quad . \tag{3.40'}$$

D. Field Operators

The creation and annihilation operators are not Hermitian and thus do not correspond to observables. In order to connect the formalism of the previous subsection to experiments, we must find Hermitian operators that are observables. We begin with

$$N_i = a_{k_i\lambda_i}^\dagger \, a_{k_i\lambda_i} \quad . \tag{3.41}$$

N is Hermitian and it satisfies the eigenvalue equation

$$N_i |n_1, \ldots, n_i, \ldots> = n_i |n_1, \ldots, n_i> \qquad . \quad (3.42)$$

The eigenvalues of N_i are just the numbers n_i of photons in the state specified by k_i and λ_i; N_i is therefore called the <u>number operator</u>. Since the energy of a photon with momentum $\vec{k}$ is given by $\omega(k) = |\vec{k}|$, the operator

$$H_\gamma = \sum_i N_i \, \omega(k_i) \qquad\qquad (3.43)$$

obeys the equation

$$H_\gamma |n_1 \ldots> = W |n_1 \ldots> \qquad\qquad (3.44)$$

with

$$W = \sum_i n_i \, \omega(k_i) \qquad . \qquad\qquad (3.45)$$

H_γ can thus be considered the Hamiltonian of the free radiation field. If the eigenvalues of the Hamiltonian are to be positive definite, there must be a state of lowest energy, the "<u>vacuum</u>," defined by

$$a_{k\lambda} |0> = 0 \qquad , \qquad\qquad (3.46)$$

for all values of k and λ.   It is not possible to
destroy a particle from the vacuum because it
contains no photons and all occupation numbers are
zero.   Starting from the vacuum, states with one,
two, and more photons can be constructed.   A state
with one free (plane wave) photon of momentum k
and helicity λ is given by

$$|k,\lambda> \equiv a^{\dagger}_{k\lambda} |0> \quad . \tag{3.47}$$

To find an expression for the operator $\vec{A}_{op}$
that corresponds to the electromagnetic potential,
we note that single photons of momentum $\vec{k}$ and
helicity λ can be created and destroyed; $\vec{A}_{op}$
should hence be linear in creation and annihilation
operators:

$$\vec{A}_{op}(x) = \vec{c}_{k\lambda}(x)\, a_{k\lambda} + \vec{d}_{k\lambda}(x)\, a^{\dagger}_{k\lambda} \quad .$$

Furthermore, $\vec{A}_{op}$ must be Hermitian since the field
$\vec{A}$ is real.   With $\vec{A}^{\dagger} = \vec{A}$, we get $\vec{d}_{k\lambda} = \vec{c}^{*}_{k\lambda}$ and
hence

$$\vec{A}_{op}(x) = \vec{c}_{k\lambda}(x)\, a_{k\lambda} + \vec{c}^{*}_{k}(x)\, a^{\dagger}_{k\lambda} \quad . \tag{3.48}$$

We determine the factor $\vec{c}^{*}_{k\lambda}$ by using the correspondence principle which demands that the
expectation value of the quantum mechanical
operator coincides with the classical observable
for large quantum numbers:

$$\langle \text{class} | \vec{A}_{op}(x) | \text{class} \rangle \rightarrow \vec{A}_{k\lambda}(x) \quad . \qquad (3.49)$$

Here, $\vec{A}_{k\lambda}(x)$ is the classical vector potential (3.21) and $|\text{class}\rangle$ is a state that permits the transition to the classical limit.[17]

Next we construct the state $|\text{class}\rangle$. The first candidate is the state vector (3.34) in which the number of photons in each mode is given; it is then easy to let any or all n_i become very large. However, the expectation values of a and $a^\dagger$ vanish even for very large n so that the correspondence principle cannot be applied. We can obtain a nonvanishing expectation value of $a_{k\lambda}$ if we can construct $|\text{class}\rangle$ as an eigenstate of $a_{k\lambda}$. It is indeed possible to do so and we define $|\text{class}\rangle$ so that[18]

$$a_{k\lambda} |\text{class}\rangle = \alpha_{k\lambda} |\text{class}\rangle \quad , \qquad (3.50)$$

with the Hermitian conjugate equation

---

17. E. Schrödinger, Naturwiss. $\underline{14}$, 664 (1926). R. J. Glauber, Phys. Rev. $\underline{130}$, 2529 (1963); $\underline{131}$, 2766 (1963). P. Carruthers and M. M. Nieto, Am. J. Phys. $\underline{33}$, 537 (1965). J. R. Klauder and E. C. G. Sudarshan, Fundamentals of Quantum Optics (W. A. Benjamin, Inc., Reading, Mass., 1968).
18. No such eigenstates exist for the creation operator $a_{k\lambda}^\dagger$. (Why not?)

$$\langle \text{class}| \; a_k^\dagger \; = \; \alpha_{k\lambda}^* \; \langle \text{class}| \quad .$$

The state $|\text{class}\rangle$ is <u>not</u> an eigenstate of the number operator (3.41), but the <u>average</u> number of photons is given by

$$\bar{n}_{k\lambda} \; = \; \langle \text{class}| a_{k\lambda}^\dagger \; a_{k\lambda} | \text{class}\rangle \; = \; |\alpha_{k\lambda}|^2 \quad , (3.51)$$

if the normalization $\langle \text{class}|\text{class}\rangle = 1$ is used. The complex eigenvalue $\alpha_{k\lambda}$ is

$$\alpha_{k\lambda} \; = \; |\alpha_{k\lambda}| \; e^{i\Phi_{k\lambda}} = \sqrt{\bar{n}_{k\lambda}} \; e^{i\Phi_{k\lambda}} \quad , \quad (3.52)$$

and we get for the expectation value of the operator (3.48)

$$\langle \text{class}| \vec{A}_{op}(x) |\text{class}\rangle$$

$$= \; \sqrt{\bar{n}_{k\lambda}} \; \left\{ \vec{c}_{k\lambda}(x) \; e^{i\Phi_{k\lambda}} + \text{c.c.} \right\} \quad .$$

Comparison of this expression with the classical potential (3.21) and use of the correspondence principle gives

$$\vec{c}_{k\lambda}(x) \; e^{i\Phi_{k\lambda}} \; = \; \frac{1}{\sqrt{2\omega V}} \; \hat{\epsilon}_{k\lambda} \; e^{-ik\cdot x} \quad . \quad (3.53)$$

The operator corresponding to a superposition of
states (k,λ), with $k = (|\vec{k}|,\vec{k})$, is given by

$$\vec{A}_{op}(x) = \sum_{\vec{k},\lambda} \frac{1}{\sqrt{2\omega V}} \tag{3.54}$$

$$\times \left\{ a_{k\lambda} \hat{\epsilon}_{k\lambda} \, e^{-ik\cdot x - i\Phi_{k\lambda}} + a^{\dagger}_{k\lambda} \hat{\epsilon}^{*}_{k\lambda} \, e^{ik\cdot x + i\Phi_{k\lambda}} \right\} .$$

In the following, we will omit the subscript "op."
The form (3.54) of $\vec{A}$ expresses the quantization
of the electromagnetic field. The vector potential
is a sum over contributions from the individual
photons; each photon can be created or destroyed.
Note that only a and $a^{\dagger}$ are operators (so-called
q numbers) and all other factors are ordinary
numbers (c numbers). Furthermore, the state
vectors are time-independent; the entire time-
dependence is in the field operator. Tradition-
ally, this is called the Heisenberg representation.
    With Eq. (3.6), the expression for the oper-
ator of the electric field strength follows as

$$\vec{E}(x) = i \sum_{\vec{k},\lambda} \sqrt{\frac{\omega}{2V}} \tag{3.55}$$

$$\times \left\{ a_{k\lambda} \hat{\epsilon}_{k\lambda} \, e^{-ik\cdot x} \, e^{-i\Phi_{k\lambda}} - a^{\dagger}_{k\lambda} \hat{\epsilon}^{*}_{k\lambda} \, e^{ik\cdot x} \, e^{i\Phi_{k\lambda}} \right\}.$$

Similarly, the magnetic field strength operator
can be found. It is then straightforward to
show that the operators $\vec{E}$ and $\vec{B}$ satisfy Maxwell's
equations.

At this point, we can understand why we could
not use the number representation to perform the
transition to the classical limit. The field
operators $\vec{A}$, $\vec{E}$, and $\vec{B}$ do not commute with the
number operator:

$$[N_i, \vec{A}] \neq 0, \quad [N_i, \vec{E}] \neq 0 \quad , \quad [N_i, \vec{B}] \neq 0. \quad (3.56)$$

Hence a state of any <u>definite</u> number of photons
does not have a well-defined field strength and
vice-versa. It is possible to measure either
the number of photons or the field strength
precisely, but not both simultaneously. This
fact is sometimes expressed by the uncertainty
relation

$$\Delta n_{k\lambda} \, \Delta \Phi_{k\lambda} > 1 \quad , \quad (3.57)$$

where $\Delta n_{k\lambda}$ is the uncertainty in the number of
photons in the state $k\lambda$ and $\Delta \Phi_{k\lambda}$ is the uncertainty in the corresponding phase. However, the
usual proof of this relation is incorrect as was
shown by Susskind and Glogower;[19] the relation
(3.57) holds for large quantum numbers but must

_____

19. L. Susskind and J. Glogower, Physics 1, 49
     (1964).

be modified for small ones.[20)] If we deal with
single photons, as we will nearly always do, the
phase has no physical meaning and we omit it in
future expressions for the field operators.

The linear momentum operator for the electromagnetic field is obtained by starting from the
Poynting vector

$$\vec{P}_{class} = \int d^3x \, \vec{E} \times \vec{B} \quad , \qquad (3.58)$$

casting it into a Hermitian form

$$\vec{P} = \frac{1}{2} \int d^3x \, (\vec{E} \times \vec{B} - \vec{B} \times \vec{E}) \quad , \qquad (3.59)$$

and inserting the operators for $\vec{E}$ and $\vec{B}$. Using
the transversality condition, Eq. (3.16), we get

$$\vec{P} = \sum_{\vec{k},\lambda}{}' \, (a_{k\lambda}^{\dagger} \, a_{k\lambda} + \frac{1}{2})\vec{k} \quad . \qquad (3.60)$$

The term $\frac{1}{2}$ represents an old problem in quantum
field theories. It is a vacuum contribution;
summed over all possible k,λ, it vanishes. (We
could have found similar terms in the occupation
number N and the energy E if we had used a sym-

---

20.   P. Carruthers and M. M. Nieto, Phys. Rev.
      Letters 14, 387 (1965); Rev. Mod. Phys. 40,
      411 (1968).

metrized number operator, $N = \frac{1}{2}(a^{\dagger}a + aa^{\dagger})$.
These terms are infinite, have no measurable
consequences, and are omitted.)[21]   The expression for the momentum operator of the electromagnetic field is thus defined to be

$$\vec{p} = \sum_{\vec{k},\lambda} a_{k\lambda}^{\dagger} a_{k\lambda} \vec{k} = \sum_{\vec{k},\lambda} N_{k\lambda} \vec{k} \quad . \tag{3.61}$$

The spin operator can be computed similarly;
inserting Eqs. (3.54) and (3.55) into Eq. (3.25),
using Eq. (3.20), and again dropping the vacuum
term, we get

$$\vec{S} = \sum_{\vec{k},\lambda} N_{k\lambda} \, \lambda \hat{k} \quad . \tag{3.62}$$

In the expansions of the field operators, for
instance Eq. (3.54), the quantization volume V
appears explicitly. For large values of V, all
physically meaningful expressions must converge
towards a finite value. This fact can often be
expressed by replacing the sums over allowed states
by integrals. The momentum k becomes a continuous

---

21.   In general, the arrangement of non-commuting
      factors in a quantum mechanical operator is
      not completely determined by the requirement
      that the operator must be Hermitian. However,
      physically measurable effects are unaffected
      by this indeterminacy. (C. P. Enz and A.
      Thellung, Helv. Phys. Acta 33, 839 (1960)).

variable and the creation and annihilation operators become functions of k, written for instance
as $a_\lambda(k)$. To find the substitution rule, we use
Eq. (3.33) which gives the number of allowed states
in the momentum volume element d^3k

$$\Delta^3\nu = \Delta\nu_1\Delta\nu_2\Delta\nu_3 = \left(\frac{L}{2\pi}\right)^3 dk_1 dk_2 dk_3$$

$$= \frac{V}{(2\pi)^3} d^3k \quad . \tag{3.63}$$

If we let the volume V go to infinity, we can make
the substitution

$$\lim_{V \to \infty} \frac{1}{V} \sum_{\vec{k}} \to \frac{1}{(2\pi)^3} \int d^3k \tag{3.64}$$

without changing the number of allowed states.
The Kronecker symbols $\delta_{\vec{k},\vec{k}'}$ must be replaced by
Dirac delta functions $\delta^3(\vec{k}-\vec{k}')$. With the
relations

$$\sum_{\vec{k}} \delta_{\vec{k},\vec{k}'} = 1 \quad , \quad \int d^3k \, \delta^3(\vec{k}-\vec{k}') = 1 \quad ,$$

we note that we must use the substitution

$$\delta_{\vec{k},\vec{k}'} \to \frac{(2\pi)^3}{V} \delta^3(\vec{k}-\vec{k}') \tag{3.65}$$

together with Eq. (3.64).

E. Single Photons

The operators of energy, momentum, and spin commute with each other and with the number operator, $N = \Sigma\, N_i$,

$$[H_\gamma,N] = [\vec{p},N] = [\vec{S},N] = 0 \quad, \tag{3.66}$$

$$[H_\gamma,\vec{p}] = [H_\gamma,\vec{S}] = [\vec{p},\vec{S}] = 0 \quad.$$

It is therefore possible to find simultaneous eigenvalues of the four operators. The one-photon state, Eq. (3.47) is such a state and it satisfies

$$N \quad |k,\lambda\rangle = |k,\lambda\rangle \quad,$$

$$H_\gamma \quad |k,\lambda\rangle = \omega|k,\lambda\rangle \quad, \tag{3.67}$$

$$\vec{p} \quad |k,\lambda\rangle = \vec{k}|k,\lambda\rangle \quad,$$

$$\vec{S} \quad |k,\lambda\rangle = \lambda\hat{k}|k,\lambda\rangle \quad, \quad \lambda = \pm 1 \quad.$$

A single photon can be characterized by its energy, its momentum, and its helicity; each of these observables can be sharp. Moreover, the helicity assumes only the values +1 and -1. However, single photons do not have to be in eigenstates of $\vec{p}$ and $\vec{S}$. The state $c_1|k,+\rangle + c_2|k,-\rangle$ still describes a single photon, but it is an eigenstate of the helicity operator only if either c_1 or c_2 is zero.

The most general one-photon state is a superposition of states with sharp momenta and helicities,

$$|1> = \sum_{\vec{k},\lambda} \hat{\epsilon}^*_{k\lambda} \, f_{k,\lambda} \, e^{i\omega t} \, a^{\dagger}_{k,\lambda} |0> \quad .$$

Such a one-photon state is still an eigenstate of
N and H_{γ}, but not of $\vec{S}$ and $\hat{p}$; it cannot be local-
ized in space. States with a definite transverse
polarization can also be formed. However, they
do not correspond to a definite number of photons.
    The expectation value of the electromagnetic
field of a single photon (or actually of any
definite number of photons) vanishes,

$$<k\lambda|\vec{A}|k\lambda> = 0 \quad . \tag{3.68}$$

The electric or magnetic field of a single photon
is hence not a meaningful quantity. The vanishing
of the expectation value, however, indicates only
that the average field is zero; the field can still
fluctuate. Indeed we find for the vacuum

$$<0|\vec{A}^2|0> = \sum_{\vec{k}\lambda} \frac{1}{2\omega V} \neq 0 \quad ,$$

and for one photon

$$<k\lambda|\vec{A}^2|k\lambda> = \frac{1}{\omega V} + <0|A^2|0> \quad .$$

So far, we have treated the properties of single photons. Assume now that we deal with a situation in which no photon is present in the initial state and one photon with four-momentum k_β and helicity λ_β is present in the final state. The only nonvanishing matrix element of the vector potential operator (3.54) is then

$$<k\lambda|\vec{A}|0> = \frac{1}{\sqrt{2\omega V}} \; \hat{\epsilon}^*_{k\lambda} \; e^{ik\cdot x} \quad . \qquad (3.69)$$

Similarly we find for the matrix element appropriate for the annihilation of a photon

$$<0|\vec{A}|k\lambda> = \frac{1}{\sqrt{2\omega V}} \; \hat{\epsilon}_{k\lambda} \; e^{-ik\cdot x} \quad . \qquad (3.70)$$

The expression (3.70) can be considered to be the wave function of a free photon with momentum k and helicity λ.[22]

3.4   PARTICLES AND ANTIPARTICLES - DESCRIPTIVE

The dream of alchemists in its purest form is the creation of matter out of nothing. Nature does not permit this dream to come true, but energy can be transformed into matter and matter into energy. These transformations are linked intimately with the existence of antiparticles. The antiparticle

---

22.   L. Landau and R. Peierls, Z. Physik $\underline{62}$, 188 (1930). R. H. Good, Jr., Phys. Rev. $\underline{105}$, 1914 (1957).

concept was revolutionary when it was introduced
by Dirac and today retains its fascination. We
retrace in the present section some of the ideas
that led to the antiparticle hypothesis and outline
some of the experiments that verify the existence
of such particles. The discussion is descriptive
and we fill in many of the gaps in later sections.

A.   Dirac's Hole Theory

> "A hole, if there were one, would be a new
> kind of a particle, unknown to experimental
> physics, having the same mass and opposite
> charge to an electron. We may call such
> a particle an anti-electron."
>
>   P. A. M. DIRAC [Proc. Roy.
>   Soc. (London) 133, 60
>   (1931)]

The energy of a particle with restmass m and
momentum $\vec{p}$ is, according to Eq. (2.2), given by

$$E^{\pm} = \pm\sqrt{\vec{p}^2 + m^2} \quad . \tag{3.71}$$

Formally, then, the energy of a free particle
with given momentum and mass can be positive or
negative. Can we find a physical reason for
excluding the negative solution or do we have to
cope with it? Are there real effects that are
caused by the negative solution? In the present
subsection we outline Dirac's interpretation of

the negative solution.[23]

In classical physics, the negative energy solution can be avoided. If only positive energy states are occupied initially, the continuity of classical physics prevents transitions to negative energy states. At one time Schrödinger suggested a similar procedure for handling the negative energy states in relativistic quantum mechanics, but his approach encountered two insurmountable difficulties: There is no a-priori reason in quantum mechanics to exclude transitions to negative energy states, and the positive energy eigenstates alone do not form a complete set. We will argue the former feature below.

The energy spectrum of a _free_ particle with mass m is shown in Fig. 3.2. The positive energy states form a continuum from E = +m to +∞; the negative energy states occupy a similar continuum from E = -m to -∞. Between E = +m and E = -m, no energy levels are available. Now consider particles in positive energy states. If only two particles interact, for instance in a collision, the conservation laws of energy and momentum guarantee that they cannot go into states of negative energy as a result of the collision. If,

23. P. A. M. Dirac, Proc. Roy. Soc. (London)
     A 126, 360 (1930) and Rapport du 7ème
     Conseil Solvay de Physique, p. 203 (1934)
     (reprinted in Quantum Electrodynamics, J.
     Schwinger, ed. (Dover Publications, New York,
     1958)).

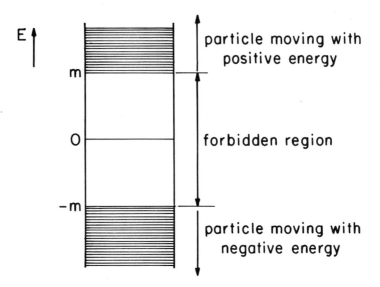

Fig. 3.2.   Positive and negative energy states
            and forbidden region for a free Dirac
            particle.

however, three or more particles interact, this
restriction is no longer present. One of the
particles can fall into a negative energy state,
with the other two absorbing the necessary energy
and momentum. Such a process is shown schemati-
cally in Fig. 3.3. The energy E can be carried
away by photons. Oppenheimer and Tamm independently computed the lifetime of electrons against
such transitions and found it to be smaller than
about 10^{-10}sec.[24)] Since electrons still exist,

24.   J. R. Oppenheimer, Phys. Rev. 35, 939 (1930).
      I. Tamm, Z. Physik 62, 545 (1930).

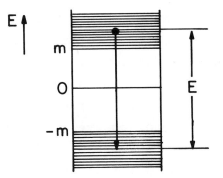

Fig. 3.3. Transition of a Dirac particle into a
negative energy state with emission of
photons of total energy E.

some mechanism must exist that forbids rapid transitions to negative energy states. Attempts to
find a simple explanation were unsuccessful[25] so
that Dirac finally cut the Gordian knot by introducing two new assumptions:[23]

   I. All negative energy states are filled
       with particles.

   II. The particles in the negative energy
       states do not produce an external field
       and do not contribute to the charge,
       energy, momentum, and spin of the system.

25. W. Pauli, "Die allgemeinen Prinzipien der
     Wellenmechanik," Handbuch der Physik (Springer
     Verlag, Berlin, 1933), Second ed., Vol. 24/1,
     §B5.

If the particles under consideration obey the
Pauli exclusion principle, the filled negative
energy levels cannot accept more particles;
transitions from the positive to the negative
energy states are thus forbidden. The stability
of electrons and protons -- the only particles
known around 1930 -- is thus guaranteed. The
assumptions I and II were, however, invented to
explain the stability. Assumptions made to
explain one fact are unsatisfactory unless they
predict new features. Dirac's assumptions led
to other observable processes. If by some means,
for instance by the action of an external field,
a particle is removed from a negative and brought
to a positive energy state, the situation shown
in Fig. 3.4 can occur. The final state consists
of a particle in a positive energy state and a hole.

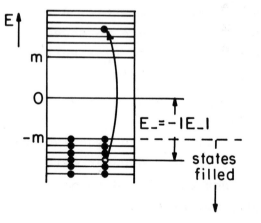

Fig. 3.4. The removal of a particle from a nega-
           tive energy state creates a particle
           in a positive energy state and a hole.

We denote with E_-, $\vec{p}_-$, $q_- = -e$, and $\vec{s}_-$ the energy, momentum, charge, and spin, respectively, of the particle while it is in its negative energy state; the corresponding quantitites for the hole are E_+, $\vec{p}_+$, q_+, and $\vec{s}_+$. By Dirac's assumption II, before the action of the external field, the hole and the particle that fills it have total energy, total momentum, total charge, and total spin all equal to zero:

$$E_+ + E_- = 0 \qquad\qquad q_+ + q_- = 0$$

$$\tag{3.72}$$

$$\vec{p}_+ + \vec{p}_- = 0 \qquad\qquad \vec{s}_+ + \vec{s}_- = 0 \quad .$$

The hole has energy, momentum, charge, and spin opposite to the corresponding quantities the particle had while it was in the negative state. (Energy and momentum of the particle after removal depend on the particular process by which it was lifted to the positive state; we are not interested in these quantities here.) In particular, the hole has positive energy and positive charge, $q_+ = e$.

At the time that Dirac postulated his hole theory, only the electron and the proton were known, and Dirac identified the particles with electrons, the holes with protons.[23] However, Oppenheimer soon pointed out that this assumption leads to insurmountable difficulties; electrons and protons in matter would annihilate each other very quickly. Weyl showed that holes must have the same mass as

the removed electrons[26] and called them "positive
electrons". Dirac accepted this interpretation[27]
and remarked: "We should not expect to find any
of them in nature, on account of their rapid state
of recombination with electrons, but if they could
be produced experimentally in high vacuum they
would be quite stable and amenable to observation."
Pauli noted that antiprotons should exist also,
but since neither positrons nor antiprotons had
been observed, he was skeptical of Dirac's hole
theory altogether.[25]   However, both antiparticles
turned up, the positron rather quickly and the
antiproton only after a long search. The actual
existence of antiparticles can be construed as a
proof of the correctness of the existence of
negative energy states. However, we will see in
Subsection C that there is an alternate and more
satisfying way to look at antiparticles.

The concept of holes is not restricted to
free particles, it is also used in the theory of
solids and nuclei. Consider as an example a
degenerate sea of fermions filling the allowed
states in a given potential up to an energy E_{max}.
At zero temperature, all states are filled
and the gas can be considered to be the vacuum.

---

26. H. Weyl, Theory of Groups and Quantum Mech-
     anics (E. P. Dutton, New York, 1931), p. 225.
27. P. A. M. Dirac, Proc. Roy. Soc. (London)
     A 133, 60 (1931); Proc. Cambr. Phil. Soc.
     30, 150 (1933-34).

Excitation of a particle from an occupied level
to a level above E_{max} corresponds to the formation
of a hole-particle pair, in analogy to Fig. 3.4,
but the minimum energy for the creation of such
a pair is no longer 2m.

B. The Positron

In a rational world, an eager experimental
physicist would have read Dirac's paper and Weyl's
book and would have rushed to his laboratory to
discover the positron. Actually, the discovery
was accidental.[28] In 1933, C. D. Anderson
observed the tracks of cosmic rays in a Wilson
cloud chamber with magnetic field. Some tracks
had a curvature corresponding to positive particles
of about the electron mass. However, these tracks
could still have been caused by electrons traveling
the wrong way. To establish the direction, Anderson put a lead plate into the chamber. The
candidates, in traversing the lead plate, lost
energy and their radius of curvature was smaller
than before they entered the lead plate: The
direction indeed corresponded to that of positive
electrons. The identification with Dirac's
anti-electrons was made by Blackett and

---

28. C. D. Anderson, Phys. Rev. $\underline{43}$, 491 (1933);
     Am. J. Phys. $\underline{29}$, 825 (1961). See also
     D. L. Anderson, The Discovery of the Electron,
     Van Nostrand Momentum Books (Van Nostrand,
     Princeton, 1964).

Occhialini.[29)]

    After positive electrons had been found, the
question arose as to how they were produced. The
answer came from Dirac's hole theory: Blackett
and Occhialini pointed out that a gamma ray could
lift an electron from a negative energy state into
a positive energy state, thereby creating a positron (the hole) and an electron, as shown in Fig.
3.5. Pair creation was indeed observed;[30)] energy
conservation gives the relation between the gamma
ray energy and the kinetic energies of the positron
and the electron,

$$E_\gamma = 2m + E_{kin}^{(+)} + E_{kin}^{(-)}. \tag{3.73}$$

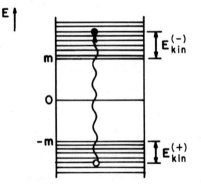

Fig. 3.5. Pair creation by radiation: The in-
           coming photon lifts an electron from
           a negative energy state to a positive
           state.

29. P. M. S. Blackett and G. P. S. Occhialini,
     Proc. Roy. Soc. (London) A 139, 699 (1933).
30. C. D. Anderson, Phys. Rev. 43, 1034 (1933).

The minimum energy required to produce an electron-positron pair is 2m = 1.02 MeV. Momentum
conservation, together with Eq. (3.73), indicates
that pair production by a single photon is not
possible without an agent, such as a nucleus, that
takes up some momentum.

The positron, once created, is presumably
just as stable as the electron as long as it moves
in free space. By using magnetic fields, positrons have been kept from colliding with matter
for a few seconds and no decays were found.[31]
Usually, however, positrons slow down in matter,
reach thermal velocity, and finally annihilate
with electrons. This annihilation corresponds to
the positive energy electron falling into the
hole, as indicated in Fig. 3.3. The energy E
is carried away by photons. Most frequently,
the annihilation of an electron-positron pair
at rest produces two photons of energy 0.51 MeV.
Experimentally, the annihilation process was
observed soon after the discovery of the
positron.[32]

31. G. Gibon, W. C. Jordan, and E. J. Lauer,
     Phys. Rev. Letters 5, 141 (1960).
32. See e.g. in M. Deutsch and S. Berko, in
     Alpha-, Beta-, and Gamma-Ray Spectroscopy,
     K. Siegbahn, ed. (North-Holland Publishing
     Co., Amsterdam, 1965), p. 1583.

C.   The Stueckelberg-Feynman Approach

"It is as though a bombardier flying
low over a road suddenly sees three
roads and it is only when two of them
come together and disappear again that
he realizes that he has simply passed
over a long switchback in a single road."
                    R. P. FEYNMAN [Phys. Rev.
            76, 749 (1949)]

We have seen that positron annihilation and
electron-positron pair creation can be described
in terms of Dirac's hole theory. Nevertheless,
the hole theory is not completely satisfactory.
Particles and antiparticles are treated in an
asymmetric way. The idea of an unobservable
sea of negative energy states is unattractive to
many people and it is applicable only to fermions.
A different way of looking at antiparticles has
been introduced by Stueckelberg[33] and Feynman:[34]
In their view the negative energy states represent
particles moving backwards in time. To make this
idea clearer, we consider solutions of a wave
equation

---

33. E. C. G. Stueckelberg; Helv. Phys. Acta 14,
     588 (1941).
34. R. P. Feynman, Phys. Rev. 74, 939 (1948); Phys.
     Rev. 76, 749 (1949).

$$i \frac{\partial \psi(\vec{x},t)}{\partial t} = H\psi(\vec{x},t) \quad , \tag{3.74}$$

where H does not depend explicitly on the time. With a solution of the form

$$\psi(\vec{x},t) = \psi_o(\vec{x}) \, e^{-iEt} \tag{3.75}$$

we get from Eq. (3.74)

$$H\psi_o = E\psi_o \quad . \tag{3.76}$$

According to Eq. (3.71), the energy E can assume positive and negative values and we get solutions belonging to E^+ and E^-. The negative energy solution has the time dependence

$$\psi^-(\vec{x},t) = \psi_o e^{-iE^- t} = \psi_o e^{+i|E^-|t} \quad ,$$

$$\tag{3.77}$$

$$E^- < 0 \quad .$$

However, we get the same time dependence if we assume that the particle has positive energy, but that the direction of time is changed.

Consider as an example the situation of Fig. 3.6. In the conventional description,

the processes shown are pair production at time
t_1 and subsequent annihilation of the antiparticle
of this pair with a particle at time t_2. Particles
are created and destroyed, but the electric charge
is conserved! This fact suggests that the description of the processes may be simpler if we
follow the charge rather than the particles.
Figure 3.6 would then represent a particle that
moves forward in time, is scattered at A and
moves backwards to B where it is scattered again.
In this approach, antiparticles are particles
moving backwards in time; the asymmetry between
particles and antiparticles present in the hole
theory has disappeared. Moreover, no sea of
negative energies need be postulated and the
particle-antiparticle concept is not restricted
to fermions. Since no negative energy states
appear, the difficulties of the catastrophic
decay of particles to negative energy levels
has disappeared. It is also clear that particles
and antiparticles are treated symmetrically; in
particular, they must have the same mass.

The description of processes in terms
of Feynman graphs, such as shown in Fig. 3.6,
has advantages in addition to the elimination
of the negative energy states. It is a convenient
way of visualizing complicated processes and it
permits the writing down of expressions for
cross sections and transition probabilities in

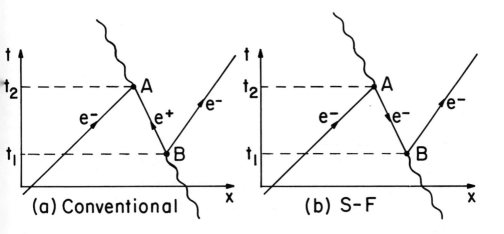

(a) Conventional (b) S-F

Fig. 3.6.   Pair production and subsequent
            particle-antiparticle annihilation.
            In the conventional description,
            the particles are the primary
            objects.   In the Stueckelberg-
            Feynman approach, the processes
            appear as motion of one charge
            forward and backward in time.

a straightforward way once the graphs have been
drawn.

    For later use, we make a remark concerning
the wave functions of free particles and anti-

particles. For a free spinless particle, we use
the time dependence given by Eq. (3.75) and write
for the wave function

$$\psi \text{ (particle)} = N e^{i(\vec{p}\cdot\vec{x} - Et)}$$

(3.78)

$$= N e^{-ip\cdot x} \quad ,$$

where N is a normalization constant and the energy
E is positive. The time part for the <u>antiparticle</u>
is given by Eq. (3.77); if the particle moves
backwards in time, its momentum is also inverted
and we get

$$\psi \text{ (antiparticle)} = N e^{-i(\vec{p}\cdot\vec{x} - Et)}$$

(3.79)

$$= N e^{ip\cdot x} \quad ,$$

where E is now also <u>positive</u>.

D.   The Antiproton

After the positron was discovered and its
role as anti-electron was clarified, the question
arose as to whether the proton and neutron also
possess antiparticles. This question is particularly important because electrons and nucleons
are fundamentally different: Electrons have only
weak and electromagnetic interactions, whereas

the proton also takes part in the hadronic interactions; electrons have a magnetic moment as
predicted by the Dirac equation, whereas nucleons
have a large anomalous magnetic moment. Nevertheless the arguments that we have given for the
existence of antiparticles in Subsections A and C
appear sufficiently general to justify a belief
in antinucleons, and the hunt for them was
started soon after the discovery of the positron.
However, even persistent search in cosmic rays
failed to provide conclusive evidence. Accelerators did not have enough energy to produce
antinucleons. Bets were made. Most famous is
one between two well-known physicists, where the
bidding reached $500 before the wives stopped
further escalation. The energy of the Berkeley
Bevatron was chosen so that it would be able to
produce antinucleons and so, finally, the antiproton was found in 1955 by Chamberlain, Segrè,
Wiegand, and Ypsilantis.[35]

The law of conservation of baryons, to be
discussed in Section 4.2, forbids the production
of single antinucleons, but for instance the
following reactions are allowed by all known
conservation laws:

---

35. O. Chamberlain, E. Segrè, C. Wiegand, and T.
    Ypsilantis, Phys. Rev. 100, 947 (1955); see
    also, E. Segrè, Am. J. Phys. 25, 363 (1957),
    Ann. Rev. Nucl. Sci. 8, 127 (1958).

$$p \; p \rightarrow p \; p \; p \; \bar{p}$$

$$\pi^- p \rightarrow p \; n \; \bar{p} \tag{3.80}$$

$$\gamma \; p \rightarrow p \; p \; \bar{p} \; .$$

In the discovery of the antiproton, the first of these reactions was used. In the c.m. system, it requires a threshold kinetic energy of $2m_p = 1.9$ GeV. Equation (2.60) gives for the corresponding kinetic energy of the incident proton in the laboratory $6m_p = 5.6$ GeV. Fortunately, if the target is a nucleus rather than a free proton, one gets help from the internal motion of the nucleons in the nucleus: nucleons in nuclei are not stationary, but they possess considerable "zero-point" energy. In a simple model, the free Fermi gas, nucleons inside typical nuclei have average kinetic energies of about 25 MeV. In the most helpful case, a head-on collision occurs and the minimum energy required for production of the antiproton is reduced considerably.[36] In the reaction $p + Cu \rightarrow \bar{p} +$ residue, antiproton production has been observed at proton kinetic energies as low as 2.9 GeV.[37]

---

36. T. Elioff, L. Agnew, O. Chamberlain, H. M. Steiner, C. Wiegand, and T. Ypsilantis, Phys. Rev. 128, 869 (1962).
37. D. E. Dorfan, J. Eades, L. M. Lederman, W. Lee, C. C. Ting, P. Piroué, S. Smith, J. L. Brown, J. A. Kadyk, and G. H. Trilling, Phys. Rev. Letters 14, 995 (1965).

To detect and identify antiprotons, their
charge (-e), their mass (m_p) and their stability
must be observed. The mass is determined by
measuring momentum and velocity; the mass then
follows from Eq. (2.58). The momentum and the
charge are found from the trajectories of the
particles in bending magnets. The velocity is
determined by time-of-flight and by Cerenkov
counters. The magnitude of the charge is also
found from the signal heights in scintillation
counters; the energy loss in a counter is proportional to $(Ze)^2$, where Ze is the charge of the
incoming particle. The stability is studied by
varying the distance traveled by the antiproton
and establishing that no antiprotons are lost
through decay. The most important parts of the
original set-up used by the Berkeley group are
shown in Fig. 3.7. The quadrupole magnets used
to focus the antiproton beam are omitted. The
results of the Berkeley experiment clearly
indicate that the antiproton exists and has the
predicted properties.

In many high-energy experiments, the Cerenkov
counter plays an important role and a few words
about its working principle are in order here.

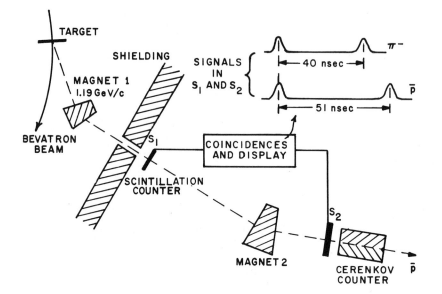

Fig. 3.7. Detection and identification of anti-
           protons. The figure shows only the
           most important elements of the set-up
           used by the Berkeley group.[35)]

E. The Cerenkov Counter[38,39)]

A charged particle moving with uniform velocity v through a medium with index of refraction

38. P. A. Cerenkov, C. R. Acad. Sci. USSR 8, 451
     (1934); 12, 413 (1936); 14, 102 (1937).
39. I. Frank and I. Tamm, C. R. Acad. Sci. USSR
     14, 109 (1937); I. Tamm, J. Phys. USSR 1,
     439 (1939).

n can emit Cerenkov radiation if v is greater than the light velocity 1/n in the medium.   This wave is essentially an electromagnetic shock wave and Huyghens' construction gives for the angle of emission of the radiation with respect to the particle momentum

$$\cos\theta_C = (n\ v)^{-1} .$$
(3.81)

The intensity of the emitted light can be calculated using classical electromagnetic theory.

Equation (3.81) implies that Cerenkov radiation permits construction of velocity-sensitive detectors: If the angle of emission is measured and the index of refraction is known, the velocity of the particle is determined.   Thus, counters can be built that record only if the particle velocity lies within a predetermined range (differential counter), or exceeds a certain value (threshold counter).   The principle of the Cerenkov counter used in the antiproton experiment, Fig. 3.7, is shown in Fig. 3.8.   The light created by the particle beam in a Cerenkov radiator only reaches the detector, a photomultiplier, if it has the correct angle of emission.   Cerenkov counters of various designs with solids, liquids, and gases under high pressure have been constructed.   Typical values of the index of refraction are: lucite n = 1.5, quartz n = 1.46, water n = 1.33.

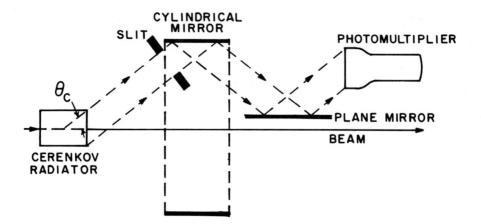

Fig. 3.8. Cerenkov counter used by the Berkeley
           group in the experiment that detected
           the antiproton.

F.   The Antineutron

     Antineutrons are created in the reaction
$pp \rightarrow ppn\bar{n}$ but these will not be monoenergetic
and therefore difficult to detect. However, once
antiprotons are available, antineutrons can be
produced by charge exchange,

$$p \, \bar{p} \rightarrow n \, \bar{n} \quad . \qquad\qquad (3.82)$$

The antiproton loses its charge to a proton and
the result is a neutron-antineutron pair. Since
the reaction is a two-body process, the antineutron emitted in a given direction has a well-

defined energy. Antineutrons were first observed
in a counter experiment,[40] using reaction (3.82).
The essential parts of the detection system are
shown in Fig. 3.9. Antiprotons, "certified" by
the system described in Subsection D, pass through
a small scintillation counter S_1 and hit a large

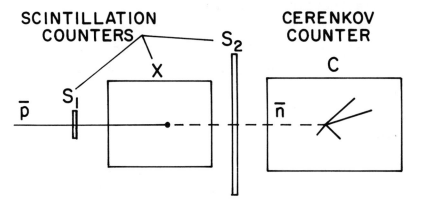

Fig. 3.9. Antineutron detection system, used by
        the Berkeley group.[40] The actual
        set-up was more complex; we only
        indicate the essential components.

scintillation counter X. If they pass through X
or undergo charge exchange according to Eq. (3.82),
they have an ionization loss of about 50 MeV in
X for the counter used by the Berkeley group. If

40. B. Cork, G. Lambertson, O. Piccioni, and W.
     Wenzel, Phys. Rev. 104, 1193 (1956).

they annihilate, the energy loss is usually much
larger than 50 MeV. If the antineutron from
a charge exchange reaction annihilates in the
Cerenkov counter C, it will not give a pulse in
the scintillation counter S_2, but will deposit
considerable energy in C. The "signature" for an
antineutron is therefore: S_1, S (<100 MeV),
$\bar{S}_2$, C (>200 MeV). The bar over S_2 indicates that
an event is only counted if S_2 is in anticoinci-
dence with the other counters, i.e., if S_2 does
not record.

G. Antibosons

In Dirac's hole theory it is hard to see
how the antiparticle concept could be valid for
bosons: Bosons do not obey the Pauli principle,
and the negative energy states can never be filled.
Bosons would decay rapidly if negative energy
states did not exist. In the Stueckelberg-Feynman
picture, no such difficulty exists. Bosons can
travel backwards in time just as well as fermions.
The question then is: Do antibosons exist? The
answer is "yes." We will encounter many examples
in later chapters and mention just one here:
The positive pion is the antiparticle of the
negative pion. Neutral bosons may, but need not,
be identical to their antibosons. The neutral
pion is its own antiparticle, but the neutral
kaon and deuterium are different from their
antiparticles. The reason for the difference
is the existence of generalized charges. We will

discuss these charges in Chapter 4.

3.5 THE DESCRIPTION OF SCALAR BOSONS

In the present section, we outline how free spinless massive bosons can be described. As prototype, we use the pion, but occasionally mention kaons.

A. The Pion

Charged pions were discovered in cosmic rays in 1947 (see Section 1.2); in 1948, they were produced artificially.[41] They were found to decay with a lifetime of about 2×10^{-8} sec; in a photographic emulsion or a bubble chamber, a typical decay looks as shown in Fig. 3.10.

Fig. 3.10. Typical pion decay, as seen in an emulsion or a bubble chamber.

---

41. E. Gardner and C. M. G. Lattes, Science 107, 270 (1948).

If the energy of the visible decay products, muon
and electron, is measured for many decays, a
striking difference between the pion and the muon
decays becomes apparent, as shown in Fig. 3.11.
The electrons from the muon decay display a

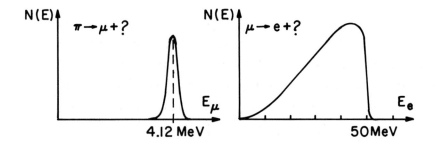

Fig. 3.11.   Energy spectrum of the decay product
             from pion and muon decay. The number
             of charged decay products with kinetic
             energy between E and E + dE is plotted
             versus E.

continuous spectrum, whereas the muons from the
pion decay are monoenergetic. The pion hence
must decay into two particles, the muon into at
least three. The unobserved particles must be
neutral; since the decay lifetimes are characteristic of weak interactions, we assume that they

are neutrinos. By now, this assumption is well
established. The spin of the pion then must be
integer; we will show later that it is zero. The
decays of the pion and the muon indeed occur
predominantly via

$$\pi \rightarrow \mu\nu \quad , \tag{3.83}$$

$$\mu \rightarrow e\bar{\nu}\nu \quad . \tag{3.84}$$

In 1950, the observation of high-energy
photons from proton-nucleon collisions was interpreted as being probably due to the decay of
neutral pions.[42] The existence of a neutral
pion was definitely established by the coincident
detection of the two gamma rays from the decay
$\pi^{0} \rightarrow 2\gamma$.[43]

We now have introduced three different pions,
π^{+}, π^{-}, and π^{0}. The mass of the neutral pion is
somewhat smaller than that of the charged ones;
nevertheless we assume that all three can be
classified by a common name. The reason for
this grouping and for the mass difference will
become clear later.

---

42. R. Bjorklund, W. E. Crandall, B. J. Moyer,
     and H. F. York, Phys. Rev. 77, 213 (1950).
43. J. Steinberger, W. K. H. Panofsky, and J.
     Stellet, Phys. Rev. 78, 802 (1950).

B.   The Schrödinger and the Klein-Gordon
     Equations

Since pions are spinless (scalar) particles,
they can be described in the nonrelativistic
approximation by the Schrödinger equation

$$i \frac{\partial \Phi}{\partial t} = H\Phi \quad . \tag{3.85}$$

Here, Φ is a one-component wave function and H
is the Hamilton operator obtained from the classical Hamiltonian $H(\vec{p},\vec{x})$ by the substitution
$\vec{p} \rightarrow -i\vec{\nabla}$. External electromagnetic fields described
by the potential $(A^0,\vec{A})$ are introduced by the
substitution (called principle of minimal electromagnetic interaction - Subsection 4.2.C),

$$p^\mu \rightarrow p^\mu - qA^\mu \tag{3.86}$$

or

$$\nabla^\mu \rightarrow \nabla^\mu + iqA^\mu \quad . \tag{3.87}$$

For a particle with charge q and mass m, H is
then given by

$$H = \frac{1}{2m}(-i\vec{\nabla} - q\vec{A})^2 + qA^0 \quad . \tag{3.88}$$

If the external fields and potentials do not
depend on time, the wave function Φ has the form

(3.75) and the energy eigenvalues E are given by
the time-independent Schrödinger equation (3.76).
The probability density

$$\rho_n = \Phi^*\Phi \qquad (3.89)$$

and the probability current density

$$\vec{j}_n = \frac{1}{2mi} (\Phi^*\vec{\nabla}\Phi - \Phi\vec{\nabla}\Phi^*) \qquad (3.90)$$

satisfy by Eq. (3.85) the continuity equation

$$\frac{\partial \rho_n}{\partial t} + \text{div } \vec{j}_n = 0 \qquad . \qquad (3.91)$$

Because the mass of the pion is only about
140 MeV, the nonrelativistic domain is severely
limited. To obtain a relativistic equation for
scalar particles, we start from the relativistic
energy-momentum connection (2.2) and use the
four-dimensional substitution $p^\mu \to i\nabla^\mu$ to get
the free-particle equation

$$(\nabla^\mu\nabla_\mu + m^2)\ \Phi = 0 \quad ,$$

or

$$(\Box^2 + m^2)\ \Phi = 0 \quad . \qquad (3.92)$$

This equation is called the Klein-Gordon equation;
it is no longer first order in time. External
electromagnetic fields are introduced again by

the substitution (3.86) or (3.87); the Klein-
Gordon equation for a particle in an external
electromagnetic field then reads

$$\{(\nabla^\mu + iqA^\mu)^2 + m^2\}\Phi = 0 \quad . \tag{3.93}$$

To interpret the solutions Φ of this equation, we require a probability density ρ and a
probability current $\vec{j}$ that satisfy the following
criteria: ρ should be real, $\int d^3x\, \rho$ should transform as a Lorentz scalar, and ρ and $\vec{j}$ should
satisfy the continuity equation (3.91). Moreover,
in the nonrelativistic limit, ρ and $\vec{j}$ should be
proportional to the corresponding relations
(3.89) and (3.90). The expressions

$$\rho = i\,(\Phi^*\frac{\partial\Phi}{\partial t} - \Phi\frac{\partial\Phi^*}{\partial t}) \tag{3.94}$$

and

$$\vec{j} = -\,i(\Phi^*\vec{\nabla}\Phi - \Phi\vec{\nabla}\Phi^*) \tag{3.95}$$

meet all these conditions; ρ and $\vec{j}$ form a four-
vector[44]

---

44. Usually, the current is written as

$$j^\mu = \frac{i}{2m}\,(\Phi^*\nabla^\mu\Phi - \Phi\nabla^\mu\Phi^*);\ \text{the nonrelativistic}$$

limit then agrees with the Schrödinger charge
density (3.89) and current (3.90). We prefer
the form (3.96) because it makes the transition to quantized fields easier.

$$j^\mu = i(\Phi^* \nabla^\mu \Phi - \Phi \nabla^\mu \Phi^*) \quad , \tag{3.96}$$

and the continuity equation (3.91) can be written as

$$\nabla^\mu j_\mu = 0 \quad . \tag{3.97}$$

When the Klein-Gordon equation was proposed in 1926, it was soon abandoned again because the probability density (3.94) is not positive definite and thus cannot be used for a single particle. In 1934, Pauli and Weisskopf[45] showed a way out of this difficulty: If the Klein-Gordon equation describes positive and negative particles together, then ρ can be taken to be the charge density and negative and positive values of this quantity are meaningful. The form Eq. (3.92) is not the most convenient one to demonstrate that the Klein-Gordon equation describes particles and antiparticles at the same time. However, it is possible to replace the second-order equation (3.92) by two first-order ones[46]; to do so, a two-component wave function

$$\Phi = \begin{pmatrix} \Phi_1 \\ \Phi_2 \end{pmatrix} \tag{3.98}$$

45. W. Pauli and V. F. Weisskopf, Helv. Phys. Acta 7, 709 (1934).
46. H. Feshbach and F. Villars, Rev. Mod. Phys. 30, 24 (1958).

is introduced. It is then straightforward to
find the two equations satisfied by Φ_1 and Φ_2
and to show that Φ_1 and Φ_2 describe positive
and negative particles, respectively.

Even though we have just pointed out that a
one-particle interpretation of the Klein-Gordon
equation is not satisfactory, it can be used to
compute the energy levels of mesonic atoms. A
mesonic atom is created if a negative pion or kaon
slows down in matter, reaches thermal velocity,
and is captured by an atom. Since the meson and
the electrons of the shell do not obey a mutual
Pauli principle, the electrons do not prevent the
meson from occupying energy levels. The excitation
energy of these levels can be calculated in a
straightforward way by inserting the Coulomb
potential ($A^0 = Ze/r$, $\vec{A} = 0$) into the Klein-Gordon
equation (3.93) and solving the eigenvalue problem.
Energy differences between levels can be measured
by observing the characteristic x rays emitted by
the meson when it cascades down from one level to
the next. Such x rays have been observed for
pions[47] and kaons[48]. Experiment and this simple

47.   M. B. Stearns, Progr. Nucl. Phys. 6, 108
      (1958); D. A. Jenkins and R. Kunselman, Phys.
      Rev. Letters 17, 1148 (1966); G. Backenstoss,
      Ann. Rev. Nucl. Sci. 20, 467 (1970).
48.   G. R. Burleson, D. Cohen, R. C. Lamb, D. N.
      Michael, and R. A. Schluter, Phys. Rev. Letters
      15, 70 (1965); C. E. Wiegand and D. A. Mack,
      Phys. Rev. Letters 18, 685 (1967); D. H.
      Wilkinson, Comments on Nuclear and Particle
      Physics 1, 112 (1967). C. E. Wiegand, Phys.
      Rev. Letters 22, 1235 (1969). G. Backenstoss
      et. al., Nucl. Phys. B73, 189 (1974).

theory agree well for the higher levels. Once the meson comes closer to the nucleus, new effects come into play. The nucleus is not a point charge and the meson interacts also hadronically with the nucleons in the nucleus. Pions and kaons can be scattered and destroyed by a hadronic interaction, e.g., $K^-p \rightarrow \Lambda^0 n$. These features are not included in the Coulomb potential and they give rise to a shift of the energy and an increase in the width of the levels.

C. Free Particle Solutions

The free-particle Klein-Gordon equation, Eq. (3.92), has solutions

$$\Phi = N \ e^{\pm ip \cdot x} \ , \ p_0^2 = E^2 = \vec{p}^2 + m^2 \quad , (3.99)$$

where N is a normalization constant. Two questions arise: What is the significance of the sign of the exponent, and what is the normalization constant N?

To answer the first question, we use the discussion in Subsection 3.4C: If we take the energy E to be always positive, then the solution with the minus sign describes particles, and the solution with the plus sign is appropriate for anti-particles.

To answer the second question and determine the normalization constant, we assume that one particle is present in the normalization volume V:

$$\int_V d^3x \; \rho(x) = 1 \quad . \tag{3.100}$$

With the probability density (3.94) and the free-particle solution (3.99), we get $2N^2EV = 1$; the normalized solution for one free particle per volume V reads

$$\Phi = \frac{1}{\sqrt{2EV}} \; e^{-ip \cdot x} \quad . \tag{3.101}$$

For the antiparticle solution, the sign in the exponent is changed.

D.   Quantization of the Free Neutral Spinless Field

We mentioned in Subsection 3.3A that we "quantize" wave functions a second time because field quantization provides a powerful tool to treat many-body problems. To quantize the solutions of the Klein-Gordon equation, we follow Section 3.3 and first define states $|n\rangle$, and annihilation and creation operators. Since we deal with scalar particles, we can omit the helicity subscript λ and specify states by the four vector $p = (E, \vec{p})$ alone.[*] Thus we get the commutation relations

---

[*]We remind the reader that, although the subscript p is used rather than $\vec{p}$, the energy E is not an independent variable.

$$[a_{op}, a^{\dagger}_{op'}] = \delta_{pp'}$$

$$[a_{op}, a_{op'}] = [a^{\dagger}_{op}, a^{\dagger}_{op'}] = 0 \quad, \tag{3.102}$$

where the subscript o denotes neutral particles.
The number operator N and the Hamiltonian H are
given by

$$N_p = a^{\dagger}_{op} a_{op} \quad, \tag{3.103}$$

$$H = \sum_{\vec{p}} a^{\dagger}_{op} a_{op} E \quad, \quad E = + \sqrt{\vec{p}^2 + m^2} \,. \tag{3.104}$$

The next step - use of the correspondence
principle - **turns out to be impossible.** Even
though pions and kaons obey Bose-Einstein statistics, the hadronic field that they mediate has
such a short range that no effects at macroscopic
distances can be observed. To find an expression
for the field operators, we use the analogy to
the electromagnetic field and expand the field
operator Φ_o in terms of annihilation and creation
operators,

$$\Phi_o(x) = \sum_{\vec{p}} \frac{1}{\sqrt{2E\,V}}$$

$$\times \left\{ a_{op} e^{-ip\cdot x} + a^{\dagger}_{op} e^{ip\cdot x} \right\}. \tag{3.105}$$

The matrix elements of this operator between single particle states are

$$<p|\Phi_o|0> = \frac{1}{\sqrt{2E\ V}}\ e^{ip\cdot x} \quad , \qquad (3.106)$$

$$<0|\Phi_o|p> = \frac{1}{\sqrt{2E\ V}}\ e^{-ip\cdot x} \quad . \qquad (3.107)$$

Comparing these matrix elements with the free-particle solution (3.101) we note that the matrix element for the destruction of a scalar boson gives the eigenfunction of the state of the particle that is destroyed; the matrix element for the creation of a scalar boson gives the complex conjugate of the eigenfunction of the state of the particle that is created.

How do we know that the operator (3.105) describes spinless neutral bosons, as claimed in the title of this subsection? The particles must be spinless because the operator Φ_o has only one component and hence cannot describe an orientation in space. To discuss the electric charge of the particles, we consider the analogy with the electromagnetic field describing neutral photons. There we know that the classical field $\vec{A}$ is real and that the field operator $\vec{A}$ correspondingly is Hermitian. In analogy we demand that Φ_o must also be Hermitian, $\Phi_o^\dagger = \Phi_o$. We see that Φ_o then describes neutral particles by considering the current, Eq. (3.96). In Eq. (3.96), Φ denotes a wave function. Now, however, Φ_o has become an operator and j^μ hence is also an operator which we write as[49]

$$j^\mu = i(\Phi^\dagger \nabla^\mu \Phi - \Phi \nabla^\mu \Phi^\dagger) \quad . \qquad (3.108)$$

For a Hermitian field, this operator vanishes; in particular the operator of the total charge,

$$Q = e \int d^3x \ j^0(x)$$

$$= ie \int d^3x \left(\Phi^\dagger \frac{\partial \Phi}{\partial t} - \Phi \frac{\partial \Phi^\dagger}{\partial t} \right) \qquad (3.109)$$

is zero. A Hermitian field describes neutral particles.

E.   Quantization of the Free Charged Spinless Field

We have seen in the previous subsection that a Hermitian field describes neutral particles and suspect that a non-Hermitian field is required for the description of charged particles. Before

---

49.   Actually, since we deal with one-component operators here, we could write Φ^* instead of $\Phi^\dagger$. (See Eq. (3.120).) However, we use the dagger for consistency.

confirming this suspicion, we treat the quantization of a non-Hermitian field. Such a field
can always be decomposed into two Hermitian
fields Φ_1 and Φ_2,

$$\Phi = \frac{1}{\sqrt{2}} (\Phi_1 - i\Phi_2) \quad ,$$

(3.110)

$$\Phi^\dagger = \frac{1}{\sqrt{2}} (\Phi_1 + i\Phi_2) \quad .$$

Since Φ_1 and Φ_2 are Hermitian, we know how to
expand them in terms of creation and annihilation
operators. Using Eq. (3.105), we write

$$\Phi_r = \sum_{\vec{p}} \frac{1}{\sqrt{2E\ V}} \left\{ a_{pr} e^{-ip\cdot x} + a_{pr}^\dagger e^{ip\cdot x} \right\} \quad (3.111)$$

with the commutation relations

$$\left. \begin{array}{c} [a_{pr}, a_{p's}^\dagger] = \delta_{p,p'} \delta_{r,s} \\[2mm] [a_{pr}, a_{p's}] = [a_{pr}^\dagger, a_{p's}^\dagger] = 0 \\[2mm] r,s = 1,2. \end{array} \right\} \quad (3.112)$$

If we introduce new operators a_p and $a_{\bar{p}}$ through
the definitions

$$a_p = \frac{1}{\sqrt{2}} (a_{p1} - ia_{p2}) \quad ,$$

(3.113)

$$a_{\bar{p}} = \frac{1}{\sqrt{2}} (a_{p1} + ia_{p2}) \quad ,$$

they satisfy by virtue of Eq. (3.112) the commutation relations

$$[a_p, a_p^{\dagger}] = [a_{\bar{p}}, a_{\bar{p}}^{\dagger}] = \delta_{p,p'}$$

(3.114)

all other $[,] = 0$.

With Eqs. (3.110), (3.111), and (3.113) we get the desired expansions

$$\Phi(x) = \sum_{\vec{p}} \frac{1}{\sqrt{2E\ V}}$$

$$\times \left\{ a_p e^{-ip\cdot x} + a_{\bar{p}}^{\dagger} e^{ip\cdot x} \right\} .$$

(3.115)

$$\Phi^{\dagger}(x) = \sum_{\vec{p}} \frac{1}{\sqrt{2E\ V}}$$

$$\times \left\{ a_{\bar{p}} e^{-ip\cdot x} + a_p^{\dagger} e^{ip\cdot x} \right\} .$$

We now turn to the interpretation of the operators a_p and $a_{\bar{p}}$. Since all a_p and $a_p^\dagger$ commute with all $a_{\bar{p}}$ and $a_{\bar{p}}^\dagger$, we can assume that the two types of operators refer to two different kinds of particles (+ and -). Since the operators satisfy the same commutation relations as the annihilation and creation operators of the Hermitian field, we interpret them as the annihilation and creation operators of the particle and anti-particles, respectively. As in Subsection D, we then introduce number operators

$$N_p = a_p^\dagger a_p \quad , \quad N_{\bar{p}} = a_{\bar{p}}^\dagger a_{\bar{p}} \quad . \tag{3.116}$$

These operators have the eigenvalues 0, 1, 2 ... and give the numbers of particles (respectively anti-particles) of momentum p. The Hamiltonian of a system of particles is given by

$$H = \sum_{\vec{p}} (N_p + N_{\bar{p}}) E \quad . \tag{3.117}$$

One essential property of the particles and the antiparticles becomes clear by introducing the fields (3.115) into the operator of the total electric charge (3.109). A short computation gives

$$Q = e \sum_{\vec{p}} (N_p - N_{\bar{p}}) \quad , \tag{3.118}$$

where we have used the commutation relations
(3.114). The charges of the two single-particle
states

$$|p> \equiv a_p^\dagger |0> \quad \text{and} \quad |\bar{p}> \equiv a_{\bar{p}}^\dagger |0>$$

then are +e and -e, since

$$Q|p> = e|p>$$

$$Q|\bar{p}> = -e|\bar{p}> \quad .$$

The operators $a_p^\dagger$ and $a_{\bar{p}}^\dagger$ create singly charged
particles of the same momentum, but with opposite
signs of the charge. We call $a_p^\dagger$ the creation
operator for particles and $a_{\bar{p}}^\dagger$ the creation operator for antiparticles. Similarly, a_p and $a_{\bar{p}}$
annihilate particles with the same momentum, but
with opposite charge; and we call a_p the annihilation operator for particles, $a_{\bar{p}}$ the annihilation
operator for antiparticles.

The formalism given here no longer contains
negative energy states; the Hamiltonian (3.117)
is positive definite. Moreover, it is symmetric
in the positive and negative charges and it is a
matter of definition which one we call particle
and which one antiparticle.

We note that the operators $a_p^\dagger$ and $a_{\bar{p}}^\dagger$ describe the creation of particles, respectively
antiparticles, in the number representation. To
describe the fields in momentum space, we require

the field operators Φ and $\Phi^\dagger$. The meaning of
these two operators is now also clear: The field
Φ can either describe the destruction of a particle or the creation of an antiparticle, and
similarly $\Phi^\dagger$ can describe the creation of a
particle or the destruction of an antiparticle.
Which of these processes takes place in a
particular situation depends on the states on
which the operators act. If, for instance, the
initial state is the vacuum and the final state
contains one antiparticle of momentum p, the
matrix elements of Φ and $\Phi^\dagger$ are

$$<\overline{p}|\Phi|0> = \frac{1}{\sqrt{2E\ V}}\ e^{ip\cdot x}$$

$$(3.119)$$

$$<\overline{p}|\Phi^\dagger|0> = 0\quad.$$

So far, we have treated neutral and charged
spinless bosons separately. Such a procedure is
not very satisfying if we assume that, for
instance, all pions are members of one family.
We will return to this point in Section 4.4 and
we will show there how to unify the description
by introducing isospin.

3.6   THE DESCRIPTION OF FREE SPIN $\frac{1}{2}$ FERMIONS

The particles that are most conspicuous in
atomic and nuclear physics are the proton, the
neutron, and the electron. These three particles

are fermions, possess spin $\frac{1}{2}$, and have antiparticles. Other spin $\frac{1}{2}$ fermions exist, for
instance the muon and the neutrinos. Although all
these particles are spin $\frac{1}{2}$ fermions, they differ
in their masses and in the interactions. In the
present section, this fact does not disturb us
since we only deal with free particles. Later,
we will have to take the various interactions into
account.

In the present section, we first outline the
nonrelativistic description of spin $\frac{1}{2}$ fermions,
then sketch the Dirac theory, and finally outline
the quantization of the Dirac field.

A. Pauli Spinors and Pauli Matrices

The Stern-Gerlach experiment shows that the
eigenvalues of the spin operator $\vec{s}$ of a spin $\frac{1}{2}$
particle can be $+\frac{1}{2}$ or $-\frac{1}{2}$ along any given direction[5]. Pauli attributed this splitting to a
"classically nondescribable two-valuedness,"[50]
and Goudsmit and Uhlenbeck introduced the idea
of a spinning electron[7]. In any case, the
spatial wave function splits into two components,
one belonging to spin "up," the other one to spin
"down." To describe this doubling, Pauli introduced a two-component wave function[51]

---

50. W. Pauli, Z. Physik 31, 373 (1925).
51. W. Pauli, Z. Physik 43, 601 (1927).

$$\psi(x,s) = \begin{pmatrix} \psi_1(x,s) \\ \psi_2(x,s) \end{pmatrix} \quad , \quad \psi^* = \begin{pmatrix} \psi_1^* \\ \psi_2^* \end{pmatrix} \quad ,$$

$$\psi^\dagger = (\psi_1^*, \psi_2^*) \quad . \tag{3.120}$$

To represent the spin, he used an operator $\vec{s}$ satisfying the commutation relations (3.12) characteristic for angular momentum operators, and obeying the eigenvalue equations (3.13) with $s = \frac{1}{2}$:

$$\vec{s}^2 | \frac{1}{2} , m> = \left(\frac{1}{2}\right) \left(\frac{3}{2}\right) | \frac{1}{2} , m> \quad ,$$

$$s_z | \frac{1}{2} , m> = m | \frac{1}{2} , m> \quad , \quad m = \pm\frac{1}{2} \quad . \tag{3.121}$$

An explicit representation of the spin operator $\vec{s}$ for spin $\frac{1}{2}$ is obtained conveniently by introducing the Pauli spin operator

$$\vec{\sigma} = 2\vec{s} \quad , \tag{3.122}$$

written in matrix form and in components in the standard representation as

$$\sigma_x = \begin{pmatrix} 0 & 1 \\ 1 & 0 \end{pmatrix} \quad , \quad \sigma_y = \begin{pmatrix} 0 & -i \\ i & 0 \end{pmatrix} \quad ,$$

$$\sigma_z = \begin{pmatrix} 1 & 0 \\ 0 & -1 \end{pmatrix} \quad . \tag{3.123}$$

The Pauli matrices (3.123) satisfy the relations

$$\sigma_x \sigma_y = -\sigma_y \sigma_x = i\sigma_z \quad , \quad \text{cycl.}$$

$$\tag{3.124}$$

$$\sigma_x^2 = \sigma_y^2 = \sigma_z^2 = I \quad ,$$

and the identity

$$(\vec{\sigma} \cdot \vec{A})(\vec{\sigma} \cdot \vec{B}) = \vec{A} \cdot \vec{B} + i\vec{\sigma} \cdot \vec{A} \times \vec{B} \quad , \tag{3.125}$$

where $\vec{A}$ and $\vec{B}$ are any two vectors that commute with $\vec{\sigma}$, but not necessarily with each other.

The component of $\vec{s}$ along a direction given by the unit vector $\hat{n}$ is $\vec{s} \cdot \hat{n}$; the eigenvalue equation for this component reads

$$(\vec{s} \cdot \hat{n}) \, \psi = m_s \psi \quad ,$$

or in terms of $\vec{\sigma}$

$$(\vec{\sigma} \cdot n) \, \psi = 2m_s \psi \quad . \tag{3.126}$$

Applying the operator $(\vec{\sigma}\cdot\hat{n})$ again and using Eq. (3.125) gives $(2m_s)^2 = 1$ or $m_s = \pm\frac{1}{2}$. The Pauli formalism hence yields the two possible states, "up" and "down," along any direction $\hat{n}$, and the operator $\frac{1}{2}\vec{\sigma}$ is an acceptable spin operator for nonrelativistic spin $\frac{1}{2}$ particles.

For free particles, the space and the spin dependence of the wave function $\psi(x,s)$ can be separated,

$$\psi(x,s) = N \, e^{-ip\cdot x} \, \phi(s) \quad , \tag{3.127}$$

where $\phi(s)$ is now a two-component Pauli spinor that no longer depends on x. To determine the constant N, we normalize to one-particle-per-volume V,

$$\int_V \psi^\dagger \psi \, d^3x = |N|^2 \phi^\dagger \phi \int_V d^3x = 1 \quad . \tag{3.128}$$

If we normalize the Pauli spinors according to

$$\phi^\dagger \phi = 1 \tag{3.129}$$

we get for the nonrelativistic free particle solution

$$\psi(x,s) = \frac{1}{\sqrt{V}} \, e^{-ip\cdot x} \, \phi(s) \quad . \tag{3.130}$$

B.   The Polarization of Nonrelativistic Spin $\frac{1}{2}$ Particles

The spinor ϕ contains the information about the direction of the spin. The direction is defined as that given by the polarization vector, Eq. (3.14). For spin $\frac{1}{2}$ particles, the polarization vector $\vec{P}$ is the expectation value of the Pauli spin vector $\vec{\sigma}$ in the normalized state ϕ:

$$\vec{P} = \phi^\dagger \vec{\sigma} \phi \equiv <\vec{\sigma}> \qquad . \tag{3.131}$$

As an example we find for the state $\phi = \begin{pmatrix} 1 \\ 0 \end{pmatrix}$:

$$P_x = (1,0) \begin{pmatrix} 0 & 1 \\ 1 & 0 \end{pmatrix} \begin{pmatrix} 1 \\ 0 \end{pmatrix} = (1,0) \begin{pmatrix} 0 \\ 1 \end{pmatrix} = 0;$$

$$P_y = 0; \; P_z = 1 \qquad .$$

The magnitude of $\vec{P}$ is called the <u>degree of polarization</u>; it is given by

$$|\vec{P}| = \sqrt{<\sigma_x>^2 + <\sigma_y>^2 + <\sigma_z>^2} \qquad . \tag{3.132}$$

For a normalized spinor $\begin{pmatrix} a \\ b \end{pmatrix}$ we have $a^* a + b^* b = 1$ and find by a straightforward computation $|\vec{P}| = 1$. To interpret $\vec{P}$ physically we note that if we have found the polarization vector $\vec{P}$ for the state

ϕ, the equation

$$\vec{\sigma} \cdot \vec{P} \, \phi = +\phi \tag{3.133}$$

holds. Comparison of Eq. (3.133) with Eq. (3.126) justifies the assertion that the electron spin points along $\vec{P}$.

The name "polarization <u>vector</u>" is somewhat misleading. Using the transformation properties of the spinor ϕ, it can be shown that $\vec{P}$ transforms like a pseudo- or axial-vector and not like a polar vector. This property will be important in the discussion of parity nonconservation.

For a moving particle, we can refer the polarization to the direction of the momentum. We can in particular set $\hat{n}$ in Eq. (3.126) equal to $\hat{p}$, the unit vector in the direction of the momentum. The operator $\vec{\sigma} \cdot \hat{p}$ is called <u>helicity</u> operator; its eigenvalues are again ± 1. We call the state with eigenvalue $+1$ right-handed and denote it by a superscript R; we call the state with eigenvalue -1 left-handed and denote it with a superscript L:

$$\vec{\sigma} \cdot \hat{p} \, \phi^R = + \, \phi^R \quad ,$$
$$\vec{\sigma} \cdot \hat{p} \, \phi^L = - \, \phi^L \quad . \tag{3.134}$$

In order to find the explicit eigenfunctions of the helicity operator, we use the representation (3.123) of the Pauli matrices and denote the components of $\hat{p}$ by α, β, and γ. We then have

$$\vec{\sigma} \cdot \hat{p} = \begin{pmatrix} \gamma & \alpha - i\beta \\ \alpha + i\beta & -\gamma \end{pmatrix} \qquad (3.135)$$

and find the normalized eigenfunctions from Eq. (3.134) to be

$$\phi^R(\hat{p}) = \frac{1}{\sqrt{2(\gamma + 1)}} \begin{pmatrix} \gamma + 1 \\ \alpha + i\beta \end{pmatrix} \, ,$$

$$\phi^L(\hat{p}) = \frac{1}{\sqrt{2(\gamma + 1)}} \begin{pmatrix} -\alpha + i\beta \\ \gamma + 1 \end{pmatrix} \, . \qquad (3.136)$$

These eigenfunctions form a complete orthonormal set:

$$\phi^{R\dagger} \phi^R = \phi^{L\dagger} \phi^L = 1 \quad ,$$

$$\phi^{R\dagger} \phi^L = \phi^{L\dagger} \phi^R = 0 \quad . \qquad (3.137)$$

In the special case where the particle moves along the z axis, the Eqs. (3.136) go over into

$$\phi^R(\gamma) = \begin{pmatrix} 1 \\ 0 \end{pmatrix} \, , \quad \phi^L(\gamma) = \begin{pmatrix} 0 \\ 1 \end{pmatrix} \, .$$

A favorite examination problem is to ask for
the spinor describing an unpolarized nonrelativistic spin $\frac{1}{2}$ beam. Most attempts to solve the
problem lead to disaster: The answer is usually
of the form

$$\phi = \begin{pmatrix} a \\ b \end{pmatrix} \quad , \quad a^*a + b^*b = 1 \quad .$$

We have already said that the degree of polarization of such a spinor is one, the spinor ϕ hence
describes a fully polarized beam, regardless of
the choice of a and b. The components a and b
only determine the direction of $\vec{P}$, but not its
magnitude. The result is due to the fact that a
wavefunction describes a <u>pure</u> state: the spinor
ϕ describes a <u>coherent</u> superposition of the two
states with spin up and spin down. A partially
polarized or an unpolarized beam must be described as an <u>incoherent</u> superposition of two
spin states or a so-called <u>mixed</u> state;[52] the
polarization is found by averaging over the
polarizations of the contributing pure states:[53]

$$\vec{P} = \overline{\langle \vec{\sigma} \rangle} \quad . \tag{3.138}$$

52. U. Fano, Rev. Mod. Phys. <u>29</u>, 74 (1957).
53. We denote the expectation value for a pure
     state by < > and the ensemble average by a
     bar.

This prescription is straightforward, but inconvenient. An elegant and concise way to handle incoherent superpositions is by means of the density matrix. Such a description is very general and applies to particles of any spin.

To introduce the density matrix, we first assume that the system under consideration is in a pure state $|\alpha>$. The expectation value of any operator σ in the state $|\alpha>$ is then given by $<\alpha|\sigma|\alpha>$. If the system is in a mixed state, it cannot be described by a single state vector $|\alpha>$, but must be given as an incoherent superposition of pure states $|\alpha>$ with weights (probabilities) g_α. The expectation values of σ for the ensemble is then obtained by averaging over the expectation values for the contributing pure states, taking into account the proper weights:

$$\overline{<\sigma>} = \sum_\alpha{}' g_\alpha <\alpha|\sigma|\alpha> \quad , \tag{3.139}$$

with

$$g \text{ real}, \sum_\alpha{}' g_\alpha = 1 \quad , \quad 1 \geq g_\alpha \geq 0 \quad . \tag{3.140}$$

The prescription (3.139) is straightforward, but it often involves tedious computations. To avoid the explicit averaging, we introduce a conveniently chosen complete set of eigenfunctions $|m>$. The expectation value for a pure state $|\alpha>$ can

then can be written with the help of the expansion
(3.28) as

$$\langle\alpha|\hat{\sigma}|\alpha\rangle = \sum_{m,m'} \langle\alpha|m'\rangle\langle m'|\hat{\sigma}|m\rangle\langle m|\alpha\rangle \qquad (3.141)$$

and the expectation value for the ensemble becomes

$$\overline{\langle\hat{\sigma}\rangle} = \sum_{\alpha,m,m'} \langle m'|\hat{\sigma}|m\rangle\langle m|\alpha\rangle g_\alpha\langle\alpha|m'\rangle \; . \qquad (3.142)$$

The factor $\langle m'|\hat{\sigma}|m\rangle$ in Eq. (3.142) depends on the
operator $\hat{\sigma}$, but not on the states $|\alpha\rangle$, whereas the
other factors are characteristic of the states
under consideration. By defining a density operator ρ,

$$\rho = \sum_{\alpha} |\alpha\rangle g_\alpha\langle\alpha| \qquad (3.143)$$

with matrix elements

$$\langle m|\rho|m'\rangle = \sum_{\alpha} \langle m|\alpha\rangle g_\alpha\langle\alpha|m'\rangle \; , \qquad (3.144)$$

we make the separation more obvious. With the
closure relation (3.30) the ensemble expectation
value becomes

$$\overline{<\sigma>} = \sum_{m,m'} <m'|\sigma|m><m|\rho|m'>$$

$$= \sum_{m'} <m'|\sigma\rho|m'> \qquad\qquad (3.145)$$

or

$$\overline{<\sigma>} = \text{Tr}(\sigma\rho) \quad . \qquad\qquad (3.146)$$

Tr denotes the trace, or sum of diagonal elements, of the matrix $\sigma\rho$ or $\rho\sigma$. The order of ρ and σ is immaterial,

$$\text{Tr}(\rho\sigma) = \text{Tr}(\sigma\rho) \quad ; \qquad\qquad (3.147)$$

the trace is invariant under a unitary transformation,

$$\text{Tr}(\rho\sigma) = \text{Tr}(S^{-1}\rho\sigma S) \quad ,$$

$$S^\dagger S = SS^\dagger = 1 \quad ; \qquad\qquad (3.148)$$

the trace of the density operator is unity,

$$\text{Tr}\rho = 1 \quad ; \qquad\qquad (3.149)$$

and the density matrix is Hermitian,

$$\langle m|\rho|m'\rangle^* = \langle m'|\rho|m\rangle \quad . \tag{3.150}$$

The probability $W(m)$ of finding the ensemble in the

state $|m\rangle$ is given by $\sum_\alpha g_\alpha \langle\alpha|m\rangle\langle\alpha|m\rangle^*$, which is

just the diagonal element of the density matrix:

$$W(m) = \langle m|\rho|m\rangle \quad . \tag{3.151}$$

For a pure state, the density matrix is hence
diagonal, with a single nonzero element.

As a simple application of the density matrix
formalism we describe nonrelativistic spin $\frac{1}{2}$ particles. As a first example, we consider the
incoherent mixture of two beams of equal intensity

$(g_1 = g_2 = \frac{1}{2})$, one with spins up, $|1\rangle = \begin{pmatrix} 1 \\ 0 \end{pmatrix}$,

and one with spins down, $|2\rangle = \begin{pmatrix} 0 \\ 1 \end{pmatrix}$. The density

operator according to Eq. (3.143) is

$$\rho = \frac{1}{2}\begin{pmatrix} 1 \\ 0 \end{pmatrix}(1,0) + \frac{1}{2}\begin{pmatrix} 0 \\ 1 \end{pmatrix}(0,1) = \frac{1}{2}\begin{pmatrix} 1 & 0 \\ 0 & 1 \end{pmatrix}$$

or

$$\rho = \frac{1}{2} I \quad .$$

To calculate the polarization we need the traces
of the Pauli matrices. With Eqs. (3.124) and (3.147)
we get

$$\text{Tr } \sigma_i = 0 \quad ,$$

$$\text{Tr}(\sigma_i \sigma_k) = 2\delta_{ik} \quad , \tag{3.152}$$

$$i,k = x,y,z \quad ,$$

and hence

$$\vec{P} = \text{Tr}(\rho \, \vec{\sigma}) = \frac{1}{2} \text{Tr}(\vec{\sigma}) = 0 \quad .$$

The <u>incoherent</u> mixture of the two states

$\begin{pmatrix} 1 \\ 0 \end{pmatrix}$ and $\begin{pmatrix} 0 \\ 1 \end{pmatrix}$ indeed represents an unpolarized
beam.

 To find the density operator for a beam with
polarization $\vec{P}$, with $0 \leq |\vec{P}| \leq 1$, we remark that
only four linearly independent two-by-two matrices
exist. The three Pauli matrices and I are linearly
independent and hence form a complete set; we
therefore expand $\rho = aI + b_1\sigma_x + b_2\sigma_y + b_3\sigma_z$. The
condition (3.149) with Eq. (3.152) yields $a = \frac{1}{2}$.
The condition $P_i = \text{Tr}(\rho\sigma_i)$ and Eq. (3.152) give
$b_i = \frac{1}{2} P_i$. The density operator hence becomes

$$\rho = \frac{1}{2} (I + \vec{\sigma}\cdot\vec{P}) \quad . \tag{3.153}$$

Actually, we could have found this form more quickly
by noting that the density matrix is a scalar; the
only linear combination of I and the Pauli matrices
that is a scalar is of the form

$$\rho = aI + \vec{b} \cdot \vec{\sigma} \quad .$$

The conditions $\mathrm{Tr}\rho = 1$, $\mathrm{Tr}(\rho\vec{\sigma}) = \vec{P}$ then yield
immediately Eq. (3.153). The density matrix for
particles of arbitrary spin S can be constructed
in a similar way; it is specified by $(2S + 1)^2 - 1$
real parameters.

C.   The Dirac Equation

In 1928, nonrelativistic quantum mechanics
was fully developed, but the special theory of
relativity had no place in it and some phenomena,
particularly the electron spin, could not be
understood within its framework. Then, in one
miraculous step, Dirac wedded quantum mechanics
and the special theory of relativity and in the
process solved a number of outstanding problems.[54]
We justify the Dirac equation here in a superficial
way.

We first consider free electrons. We know
from experiment that a free electron can be characterized by its mass, its four-momentum and its

54.   P. A. M. Dirac, Proc. Roy. Soc. (London) A117,
        610 (1928); A118, 351 (1928).

spin. What, then, is the simplest equation that
can describe free electrons? In analogy to the
successful Schrödinger equation (3.85), we want
this equation to be of the Hamiltonian form

$$i \frac{\partial \psi}{\partial t} = H\psi \qquad . \qquad\qquad (3.154)$$

Since the equation should describe relativistic
electrons, it should be covariant, and only
first order spatial derivatives can occur in H.
To be manifestly covariant, the equation should
contain the time and the spatial derivatives in
the form of the four-vector ∇^μ. If we now try to
write a covariant equation with ∇^μ (or p^μ) and the
mass m, we fail because we cannot form a scalar
with one four-vector alone. We need a second
four-vector and since we have no obvious candidate,
we invent one and call it γ, with $\gamma^\mu = (\gamma^0, \vec{\gamma})$. We
will find the physical meaning of this four-vector
below. With m, p, and γ, the simplest equation is

$$(\gamma^\mu p_\mu)\psi = m\psi \qquad ,$$

or with the substitution (2.16)

$$i\gamma^\mu \frac{\partial \psi}{\partial x^\mu} - m\psi = 0 \quad \text{or} \quad (i\gamma^\mu \nabla_\mu - m)\psi = 0 \ . (3.155)$$

This equation can only be correct if the wave
function ψ has at least four components. This
fact was first discovered by Dirac and it was
a surprising result then. Today, with hindsight

and with more experimental information we can
understand it physically: We know that electrons
possess antiparticles, the positrons, and that
each electron and each positron has two spin
states. We therefore require two components to
describe particles and antiparticles, just as in
Eq. (3.98) in the case of relativistic scalar
particles. Each of these components is again a
two-component spinor. The total spinor describing
relativistic spin $\frac{1}{2}$ particles is hence of the
form

$$\psi = \begin{pmatrix} \Phi_1 \\ \Phi_2 \end{pmatrix},$$

$$\psi^* = \begin{pmatrix} \Phi_1 \\ \Phi_2 \end{pmatrix}^*, \qquad (3.156)$$

$$\psi^\dagger = (\Phi_1^\dagger, \Phi_2^\dagger),$$

where Φ_1 and Φ_2 are two-component Pauli spinors.
    The simplicity of the Dirac equation as
written in Eq. (3.155) is misleading. First of
all, as we have just stated, ψ is a four-component
spinor. Secondly, each of the four components of
γ^μ can still be an operator in spin space, acting
on ψ just as the two-by-two Pauli spin matrices
act on two-component Pauli spinors. Each component of γ^μ can thus be a four-by-four matrix.

In order to find out more about the four-vector operator γ^μ, we note that the Klein-Gordon equation (3.92) was obtained from the general equation (2.2) with the substitution (2.16); it should hence be satisfied by each component of the wave function of any relativistic particle, not just scalar ones. Each component of the Dirac spinor ψ hence should also be a solution of the Klein-Gordon equation. This condition severely restricts the choice of the vector γ, as we will see now. In order to arrive at the Klein-Gordon equation, we must make first-order terms in the Dirac equation disappear. To do this, we multiply Eq. (3.155) by the operator $(i\gamma^\nu \nabla_\nu + m)$ and get

$$\left[\gamma^\nu \gamma^\mu \nabla_\nu \nabla_\mu + im(\gamma^\nu \nabla_\nu - \gamma^\mu \nabla_\mu) + m^2 \right] \psi = 0 \quad .$$

Since the equation must be symmetric in the indices μ and ν, we write a second equation with $\mu \overset{\leftarrow}{\to} \nu$; adding the two then yields

$$\left[(\gamma^\mu \gamma^\nu + \gamma^\nu \gamma^\mu) \nabla_\nu \nabla_\mu + 2m^2 \right] \psi = 0 \quad .$$

We obtain the Klein-Gordon equation if the operators γ satisfy the relation

$$(\gamma^\mu \gamma^\nu + \gamma^\nu \gamma^\mu) = 2g^{\mu\nu} I \quad , \qquad (3.157)$$

where I is the unit operator. The left-hand side of Eq. (3.157) is called the anticommutator of

γ^μ and γ^ν; we use the notation

$$(ab + ba) = \{a,b\} \quad . \tag{3.158}$$

With Eq. (3.157), we can now cast the Dirac equation (3.155) into the Hamiltonian form (3.154); the Hamiltonian H is then given by

$$H = -i\ \gamma^0\vec{\gamma}\cdot\vec{\nabla} + \gamma^0 m = \gamma^0\vec{\gamma}\cdot\vec{p} + \gamma^0 m \quad . \tag{3.159}$$

To insure the Hermiticity of the Hamiltonian, we demand (with $\vec{p}^{\dagger} = \vec{p}$)

γ^0, $i\vec{\gamma}$ Hermitian, or

$$(\gamma^0)^{\dagger} = \gamma^0, \ \vec{\gamma}^{\dagger} = -\vec{\gamma} \quad . \tag{3.160}$$

The Hermitian properties can be summarized by the expression

$$\gamma^{\mu\dagger} = \gamma^0\gamma^\mu\gamma^0 \quad . \tag{3.161}$$

The Hermitian conjugate of the Dirac equation (3.155) is

$$i\ \frac{\partial\psi^{\dagger}}{\partial x^\mu}\ \gamma^{\mu\dagger} + m\psi^{\dagger} = 0 \quad .$$

Because this form involves the matrices $\gamma^{\mu\dagger}$, it is inconvenient to use. It is customary to introduce the adjoint spinor $\bar{\psi}$ by

$$\overline{\psi} = \psi^{\dagger}\gamma^0 \quad . \tag{3.162}$$

With Eq. (3.161), the Dirac equation for $\overline{\psi}$ becomes[55]

$$i\frac{\partial\overline{\psi}}{\partial x^{\mu}} \gamma^{\mu} + m\overline{\psi} = 0 \quad . \tag{3.163}$$

Dirac equations can be written even more succinctly by introducing the Feynman slash notation.[56]   If A is a four-vector, A-slash is defined by

$$\rlap{/}{A} = \gamma^{\mu}A_{\mu} \quad . \tag{3.164}$$

In particular we get (with Table 2.3)

$$\rlap{/}{p} \equiv \gamma^{\mu}p_{\mu} \equiv \gamma^0 E - \vec{\gamma}\cdot\vec{p} \quad , \tag{3.165}$$

$$\rlap{/}{\nabla} \equiv \gamma^{\mu}\nabla_{\mu} \equiv \gamma^0\frac{\partial}{\partial t} + \vec{\gamma}\cdot\vec{\nabla} \quad . \tag{3.166}$$

---

55.   The Hermitian conjugate of the Dirac equation in the form $\gamma^{\mu}p_{\mu}\psi = m\psi$ is, with $p_{\mu}^{\dagger} = p_{\mu}$, given by $\psi^{\dagger}\gamma^{\mu\dagger}p_{\mu} = m\psi^{\dagger}$. With Eqs. (3.161) and (3.162), the equation for $\overline{\psi}$ is $\overline{\psi}\gamma^{\mu}p_{\mu} - m\overline{\psi} = 0$. At first sight, this expression appears to differ by a minus sign from Eq. (3.163). However, the operator p_{μ} now stands to the right of $\overline{\psi}$. In the coordinate representation, a partial integration is needed to bring the operator p_{μ} to the other side and this partial integration supplies the missing minus sign.

56.   R. P. Feynman, Phys. Rev. 76, 749 (1949).

With the slash notation, Eqs. (3.155) and (3.163)
become

$$(i\slashed{\partial} - m)\psi = 0 \quad , \tag{3.167}$$

$$\bar{\psi}(i\overleftarrow{\slashed{\partial}} + m) = 0 \quad . \tag{3.168}$$

The second equation here is ambiguous. The operator acts to the left and its action is defined by
Eq. (3.163). In case of doubt it is safest to
return to the explicit expression (3.163). Defining a probability current density j^μ by

$$j^\mu = \bar{\psi} \, \gamma^\mu \, \psi \quad , \tag{3.169}$$

we find with the Dirac equations that it satisfies
the continuity equation

$$\nabla_\mu j^\mu = 0 \quad . \tag{3.170}$$

Here the physical meaning of the four-vector γ
becomes apparent: if sandwiched between spinors,
it gives an expression that can be interpreted as
probability current. The charge density

$$j^0 = \bar{\psi} \, \gamma^0 \, \psi = \psi^\dagger \psi = \sum_\alpha |\psi_\alpha|^2 \quad ,$$

where the sum extends over the four components, is
positive definite. It follows from the continuity

equation (3.169) that the total charge

$$Q = -e \int d^3x \ j^0(x)$$

is a constant of the motion, if the current $\vec{j}(x)$
properly vanishes at infinity.   The minus sign in
Q is required because the electron is treated as
a particle and not as an antiparticle.   The constant charge and the positive definite probability
density make a one-particle interpretation of the
Dirac equation possible.   However, the price paid
for such an interpretation is the unavoidable
appearance of negative energy states.   It is more
satisfactory to assume that the Dirac equation
describes many particles, just as the Klein-Gordon
equation does.   The difficulties with the negative
energies then disappear.   We outline the many-
particle approach in Subsection F.

D.   Dirac Operators

We have introduced four contravariant Dirac
operators by the definition (3.157).   The corresponding covariant operators are given by

$$\gamma_\mu = g_{\mu\nu}\gamma^\nu \quad . \tag{3.171}$$

A fifth independent Dirac operator is defined by

$$\gamma^5 \equiv i \ \gamma^0\gamma^1\gamma^2\gamma^3 \quad ; \tag{3.172}$$

it is Hermitian and anticommutes with the other
Dirac operators:

$$\left\{\gamma^5, \gamma^\mu\right\} = 0 \quad .$$

(3.173)

Note, however, that γ^5 is <u>not</u> a component of a
four-vector and the covariant form γ_5 is <u>not</u> given
by Eq. (3.171), but is <u>defined</u> by

$$\gamma_5 \equiv \gamma^5 = - i \; \gamma_0 \gamma_1 \gamma_2 \gamma_3 = \gamma^{5\dagger} \quad .$$

(3.172')

Most calculations involving the Dirac equation
can be performed by using the algebraic properties
of the Dirac operators. Occasionally, however,
it is convenient to represent each operator by a
matrix. By trial, it is easy to see that the
condition (3.157) cannot be satisfied by two-by-two
or three-by-three matrices, but that at least
four-by-four matrices are required. We thus see
that our earlier statement - that the Dirac
spinors must have four components -- is verified.
The choice of the Dirac matrices is not unique;
we use mostly the <u>standard representation</u> (Pauli
representation) in which γ^0 is taken to be diagonal.
We write this representation with the help of the
Pauli matrices (3.123) and the two-by-two unit
Matrix I:

$$\gamma^0 = \begin{pmatrix} I & 0 \\ 0 & -I \end{pmatrix} \,, \quad \vec{\gamma} = \begin{pmatrix} 0 & \vec{\sigma} \\ -\vec{\sigma} & 0 \end{pmatrix} \,,$$

$$\gamma^5 = \begin{pmatrix} 0 & I \\ I & 0 \end{pmatrix} \,. \tag{3.174}$$

These matrices satisfy the conditions (3.157), (3.160), and (3.173). Sometimes, all sixteen elements of the Dirac matrices are displayed explicitly. However, computations are more transparent if the Dirac matrices are written in terms of Pauli matrices, as in Eq. (3.174).

So far, we have introduced six independent Dirac operators, I, γ^0, γ^1, γ^2, γ^3, γ^5, which can be represented by four-by-four matrices. By multiplying two or more of these, we can create new operators or matrices that are independent of the six original ones, i.e., cannot be written as linear combinations of them. Since we are dealing with four-by-four matrices, we expect sixteen independent Dirac operators to exist. This expectation is indeed correct[57] and the sixteen operators can be arranged in five groups, as shown in Table 3.1. The operator $\sigma^{\mu\nu}$ is given by

57. W. Pauli, Ann. Inst. Henri Poincare 6, 109 (1936). R. H. Good, Jr., Rev. Mod. Phys. 27, 187 (1955).

$$\sigma^{\mu\nu} = \frac{i}{2}(\gamma^{\mu}\gamma^{\nu}-\gamma^{\nu}\gamma^{\mu}) = \frac{i}{2}\left[\gamma^{\mu},\gamma^{\nu}\right] \quad . \qquad (3.175)$$

Table 3.1   Transformation properties of the Dirac
bilinear covariants.

i	Symbol	Operator σ^i	Transformation property of $\bar{\psi}\sigma^i\psi$.
1	S	I	Scalar
2	V	γ^{μ}	Vector
3	T	$\sigma^{\mu\nu}$	Tensor
4	A	$\gamma^{\mu}\gamma^5$	Axial vector
5	P	γ^5	Pseudoscalar

The groups in Table 3.1 are characterized by the
transformation properties of the corresponding
"Dirac bilinear covariant," $\bar{\psi}\sigma^i\psi$:   Under a coordinate transformation, $\bar{\psi} I \psi$, for instance, remains
unchanged and $\bar{\psi}\gamma^{\mu}\psi$ behaves like a four-vector.[57]
These transformation properties are essential in
the treatment of form factors and of weak interactions.

E. Free Particle Solutions

The free-particle solutions of the Dirac equation are important for many later applications. To solve the free-particle Dirac equation (3.167), we describe a free electron with momentum $\vec{p}$ and energy $E = p^0$ by the plane wave

$$\psi = \psi_o(p) \, e^{i(\vec{p} \cdot \vec{x} - Et)} = \psi_o(p) \, e^{-ip \cdot x}, \quad (3.176)$$

where ψ_o is a four-component Dirac spinor of the form (3.156). For a free particle, ψ_o depends only on p, but not on x; the operator ∇^μ hence acts only on the exponential. After inserting (3.176) into Eq. (3.167), we find the time-independent Dirac equation

$$(\not{p} - m)\psi_o(p) = 0 \quad . \quad (3.177)$$

It represents four simultaneous homogeneous linear equations and hence has four independent solutions. The condition that solutions exist is

$$\det(\not{p} - m) = 0 \quad .$$

With the standard representation (3.174) and with Eq. (3.164), the determinant becomes

$$\det(\not{p} - m) = \det(\gamma^0 E - \vec{\gamma} \cdot \vec{p} - m)$$

$$= \begin{vmatrix} E-m & , & -\vec{\sigma} \cdot \vec{p} \\ & & \\ \vec{\sigma} \cdot \vec{p} & , & -(E+m) \end{vmatrix}$$

$$= -E^2 + m^2 + \vec{\sigma} \cdot \vec{p} \; \vec{\sigma} \cdot \vec{p} \quad ,$$

or with Eq. (3.125)

$$\det(\not{p} - m) = -E^2 + \vec{p}^2 + m^2 = 0 \quad .$$

The plane wave is a solution of the Dirac equation if its energy and momentum satisfy the relation (2.2). For a given rest mass m and momentum $\vec{p}$, the energy can have the two values

$$E^{\pm} = \pm \sqrt{\vec{p}^2 + m^2} \quad .$$

Here we encounter in more explicit form the problem that we have stated in connection with Eq. (3.71) in Section 3.4, namely the occurrence of negative energy states. We have discussed the interpretation of such states in Section 3.4 and we know that they describe antiparticles with positive energy. The general solution of the Dirac equation is a sum of four terms, two corresponding to particles and two to antiparticles. Taking the energy

to be always positive from now on, we write the
general solution as a superposition of four independent solutions

$$\psi = (u_R(p) + u_L(p))\, e^{-ip\cdot x}$$
$$+ (v_R(p) + v_L(p))\, e^{ip\cdot x} \quad . \tag{3.178}$$

The arguments given in Subsection 3.4C show that
the first two terms describe particles, the
second two antiparticles. We will discuss the
differences between u_R and u_L, and v_R and v_L below
and write in the following only u and v. Inserting
Eq. (3.178) into the Dirac equation then gives the
following time-independent equations for the
particle spinors u and the antiparticle spinors v:

$$(\not{p} - m)\, u(p) = 0 \quad , \quad \bar{u}(p)(\not{p} - m) = 0 \quad , \tag{3.179}$$

$$(\not{p} + m)\, v(p) = 0 \quad , \quad \bar{v}(p)(\not{p} + m) = 0 \quad . \tag{3.180}$$

For particles at rest, the Dirac equation is
simply:

$$(\gamma^0 - 1)\, u(\vec{p} = 0) = 0$$
$$(\gamma^0 + 1)\, v(\vec{p} = 0) = 0 \quad . \tag{3.181}$$

To find explicit expressions for the spinors u and
v, we must select a specific representation. With

the standard representation (3.174), and with

spinors of the form $\begin{pmatrix} \phi \\ \chi \end{pmatrix}$, where ϕ and χ are two-component Pauli spinors, we find for particles at rest

$$u(\vec{p} = 0) = N \begin{pmatrix} \phi \\ 0 \end{pmatrix}, \quad v(\vec{p} = 0) = N \begin{pmatrix} 0 \\ \chi \end{pmatrix}.$$

N is a normalization factor which we will determine later. In the standard representation and for particles at rest, the upper two components of the Dirac spinor describe particles, the lower two antiparticles. For particles with arbitrary momentum, we again use the standard representation; Eq. (3.179) for u then reads

$$(E - m) \phi - \vec{\sigma} \cdot \vec{p} \chi = 0 \quad,$$

$$\vec{\sigma} \cdot \vec{p} \phi - (E + m) \chi = 0 \quad.$$

For $\vec{p} \rightarrow 0$ we want to regain the spinor $\begin{pmatrix} \phi \\ 0 \end{pmatrix}$ and

we hence eliminate χ; from the second equation we get

$$\chi = \frac{\vec{\sigma} \cdot \vec{p}}{E + m} \phi$$

and find for the particle solution

$$u(p) = N \begin{pmatrix} \phi \\ \dfrac{\vec{\sigma} \cdot \vec{p}}{E + M} \phi \end{pmatrix} . \qquad (3.182)$$

Similarly the antiparticle solution is

$$v(p) = N \begin{pmatrix} \dfrac{\vec{\sigma} \cdot \vec{p}}{E + m} \chi \\ \chi \end{pmatrix} . \qquad (3.182')$$

In Eq. (3.178) we expressed the fact that u and v each are twofold degenerate. In order to characterize a solution uniquely, we must introduce another observable. As expected, the spin provides such an observable. The general description of the polarization of a relativistic electron is somewhat tricky, but the helicity can be introduced easily. Both u and v can be chosen to be either right-handed or left-handed and four independent solutions, u_R, u_L, v_R, and v_L are thus obtained.

Finally, we return to the normalization of the spinors u and v. In the nonrelativistic region, we have normalized the Pauli spinors to unity by Eq. (3.129). The corresponding normalization for Dirac spinors,

$$u^\dagger u = \bar{u} \gamma^0 u = 1 \quad ,$$

is possible, but it is not Lorentz invariant: The right-hand side of this equation is a scalar, while according to Eq. (3.169) or Table 3.1 the left-hand side transforms like the zeroth component of a four-vector. We obtain a covariant normaliz- ation if both sides transform in the same way; to achieve this goal $u^{\dagger}u$ is set equal to the zeroth component of some four vector, for instance[58]

$$u^{\dagger}u = v^{\dagger}v = 2E \quad , \; E > 0 \quad . \tag{3.183}$$

The corresponding normalization for $\bar{u}u$ is found by writing the Dirac equation in the form

$$(\gamma^0 E - \vec{\gamma} \cdot \vec{p} - m) \, u = 0 \quad .$$

Multiplication from the left with $u^{\dagger}$,

$$u^{\dagger} (\gamma^0 E - \vec{\gamma} \cdot \vec{p} - m) \, u = 0 \quad ,$$

taking the Hermitian conjugate of this equation, and adding the two yields

$$u^{\dagger} \gamma^0 E \, u - u^{\dagger} m \, u = 0 \quad ,$$

---

58. The normalization $u^{\dagger}u = E/m$ is also used in the literature. It has the advantage that its nonrelativistic limit agrees with the conventional definition and that $\bar{u}u = 1$. Our convention has the advantage that the normal- ization is the same for fermions and bosons.

or

$$\bar{u}u = \frac{m}{E} u^{\dagger}u = 2m \quad .$$

Similarly we find $\bar{v}v = -2m$. We also note that u
and v are orthogonal to each other and that solutions belonging to different helicities are orthogonal. We summarize the orthonormality properties
as follows:

$$\bar{u}u = - \bar{v}v = 2m \tag{3.184}$$

$$\bar{u}v = \bar{u}_R u_L = \bar{v}_R v_L = 0 \quad .$$

In order to establish the closure relation
for Dirac spinors, we expand an arbitrary spinor
ψ in terms of the complete set u_i and v_i:

$$\psi = \sum_{i=R,L} (u_i c_i + v_i d_i) \quad .$$

The coefficients c_i and d_i are obtained with Eq.
(3.184) as $c_i = \bar{u}_i \psi/2m$ and $d_i = -\bar{v}_i \psi/2m$; inserting
the coefficients into the expansion for ψ and
noting that ψ is arbitrary gives the closure
relation

$$\sum_{i=R,L} (u_i \bar{u}_i - v_i \bar{v}_i) = 2mI \quad , \tag{3.185}$$

With Eqs. (3.129) and (3.184), the normalization constant N in Eq. (3.182) can be calculated
easily; we get N = $(E + m)^{1/2}$ so that the free
particle solutions are

$$
u(p) = (E + m)^{1/2} \begin{pmatrix} \phi \\ \frac{\vec{\sigma} \cdot \vec{p}}{E + m} \phi \end{pmatrix} \quad ,
$$

(3.186)

$$
v(p) = (E + m)^{1/2} \begin{pmatrix} \frac{\vec{\sigma} \cdot \vec{p}}{E + m} \chi \\ \chi \end{pmatrix} \quad ,
$$

where E is positive.

The total probability of finding the free
particle in a volume V is

$$
\int_V j^0 \, d^3x = \int_V \bar{\psi} \, \gamma^0 \psi \, d^3x
$$

$$
= \int_V \psi^\dagger \psi \, d^3x = \int_V u^\dagger u \, d^3x = 2EV \quad .
$$

The normalization (3.183) corresponds to having
2E particles in the volume V. The wave function
of one free particle per volume V thus is given by

$$\psi_1 = (2EV)^{-1/2}u(p)e^{-ip\cdot x} \quad . \qquad (3.187)$$

F. Quantization of the Dirac Field

We finally turn to the quantization of the Dirac field, i.e., the field describing spin $\frac{1}{2}$ fermions. Since fermions obey the Pauli principle, no classical fermion field can exist and the correspondence principle cannot be used to arrive at expressions for the field operators. Nevertheless we need the field operators in order to describe creation and annihilation of fermions. To arrive at the desired expressions, we introduce the number representation and creation and annihilation operators in analogy to the charged boson case in Section 3.5. However, a major difference between the boson and the fermion case must be taken into account: For fermions, the number of particles in any one state can only be 0 or 1.

We denote the annihilation and creation operators for a fermion with four-momentum p and helicity h (h = R or L) by a_{ph} and $a_{ph}^{\dagger}$, respectively. Similarly we denote the annihilation and creation operators for the corresponding anti-fermions by b_{ph} and $b_{ph}^{\dagger}$, respectively. A state with one free fermion of momentum p and helicity h is then described by

$$a_{ph}^{\dagger} |0> \quad ,$$

and a state with two fermions is given by

$$a^\dagger_{ph} \, a^\dagger_{p'h'} \, |0\rangle \quad .$$

Interchange of two fermions, according to the Pauli principle, should multiply the state by a minus sign:

$$a^\dagger_{ph} \, a^\dagger_{p'h'} = - \, a^\dagger_{p'h'} \, a^\dagger_{ph}$$

or

$$\left\{ a^\dagger_{ph} , a^\dagger_{p'h'} \right\} = 0 \quad .$$

Whereas the operators for boson fields obey commutation relations, the operators for a fermion field obey anticommutation relations! The quantization with anticommutators was first introduced by Jordan and Wigner.[59]   In order to be consistent, all commutation relations involving a, $a^\dagger$, b, and $b^\dagger$ are replaced by anticommutation relations:

$$\left.
\begin{aligned}
\left\{ a_{ph} , a^\dagger_{p'h'} \right\} &= \left\{ b_{ph} , b^\dagger_{p'h'} \right\} \\
&= \delta_{pp'} \, \delta_{hh'} \quad , \\
\text{all other anticommutators vanish.}
\end{aligned}
\right\} \qquad (3.188)$$

59.   P. Jordan and E. Wigner, Z. Physik <u>47</u>, 631
      (1928).

The anticommutation rules are in accord with the
exclusion principle: Apply, for instance, the
creation operator $a_{ph}^\dagger$ twice; the anticommutation
rules then give

$$(a_{ph}^\dagger)^2 = 0 \quad .$$

Two particles cannot be in the same state. Furthermore, if we define the number operator in analogy
to Eq. (3.116) as

$$N_{ph}^a = a_{ph}^\dagger \, a_{ph} \quad , \quad N_{ph}^b = b_{ph}^\dagger \, b_{ph} \quad , \qquad (3.189)$$

we find from the anticommutation rules

$(1 - N_{ph}^a) = a_{ph} \, a_{ph}^\dagger$ and hence, for instance,

$$N_{ph}^a (1 - N_{ph}^a) = 0 \quad .$$

The eigenvalues of the number operator are 0 and 1,
in agreement with the Pauli principle.

To get expressions for the momentum space
field operators ψ and $\bar{\psi}$ for free spin $\frac{1}{2}$ fermions,
we use the analogy to the Eqs. (3.115) and write

$$\psi(x) = \sum_{p,h} \frac{1}{\sqrt{2E\ V}}$$

$$\times (a_{ph}\ u_h(p)\ e^{-ip\cdot x} + b^\dagger_{ph}\ v_h(p)\ e^{ip\cdot x}) \quad,$$

$$\bar{\psi}(x) = \sum_{p,h} \frac{1}{\sqrt{2E\ V}} \qquad\qquad (3.190)$$

$$\times (\bar{u}_h(p)\ a_{ph}\ e^{ip\cdot x} + \bar{v}_h(p)\ b_{ph}\ e^{-ip\cdot x}) \quad.$$

The Dirac current also becomes an operator. However, in the form of Eq. (3.169) it is not symmetric in particles and antiparticles. Heisenberg suggested proper antisymmetrization by writing the current as a commutator,[60]

$$j^\mu(x) = \frac{1}{2}\left[\bar{\psi}(x), \gamma^\mu \psi(x)\right] \quad. \qquad (3.191)$$

In this form, the vacuum expectation value of j^μ vanishes:

$$<0|j^\mu(x)|0> = 0 \quad, \qquad (3.192)$$

and the operator of the total electric charge

---

60.   W. Heisenberg, Z. Physik 90, 209 (1934).

becomes

$$Q = e \int d^3x \ j^0(x) = \frac{1}{2} e \int (\psi^\dagger \psi - \psi \psi^\dagger) d^3x \quad .$$

One point can lead to confusion when we insert ψ and $\psi^\dagger$ from Eq. (3.190): The reversed ordering of the operators ψ and $\psi^\dagger$ in the term $\psi \psi^\dagger$ only applies to the annihilation and creation operators; the eigenfunctions $u^\dagger$, u, $v^\dagger$, and v remain in their proper order, for instance $u^\dagger u$. Evaluating the resulting expression with the help of Eqs. (3.183) and (3.188) gives

$$Q = e \sum_{p,h} (N^a_{ph} - N^b_{ph}) \quad . \tag{3.193}$$

This form of the current operator shows that we can interpret the various operators in terms of particles and antiparticles: The operator $a^\dagger_{ph}$ creates a particle with charge e, momentum p, and helicity h and the operator $b^\dagger_{ph}$ creates the corresponding antiparticle with charge -e, but with the same momentum p and the same helicity h.*
The momentum space field operator $\psi(x)$ can destroy particles and create antiparticles while $\psi^\dagger$ and $\bar{\psi}$ create particles and destroy antiparticles.

---

*The convention for leptons is the opposite.

3.7 CANONICAL FIELD QUANTIZATION[61]

In the previous sections, we have introduced
field quantization by defining creation and annihilation operators and by using the correspondence
principle for the electromagnetic field. The
conventional approach is different; it follows
the general procedure for quantizing the equations
of motion of a classical system. We sketch the
main steps of the "canonical" quantization procedure in the present section.

A. Classical Fields

We first consider a one-component real classical field $\Phi(x)$ and describe this field by a
Lagrangian density which we assume to be a function
of Φ and $\nabla^\mu \Phi$ only:

$$\mathcal{L} = \mathcal{L}(\Phi, \nabla^\mu \Phi) \quad . \tag{3.194}$$

The total Lagrangian of the field is given by

$$L = \int d^3x \, \mathcal{L} . \tag{3.195}$$

The field equations are determined by Hamilton's
principle of least action,

---

61. W. Heisenberg and W. Pauli, Z. Phys. 56, 1
     (1929); 59, 168 (1930).

$$\delta \int_{t_1}^{t_2} L dt = \delta \int \mathscr{L} d^4 x = 0 \quad , \tag{3.196}$$

where the variation is subject to the restrictions

$$\delta \Phi(\vec{x}, t_1) = \delta \Phi(\vec{x}, t_2) = 0 \quad .$$

This variational principle leads to the Euler-Lagrange field equations

$$\frac{\partial \mathscr{L}}{\partial \Phi} - \nabla^\mu \left(\frac{\partial \mathscr{L}}{\partial \nabla^\mu \Phi} \right) = 0 \quad . \tag{3.197}$$

$\mathscr{L}$ is called the Lagrangian density of the field Φ if Eq. (3.197) yields the proper field equations. Note, however, that the choice of the Lagrangian density is not unique.

The momentum density of the field Φ is defined by

$$\pi = \frac{\partial \mathscr{L}}{\partial \dot{\Phi}} \quad , \tag{3.198}$$

and the Hamiltonian H and the Hamiltonian density $\mathscr{H}$ are given by (compare Eq. (2.24))

$$H = \int d^3 x \ (\pi \dot{\Phi} - \mathscr{L}) \equiv \int d^3 x \ \mathscr{H}, \tag{3.199}$$

$$\mathscr{H} = \pi \dot{\psi} - \mathscr{L} \quad . \tag{3.200}$$

If a field ψ has more than one component, each
component is varied independently and an equation
of the type (3.197) obtains for each component.
If, for instance, ψ is complex, we write it in
terms of two real fields ψ_1 and ψ_2 ,

$$\psi = \frac{1}{\sqrt{2}}(\psi_1 + i\psi_2), \ \psi^\dagger = \frac{1}{\sqrt{2}} (\psi_1 - i\psi_2) \ , (3.201)$$

and vary ψ_1 and ψ_2 independently. In Table 3.2
we list two Lagrangians and the corresponding
field equations. Note that we consider Φ and ψ
to be classical fields and not wave functions.

Table 3.2 Lagrangians and Field Equations

Lagrangian	Field Equation	
$\frac{1}{2}(\nabla^\mu\Phi\nabla_\mu\Phi - m^2\Phi^2)$	$(\Box^2+m^2)\Phi = 0$	Klein-Gordon equation
$\bar{\psi}(i\slashed{\partial} - m)\psi$	$(-i\slashed{\partial} +m)\psi = 0$	Dirac equation

B.   Quantization

To quantize the real Klein-Gordon field $\Phi_0(x)$,
we use the analogy with Eq. (3.32) and postulate
commutation relations between $\Phi_0(x)$ and the conjugate momentum density field $\pi_0(x)$. Since we now

deal with continuous functions of x, we replace
the Kronecker symbol δ_{ij} by the Dirac delta function
and get the commutation relations for the field
operators Φ_0 and π_0 at equal times

$$[\Phi_0(\vec{x},t), \Phi_0(\vec{x}',t)] = [\pi_0(\vec{x},t), \pi_0(\vec{x}',t)] = 0,$$

$$[\Phi_0(\vec{x},t), \pi_0(\vec{x}',t)] = i \, \delta^3(\vec{x}-\vec{x}') \qquad . \tag{3.202}$$

To establish a connection between the canonical quantization procedure and our earlier
approach, we will show that the field operators
given in Section 3.5 satisfy the commutation
relations Eq. (3.202). The field operator Φ_0 for
the real Klein-Gordon field is given by Eq. (3.105);
the conjugate momentum is obtained with Eq. (3.198)
and with Tables 3.2 and 2.3 as

$$\pi_0 = \frac{\partial \mathcal{L}}{\partial \dot{\Phi}} = \dot{\Phi}_0 = -i \sum_{\vec{p}} \frac{E}{\sqrt{2E\,V}}$$

$$\times \left\{ a_{op} e^{-ip\cdot x} - a_{op}^{\dagger} e^{ip\cdot x} \right\} . \tag{3.203}$$

With the expansions (3.105) and (3.203) and the
commutation rules Eq. (3.102) for the operators
a_p and $a_p^{\dagger}$, we find for the equal-time commutator
of Φ_0 and π_0

$$[\Phi_o(\vec{x},t), \ \pi_o(\vec{x},t)] \ =$$

$$\frac{i}{2V} \sum_{\vec{p}} \left\{ e^{i\vec{p}\cdot(\vec{x}'-\vec{x})} + e^{i\vec{p}\cdot(\vec{x}-\vec{x}')} \right\} \ .$$

We perform the transition to the continuous limit
with the help of the substitution Eq. (3.64);
since

$$\frac{1}{(2\pi)^3} \int d^3p \ e^{i\vec{p}\cdot(\vec{x}'-\vec{x})} = \delta^3(\vec{x}'-\vec{x}) \ , \quad (3.204)$$

we find

$$[\Phi_o, \pi_o] \ = \ i \ \delta^3(\vec{x}'-\vec{x})$$

in agreement with our postulated commutation rule.
Similarly it is possible to start from the postulated commutation rules Eq. (3.202), expand the
field operators in terms of creation and annihilation operators, and find the proper commutation
relations for these operators.

The commutation rules Eq. (3.202) apply to
neutral spinless bosons which can be described
by a one-component field. For fields with more
than one component, the components are independent and the commutation relations are taken to be

$$[\Phi_\alpha(\vec{x},t), \Phi_\beta(\vec{x}',t)] = [\pi_\alpha(\vec{x},t), \pi_\beta(\vec{x}',t)] = 0$$

$$\tag{3.205}$$

$$[\Phi_\alpha(\vec{x},t), \pi_\beta(\vec{x}',t)] = i\,\delta_{\alpha\beta}\,\delta^3(\vec{x}-\vec{x}') \quad .$$

If the field Φ is complex, we write it in terms of two real fields Φ_1 and Φ_2, as in Eq. (3.201). Equation (3.198) then yields for the conjugate momentum

$$\pi = \frac{1}{\sqrt{2}}\,(\pi_1 - i\pi_2), \quad \pi^\dagger = \frac{1}{\sqrt{2}}(\pi_1 + i\pi_2). \tag{3.206}$$

The commutation relations for the field operators then follow by applying Eq. (3.203) to the two real fields Φ_1 and Φ_2:

$$[\Phi(\vec{x},t), \pi(\vec{x}',t)] = [\Phi^\dagger(\vec{x},t), \pi^\dagger(\vec{x}',t)]$$

$$= i\,\delta^3(\vec{x}-\vec{x}') \quad . \tag{3.207}$$

All other commutators vanish.

For fermion fields, the commutators are replaced by anticommutators. At equal times, these anticommutation relations are

$$\left\{\psi_\alpha(\vec{x},t), \psi_\beta(\vec{x}',t)\right\} = \left\{\psi_\alpha^\dagger(\vec{x},t), \psi_\beta^\dagger(\vec{x}',t)\right\} = 0$$

$$\left\{\psi_\alpha(\vec{x},t), \psi_\beta^\dagger(\vec{x}',t)\right\} = \delta_{\alpha\beta}\,\delta^3(\vec{x}-\vec{x}') \quad . \tag{3.208}$$

PROBLEMS

3.1   Show that the photon mass must be zero if the
      speed of light in vacuum is the same in all
      inertial frames of reference.

3.2   Write Maxwell's equations for a photon of
      rest mass m_γ. Determine the effect of such
      a mass on the magnetic field of a dipole at
      large distances. Assume that the magnetic
      field of the earth falls off, to within one
      percent, as the third power of the distance
      out to 5 earth radii. What is the corres-
      ponding limit on m_γ?

3.3   Show that the gauge transformation Eq. (3.8)
      leaves the electric and magnetic field
      strengths unaltered.

3.4   Justify Eqs. (3.23) to (3.25).

3.5   Calculate the spin $\vec{S}$, Eq. (3.25), for a lin-
      early polarized electromagnetic wave. Explain
      the result.

3.6   Discuss the Mössbauer experiment that shows
      the absence of the third component of the
      photon's spin, quantized along the direction
      of motion.

3.7   Find the operator corresponding to the mag-
      netic field strength $\vec{B}$.

3.8   Show that the operators $\vec{E}$ and $\vec{B}$ satisfy Max-
      well's equations. What is the difference of
      of the classical and the "field theoretic"
      equations?

3.9   Discuss the analogy of the quantization pro-
      cedure used in Section 3.3 with the harmonic
      oscillator. Show that for the harmonic oscil-
      lator $\langle n|\vec{x}|n\rangle = \langle n|\vec{p}|n\rangle = 0$. Calculate the
      fluctuations of $\vec{x}$ and $\vec{p}$ and compare with the
      fluctuations of $\vec{A}$.

3.10 Investigate the "coherent" states $|$class$>$, Eq. (3.50).

- (a) Construct the states out of eigenstates of the number operator.
- (b) Show that the states form a complete set, but that they are not orthogonal.
- (c) Show that the states form a "minimum wave packet."

3.11 Follow the usual proof (e.g. in Heitler) for the uncertainty relation

$$\Delta n \Delta \phi \geq 1$$

and discuss where the arguments fail.

3.12 Show that linearly polarized waves do not correspond to a definite number of photons.

3.13 Show that the conservation laws of energy and momentum prevent the transition to negative energy states in a two-body collision.

3.14 Show that pair production by a single photon $(\gamma \rightarrow e^+ e^-)$ is not possible in the absence of a catalyst to take up momentum.

3.15 Discuss the experimental production of antiprotons by high-energy photons. (W. Bertram et al., Phys. Letters 21, 471 (1966).)

- (a) Find the laboratory threshold energy for antiproton production on a nucleon at rest, and on a nucleon bound in a nucleus.
- (b) How were the antiprotons identified?
- (c) How does the observed cross section compare with the cross section for antiproton production by protons?

3.16 Discuss scintillation counters.

- (a) How does a scintillation counter work?
- (b) A parallel beam of gamma rays of 1.2 MeV energy hits a Na-I scintillator,

5 cm deep and with a diameter of 5 cm.
Estimate the efficiency for detection.
Sketch the pulse height spectrum.

(c) When measuring the Mössbauer effect of
the ^{57}Fe 14 keV transition, a 122 keV
gamma ray disturbs the measurement.
Estimate the optimum thickness of a
Na-I scintillator used to detect the
14 keV gamma rays.

(d) A proton of momentum 2 GeV/c passes
through a 2 cm thick plastic scintillator. How much energy does the proton
lose? How many photons are produced
approximately? How many electrons will
be released at best in the photomultiplier?

3.17 How are pions, muons, and electrons distinguished in photographic emulsions and in
bubble chambers?

3.18 Find the pion-muon mass difference from
Fig. 3.11.

3.19 Recast the Klein-Gordon equation into a
two-component form. Show that one component
describes particles, the other antiparticles.

3.20 Show that the four-current Eq. (3.96) satisfies the conditions listed after Eq. (3.93).

3.21 (a) Solve the Klein-Gordon equation for the
Coulomb potential of a point nucleus.
Find the energy levels and estimate
the line width for the various levels.

(b) Describe the experimental arrangement
used to detect pionic x rays.

(c) Compare the x-ray energies and widths
from part (a) with experimentally observed values, particularly for ^{32}S,
^{59}Co, Y, Pr. Discuss deviations.

(d) Describe the eventual fate of the pion.
What is the Panofsky ratio?

3.22   Discuss experiments and experimental results
       on K-mesic atoms.

3.23   An electron at rest is described by the spinor

       $\frac{1}{\sqrt{2}}\begin{pmatrix} 1 \\ -1 \end{pmatrix}$.   Find the polarization vector $\vec{P}$.

       Calculate the degree of polarization.

3.24   Show that the polarization vector $\vec{P}$ trans-
       forms like a pseudovector.

3.25   Assume a spinor $\phi = \begin{pmatrix} a \\ b \end{pmatrix}$, $a^*a + b^*b = 1$.

       Show that Eq. (3.133) is satisfied.

3.26   (a)   Expand $e^{i\vec{\sigma}\cdot\vec{A}}$.   Express the answer in the
             form $\alpha + \vec{\beta}\cdot\vec{\sigma}$ and show that this is the
             most general form.

       (b)   Compare $e^{i\vec{\sigma}\cdot\vec{A}}\ e^{i\vec{\sigma}\cdot\vec{B}}$ and $e^{i\vec{\sigma}\cdot(\vec{A}+\vec{B})}$ if $\vec{A}$
             and $\vec{B}$ do not commute.   (Compare the
             first four terms in the expansion.)

3.27   A nonrelativistic spin $\frac{1}{2}$ particle moves along
       the z axis.   Its polarization vector lies in
       the positive quadrant of the x-z plane and
       makes an angle of 30° with the z axis.
       Calculate the helicity and a suitably chosen
       "transverse polarization."

3.28   Prove Eq. (3.136).

3.29   An electron beam, with a degree of polariza-
       tion of 50%, moves along the x axis.   Its
       polarization vector points along the z axis.
       Find the density matrix.

3.30   Show that γ^5 anticommutes with all components
       of γ^μ.   Prove that $\gamma^{5\dagger}=\gamma^5$.

3.31   Show that in the absence of a field the
       components of a solution of the Dirac equation

satisfy the Klein-Gordon equation. Show that
the converse is not true by finding a counter
example.

3.32 Show that the operator

$$\vec{\sigma}_D \cdot \hat{p} = \begin{pmatrix} \vec{\sigma} \cdot \hat{p} & 0 \\ 0 & \vec{\sigma} \cdot \hat{p} \end{pmatrix}$$

where $\vec{\sigma}$ is the Pauli spin vector, can be
used as helicity operator in the Dirac theory,
if the standard representation is used. Find
the eigenvalues and the eigenfunctions of this
operator. Express the eigenfunctions
u_R, u_L, v_R, and v_L in terms of the two-component eigenstates Eq. (3.136). Write these
eigenfunctions for the special case of a
particle moving along the +z axis.

3.33 An electron with momentum 0.51 MeV/c moves
in the positive quadrant of the x-y plane,
at 30° to the x axis. Find the explicit
expression for the spinors u_R, u_L, v_R, v_L.
Show that they are orthogonal.

3.34 (a) Verify the transformation properties
given in Table 3.1.

(b) Show that $\gamma^\mu \gamma^\nu = g^{\mu\nu} I - i\sigma^{\mu\nu}$.

3.35 Verify the orthonormality relations, Eq.
(3.184).

3.36 Use the Dirac equation in the Hamiltonian
form and show that H u = E u, H v = - E v,
where E is positive.

3.37 Show that $a^\dagger$ is a charge raising operator
and $b^\dagger$ is a charge lowering operator for
fermions in the sense that if

$$Q|\psi> = Nq|\psi> \qquad ,$$

then

$$Qa^\dagger|\psi> = (N + 1) \, q \, a^\dagger|\psi> \text{ and}$$

$$Qb^\dagger|\psi> = (N - 1) \, q \, b^\dagger|\psi> \quad .$$

3.38 Show that the variational principle Eq. (3.196) leads to the Euler-Lagrange field equations Eq. (3.197).

3.39 Show that the field equations in Table 3.2 follow from the given Lagrangians.

3.40 Find Lagrangian densities other than the ones given in Table 3.2 that lead to the Klein-Gordon and the Dirac equation.

3.41 Find a Lagrangian density that leads to the Schrödinger equation for ψ and for $\psi^\dagger$. Is your Lagrangian density Hermitian? Compute the momenta conjugate to ψ and $\psi^\dagger$. Explain the result.

ADDITIONAL REFERENCES

The history of the development of the photon concept is discussed in M. J. Klein, "Einstein's First Paper on Quanta" and "Einstein and the Wave-Particle Duality," in The Natural Philosopher, Vol. 2, 1963 and Vol. 3, 1964 (Blaisdell Publishing Company, New York). Further references and a selection of reprints can be found in Quantum and Statistical Aspects of Light (American Institute of Physics, New York, 1963).

Among the many excellent books treating Maxwell's theory and classical electromagnetism, we mention only J. D. Jackson, Classical Electrodynamics (John Wiley and Sons, Inc., New York, 1962), where the derivations of the equations used in Section 3.1 can be found.

Spin is discussed in all quantum mechanics texts; for instance, E. Merzbacher, Quantum Mechanics (John Wiley and Sons, New York, Second Edition, 1970); in R. P. Feynman, R. B. Leighton, and M. Sands, The Feynman Lectures, Vol. III (Addison-Wesley, Reading, Mass., 1965) (Sects. 11.4 and 17.4); and in K. Gottfried, Quantum Mechanics, Vol. I (W. A. Benjamin, Inc., Reading, Mass., 1966). The historical development is described by B. L. Van der Waerden, in Theoretical Physics in the 20th Century, M. Fierz and V. F. Weisskopf, eds. (Interscience Publishers, Inc., New York, 1960) and by S. A. Goudsmit, Physikalische Blätter 21, 445 (1965); J. Physique 28, 123 (1967).

Quantum Field Theory is treated in a very large number of texts and reviews. The early history is vividly described by G. Wentzel in Theoretical Physics in the Twentieth Century, M. Fierz and V. F. Weisskopf, eds. (Interscience Publishers, Inc., New York, 1960). A sketch of the history and reprints of the main papers are contained in a book by Schwinger, Quantum Electrodynamics, J. Schwinger, ed. (Dover Publications, Inc., New York, 1958). The present status of field quantization is surveyed in the report of the 12th Solvay Conference (The Quantum Theory of Fields, R. Stoops, ed. (Interscience Publishers, Inc., New York, 1962)); particularly the remarks made during the discussions at the conference make fascinating reading.

As a first introduction to quantum electro-
dynamics, the classic review by Fermi is still un-
surpassed. (E. Fermi, Rev. Mod. Phys. $\underline{4}$, 87 (1932).)
Readable introductions to field theory are given in
F. Mandl, Introduction to Quantum Field Theory
(Interscience Publishers, Inc., New York, 1959),
J. M. Ziman, Elements of Advanced Quantum Theory
(Cambridge University Press, 1969), and K. Nishijima,
Fields and Particles (W. A. Benjamin, Inc., Reading,
Mass., 1969). Nonrelativistic field theory is
treated in E. M. Henley and W. Thirring, Elementary
Quantum Field Theory (McGraw-Hill Publishing Co.,
New York, 1962).

The basic principles of quantum electrodynamics
are discussed in W. E. Thirring, Principles of
Quantum Electrodynamics (Academic Press, New York,
1958); and R. P. Feynman, Quantum Electrodynamics
(W. A. Benjamin, Inc., Reading, Mass., 1962). The
older approach is fully covered in W. Heitler,
Quantum Theory of Radiation (Oxford University Press,
Oxford, 1954). For many years, three successive
editions of Heitler's work have set the standard for
all treatments of quantum electrodynamics. Even
though the modern approach is more elegant, Heitler's
book still gives considerable physical insight into
many problems. Complete treatments of the modern
approach are contained in J. M. Jauch and F. Rohr-
lich, The Theory of Photons and Electrons (Addison-
Wesley, Reading, Mass., 1955); G. Källen, Quan-
tenelektrodynamik, Encyclopedia Physics V/1

(Springer-Verlag, New York, 1958); A. I. Akhiezer
and V. B. Berestetskii, Quantum Electrodynamics
(Interscience Publishers, Inc., New York, 1965).
All three contain the general theory and applications with rather complete calculations; the last
work gives the most detailed information.

Quantum field theory is fully treated in three
large works: N. N. Bogoliubov and D. V. Shirkov,
Introduction to the Theory of Quantized Fields
(Interscience Publishers, Inc., New York, 1959);
S. S. Schweber, An Introduction to Relativistic
Quantum Field Theory (Harper and Row, New York,
1961); J. D. Bjorken and S. D. Drell, Relativistic
Quantum Mechanics (McGraw Hill Publishing Company,
New York, 1964), Relativistic Quantum Fields
(McGraw-Hill, New York, 1965).

CHAPTER 4 NONCLASSICAL CONSERVATION LAWS

> Symmetry, as wide or as narrow as you
> may define its meaning, is one idea
> by which man through the ages has tried
> to comprehend and create order, beauty,
> and perfection.

H. WEYL, Symmetry
(Copyright 1952 by Princeton
University Press, Princeton,
New Jersey) p. 5. Reprinted
by permission of Princeton
University Press.

At present, we have no complete dynamical
theory of elementary particles and of the strong
and weak interactions. For this reason, symmetries
and conservation laws form a central part of
particle and nuclear physics; they allow us to
classify the observed phenomena, predict new
features, and they can serve as guides towards
a future dynamical theory. In this chapter, we
first survey the types of symmetries that we will

encounter and then discuss some of the symmetries
that play an important role in particle and nuclear
physics.

4.1   CONSERVATION LAWS IN QUANTUM MECHANICS

The connection between invariance principles
and conserved quantities has already appeared twice.
In Section 1.5 we have stressed the importance of
this connection for all of particle and nuclear
physics and in Section 2.3 we have sketched the
connection in classical physics. In the present
section, we extend the treatment to quantum mechanics and we will find some new and important
features. The translation of Eq. (2.26) into
quantum mechanical language is well known: If H
is the Hamiltonian of the system, and $\mathcal{O}$ a Hermitian
operator, then

$$\frac{d\mathcal{O}}{dt} = \frac{\partial \mathcal{O}}{\partial t} - i\ [\mathcal{O},H] \quad , \tag{4.1}$$

where $[\mathcal{O},H] = \mathcal{O}H - H\mathcal{O}$ is the commutator. The
observable represented by the operator $\mathcal{O}$ is a
constant of motion if $\mathcal{O}$ commutes with the Hamiltonian and does not depend explicitly on the time.
If the Hamiltonian is known, the question is: What
are the operators that correspond to conserved
quantities? To answer this question, the behavior
of the Hamiltonian or Lagrangian under various
transformations is studied. We distinguish below

two different possibilities: transformations
that form a continuous group and transformations
that are discrete.

We have mentioned here the study of symmetries
based on a Hamiltonian description. Clearly, however, symmetries do not depend on such a description; they are valid more generally.

To discuss continuous and discrete transfor-
mations, we consider an operator $\mathcal{O}$ and a unitary
transformation U for which

$$U^\dagger U = UU^\dagger = 1 \quad . \tag{4.2}$$

With U, we can rewrite any matrix element of $\mathcal{O}$ as

$$\langle b| \; \mathcal{O} \; |a\rangle = \langle b|U^\dagger U \; \mathcal{O} \; U^\dagger U|a\rangle = \langle b'| \; \mathcal{O}' \; |a'\rangle \quad .$$

$$\tag{4.3}$$

The simultaneous transformation

$$\mathcal{O}' = U \; \mathcal{O} \; U^\dagger$$

$$|a'\rangle = U|a\rangle$$

thus leaves any matrix element invariant. The
unitarity of U, expressed by Eq. (4.2), guarantees
that normalization and orthogonality of the states
is preserved. Equation (4.3) states the trivial
result that the matrix element of the operator $\mathcal{O}$
between the original states is the same as that
of the transformed operator $\mathcal{O}'$ between the transformed states; Eq. (4.3) is valid for any unitary
operator U and any linear operator $\mathcal{O}$ and has nothing

to do with symmetries. Invariance arguments come
into play if we have found an operator $\mathcal{O}$ that is
invariant under the transformation U:

$$\mathcal{O}' = U \; \mathcal{O} \; U^\dagger = \mathcal{O} \quad , \text{ or } [\mathcal{O}, U] = 0 \qquad . \qquad (4.4)$$

In this case we get

$$<b'| \; \mathcal{O} \; |a'> \; = \; <b| \; \mathcal{O} \; |a> \quad ; \qquad\qquad (4.5)$$

the matrix element of the operator $\mathcal{O}$ is the same
between the original and the transformed states.
Of particular importance is the Hamiltonian, $\mathcal{O}$ = H.
If

$$H' \equiv UHU^\dagger = H \quad \text{or} \quad [U, H] = 0 \qquad ,$$

then the dynamics of the system remains invariant
under the transformation U. If |E> is an eigen-
state of H with eigenvalue E,

$$H|E> \; = \; E|E> \quad , \qquad\qquad\qquad (4.6)$$

then the transformed state U|E> is also an eigen-
state of H with the same energy E. Since
[U,H] = 0, U is an invariant, a constant of the
motion. If U is Hermitian, it thus becomes an
observable. If U is not Hermitian, it has a
Hermitian operator associated with it which will
be a constant of the motion. Further discussion
thus depends on the properties of U.

Continuous transformations.[1] Any unitary
operator that corresponds to a continuous transformation can be written as

$$U(\varepsilon) = e^{i \varepsilon F} \quad, \tag{4.7}$$

where F is called the generator of the invariance
operator U and ε is a real parameter. The operator
F in Eq. (4.7) appears in the exponent; its action
is defined by

$$e^{i \varepsilon F} \equiv \sum_{n} (i \varepsilon F)^{n}/n! \quad. \tag{4.8}$$

The parameter ε can be made infinitesimally small
and the transformation can be built up from successive application of infinitesimal transformations,

$$U = 1 + i \varepsilon F \quad, \quad \varepsilon \ll 1 \quad. \tag{4.9}$$

With Eq. (4.2) we get from (4.9) that $F^{\dagger} = F$;
F is Hermitian. Furthermore, if U commutes with
H, F also commutes with H and we have then found
in F a candidate for an observable that is conserved under the operation U and is a constant of
motion.

---

1. G. Racah, "Group Theory and Spectroscopy,"
    Springer Tracts in Modern Phys. 37, 28 (1965).
    W. Pauli, "Continuous Groups in Quantum Mech-
    anics," Springer Tracts in Modern Phys. 37,
    85 (1965).

We will study two types of continuous transformations, gauge transformations in Section 4.2
and rotations in Sections 4.3 and 4.4. Under a
gauge transformation, the wave function $\psi(x)$ is
transformed into $\psi'(x)$, with

$$\psi'(x) = \psi(x)e^{-i\alpha} \quad , \tag{4.10}$$

where α is either a real constant or a real function of space and time. Invariance under such a
gauge transformation leads to an additive conservation law. An additive quantum number has the
property that the total quantum number of a system
is the sum of the quantum numbers of its parts.
The best known example is the electric charge.

Two rotations of importance in particle and
nuclear physics are rotation in ordinary space
and rotation in "charge space." Invariance under
rotation in ordinary space leads to conservation
of the total angular momentum. The operator $U_z(\alpha)$
that produces a rotation through the angle α about
the z axis is given by

$$U_z(\alpha) = e^{-i\alpha J_z} \quad , \tag{4.11}$$

where J_z is the z component of the angular momentum.
If the Hamiltonian is invariant under this rotation,
i.e., if it commutes with U_z, J_z is a constant of
motion. This example makes the meaning of the
real parameter $\alpha(=\varepsilon)$ clearer. The next case is
without classical analogue and we will discuss it

in considerable detail: Invariance of a system, for instance an assembly of nucleons, under rotations in the so-called charge space leads to conservation of isospin. Isospin is unobservable classically because the forces or fields associated with strongly interacting particles are of short range.

Discrete transformations. In some cases, U cannot be reduced continuously to the identity. Application of a discrete transformation twice often leads to the identity:

$$U^2 = 1 \quad .\tag{4.12}$$

From Eqs. (4.2) and (4.12) it follows immediately that $U^{-1} = U^{\dagger} = U$; the transformation operator itself is Hermitian. It will therefore be an observable, with eigenvalues ± 1. These operators lead to multiplicative quantum numbers, which have the property that the total quantum number of a system is the product of that of its parts. We treat such quantum numbers in Sections 4.5 and 4.6.

All conservation laws must be tested experimentally, no matter how convincing the underlying symmetry principle. Such tests can obviously only give upper limits on possible violations. When we call a conservation law exact, we mean that no experimental evidence against it has been found and that the theoretical arguments make it reasonable that it indeed is true. However, in every conservation law, more and better evidence as to

its validity is valuable. Of particular importance
are checks that bear on the validity for all types
of interactions.

In the following sections, we will encounter
exact and approximate conservation laws. It is a
strange fact that some conservation laws and the
underlying symmetry principles depend on the type
and strength of interactions. Parity is conserved,
as all experimental evidence indicates, in the
strong interactions. It is, however, violated in
a most dramatic fashion in the weak interactions.
All experimental studies must therefore consider
also the strength of interactions.

4.2   GENERALIZED CHARGES

A.   Generalized Charges and Quantum Numbers

Familiarity with electrical phenomena lets us
forget frequently that the elementary electric
charge is really a stranger in our world and that
we do not understand it. We do not know why it is
quantized and what determines its smallest nonvanishing value. The electric charge actually plays
two roles, as was pointed out by Wigner:[2] On the
one hand, the charge characterizes the interaction
strength of a particle with an external electro-

---

2.   E. P. Wigner, Proc. Nat. Acad. Sci. US 38,
     449 (1952).

magnetic field: In a field $\vec{E}$, the force on the particle with charge q is given by $\vec{F} = q\,\vec{E}$. On the other hand, charge can be used as an <u>additive quantum number</u>: The charge of a collection of particles is the algebraic sum of the charges of the individual particles.

In the following we call a <u>generalized charge</u> one that acts as an additive quantum number <u>and</u> as an interaction constant. In the case of the electric charge, the connection between these two properties is given by

$$q = N\ e \qquad\qquad (4.13)$$

where q, the electric charge, plays the role of the interaction constant and N is the corresponding additive quantum number (electric number).

In a closed system, the electric charge obeys a conservation law; the total charge of the system is conserved. Classically, such a conservation law is expressed by the continuity equation. Microscopically, and with the introduction of the notion of particles, we describe the conservation by an additive quantum number. Consider a reaction

$$AB \rightarrow CDE$$

and denote the electric numbers by N_A, N_B, N_C, N_D, N_E. An <u>additive</u> conservation law then states

$$N_A + N_B = N_C + N_D + N_E \qquad . \qquad (4.14)$$

The electric charge, as we know from experiment, satisfies an additive conservation law.

The notion of an additive quantum number, originally applied only to electricity, is now used for other properties. In particular, the introduction of the baryonic number comes from the observation that matter is stable. The leptonic number has been introduced to account for observed selection rules in the Fermi interaction. The muonic number can account for the absence of the decay $\mu \rightarrow e \gamma$. The hypercharge (hyper number) describes many features observed in the creation of strange particles. To the best of our present knowledge (as we will discuss in the following subsections), the electric, the baryonic, the leptonic and the muonic numbers are strictly conserved while the hyper number (hyper charge) is only partially conserved.

An interesting question arises in connection with additive quantum numbers. We pointed out that the electric charge not only counts the number of charged particles algebraically, it also measures the strength of interaction of these particles with an external electromagnetic field. Now, if a particle carries another additive quantum number A, will this quantum number be associated with a generalized charge q_A? Such a generalized charge would be proportional to A and it would measure the strength of interaction with a new field. There exist some theoretical arguments in favor of this idea.[3-5] We will see in the next sub-sections that an additive conservation law can be the result of invariance under a gauge transformation, and that, under certain conditions, such a gauge transformation suggests the existence of a new interaction. At the present time it is not known whether this idea is correct.

B. Additive Quantum Numbers

We consider a particle or nucleus character-ized by an additive quantum number N_a decaying to a state with quantum number N_b and ask: What in-variance principle expresses conservation of this additive quantum number, $N_a = N_b$? To answer this question we consider a phase transformation of the wave function, with

$$\psi' = \psi \, e^{-iN\lambda} \tag{4.15}$$

where N is the additive quantum number of the state ψ and λ is an arbitrary constant. Under the transformation (4.15) the matrix element of an operator $\mathcal{O}$ between the states a and b changes as

3. C. N. Yang and R. L. Mills, Phys. Rev. <u>96</u>, 191 (1954).
4. T. D. Lee and C. N. Yang, Phys. Rev. <u>98</u>, 1501 (1955).
5. J. J. Sakurai, Ann. Phys. <u>11</u>, 1 (1960). J. Schwinger, Phys. Rev. <u>125</u>, 397 (1962); Phys. Rev. <u>128</u>, 2425 (1962).

$$\langle b' | \, \sigma \, | a' \rangle = \langle b | \, \sigma \, | a \rangle \, e^{i\lambda(N_b - N_a)} \quad .$$

In order for the matrix element $\langle b | \, \sigma \, | a \rangle$ to re-
main unchanged under the transformation Eq. (4.15),
the exponential factor must be unity. Since λ is
arbitrary (and phases are only defined to within
multiples of 2π), we get

$$N_b = N_a \quad .$$

Invariance under the gauge transformation (4.15)
leads to conservation of the additive quantum
number N. If there are particles with quantum
numbers $N_1 \, N_2 \cdot \cdot N_i$ in the state a, we transform the
total wave function as

$$\psi' = \psi \, e^{-i\lambda \Sigma_a N_i}$$

and find

$$\sum_{\text{state a}} N_i = \sum_{\text{state b}} N_i \quad . \tag{4.16}$$

The number N in Eq. (4.13) indeed plays the role
of an <u>additive</u> quantum number.

C. Gauge Invariance, Charge Conservation, and
    Minimal Electromagnetic Interaction

    At this point a new question arises. We can
take N to be the electric number q/e. How do we
know that q then is really the electric charge
and not some other quantum number? Certainly none
of the arguments given so far helps us decide this
question because we have not yet introduced the
interaction with an external field. We will get
a relation of the form Eq. (4.13) if we introduce
a gauge transformation that depends on x, and also
use the gauge transformation of the electromagnetic
field, Eq. (3.8). A connection between the latter
gauge transformation and charge conservation exists
already in classical physics, as was shown by
Weyl[6] and this connection can be taken over into
quantum mechanics.[7]

    The interaction of a particle of charge q
with the electromagnetic field is introduced through
the substitution (3.87)

$$\nabla^\mu \rightarrow \nabla^\mu + iqA^\mu \qquad\qquad (4.17)$$

that was first used by Larmor.[8] This substitution

---

6. See W. Pauli, Theory of Relativity (Pergamon
    Press, London, 1958) § 65, (δ) and Supple-
    mentary Note 22. Original German Edition
    published 1921.
7. H. Weyl, Z. Phys. 56, 330 (1929); F. London,
    Z. Phys. 42, 375 (1927).
8. J. J. Larmor, Aether and Matter (Cambridge
    University Press, Cambridge, 1900).

is performed wherever a gradient occurs. An
example is given in Eq. (3.88). We postulate that
there is no other electromagnetic interaction
and that for instance interactions involving dipole
moments and higher moments are a consequence of
Eq. (4.17). Gell-Mann calls the substitution
(4.17) the principle of the minimal electromagnetic interaction.[9]

From classical electrodynamics and from quantum mechanics we know that the vector potential
A^{μ} in Eq. (4.17) is not directly observable;[10]
any transformation of the form Eq. (3.8) should
leave physical results unchanged. In addition
we demand that the wave function ψ can undergo
a gauge transformation without physical manifes-
tation, and we consider a gauge transformation
that is more general than Eq. (4.15): Invariance
under Eq. (4.15) means only that the overall
phase of the wave function is not observable. We
now assume that the phase between any two points
is not observable; the gauge transformation then
is of the form

9. M. Gell-Mann, Supplemento Nuovo Cimento 4,
    848 (1956).
10. The role of the electromagnetic potential in
    quantum mechanics has been discussed by Y.
    Aharonov and D. Bohm, Phys. Rev. 115, 485
    (1959); W. H. Furry and N. F. Ramsey, Phys.
    Rev. 118, 623 (1960); V. F. Weisskopf, in
    1960 Lectures in Theoretical Physics, Vol.
    III, W. E. Brittin, B. W. Downs, and J. Downs,
    Eds. (Interscience Publishers, New York, 1961)
    p. 62 ff.

$$\psi' = \psi \, e^{-iN\Lambda(x)} \quad , \tag{4.18}$$

where $\Lambda(x)$ is an arbitrary real function of x.
Introducing the two gauge transformations Eq. (3.8)
and Eq. (4.18) into the right-hand side of Eq.
(4.17), operating on ψ, we get the expression

$$\nabla^\mu \psi \rightarrow e^{-iN\Lambda(x)} \, (\nabla^\mu - iN\nabla^\mu\Lambda(x)$$

$$+ \; iq \; \nabla^\mu\chi(x) + iq \; A^\mu)\psi \quad .$$

The expressions before and after the gauge transformations are, apart from a phase factor, identical only if we set

$$N \; \nabla^\mu\Lambda(x) = q \; \nabla^\mu\chi(x) \quad ,$$

or

$$N\Lambda(x) = q\chi(x) + d \quad .$$

The additive constant d introduces only a shift in
the assignment of the quantum numbers N and can be
set equal to zero. Since $\Lambda(x)$ is an arbitrary
function, we can set

$$\Lambda(x) = e \; \chi(x)$$

and then get

$$q = N e \quad .$$

The substitution representing the principle of minimal electromagnetic interaction thus remains invariant under the simultaneous transformation

$$\psi' = \psi \, e^{-iN\Lambda(x)} \quad , \tag{4.19a}$$

$$A^{\mu}{}' = A^{\mu} + \nabla^{\mu}\Lambda(x)/e \quad . \tag{4.19b}$$

As an application, consider the Dirac equation in the presence of an external electromagnetic field,

$$(i\not{\partial} - q \not{A} - m)\psi = 0.$$

The simultaneous transformations Eq. (4.19) lead to

$$[i\not{\partial} + N\not{\partial}\Lambda(x) - q \not{A} - (q/e)\not{\partial}\Lambda(x) - m]\psi' = 0.$$

This equation has the same form as the original Dirac equation only if $q = N e$.

The combined transformation Eq. (4.19) provides an example for the existence of a <u>compensating field</u>: If the phase of the wave function changes over space and time in a manner given by the function $\Lambda(x)$, this variation is counteracted by a corresponding change in the vector potential.[3] The counteraction is such that the substitution (4.17) is always valid. We further note that we have found the connection between

the additive quantum number N in Eq. (4.18) and
the electric charge q. The terms introduced by
the transformations of the wave function and of
the electromagnetic field only cancel if q is
proportional to N.

In field theory, where ψ and A^μ are operators,
the transformation of ψ, Eq. (4.19a), is called a
gauge transformation of the first type and the
transformation of A^μ, Eq. (4.19b) is called a
gauge transformation of the second type.[11] We
note that a gauge transformation of the first type
with $\Lambda(x)$ = constant already yields an additive conservation law. In order to introduce the interaction with the electromagnetic field, the gauge
function $\Lambda(x)$ must be a function of x; the compen-
sating field is then subject to gauge transformation
of the second type.

D. Gauge Invariance and Current Conservation[11,12]

In the previous subsection we have made it
plausible that invariance of a matrix element
under a gauge transformation of the wave function
leads to conservation of an additive quantum number.
In the present subsection, we treat this connection
in a more general way, using the language of field
theory. At the same time, we clear up another
problem: The probability currents for the
Schrödinger, the Klein-Gordon, and the Dirac equa-

---

11. W. Pauli, Rev. Mod. Phys. 13, 203 (1941).
12. M. Gell-Mann and M. Levy, Nuovo Cimento 16,
     705 (1960).

tions do not look the same and the choices that
we have made appear somewhat unmotivated. Here
we will show how the three different forms follow
from one common expression.

Under a variation $\delta\psi_i$ of the fields ψ_i, the
Lagrangian density $\mathcal{L}(\psi_i, \nabla^\mu\psi_i)$ varies as

$$\delta\mathcal{L} = \frac{\partial\mathcal{L}}{\partial\psi_i}\,\delta\psi_i + \frac{\partial\mathcal{L}}{\partial(\nabla^\mu\psi_i)}\,\delta(\nabla^\mu\psi_i)$$

$$= \delta\psi_i\left(\frac{\partial\mathcal{L}}{\partial\psi_i} - \nabla^\mu\frac{\partial\mathcal{L}}{\partial\nabla^\mu\psi_i}\right) + \nabla^\mu\left(\frac{\partial\mathcal{L}}{\partial\nabla^\mu\psi_i}\,\delta\psi_i\right)$$

where we have used $\nabla^\mu(\delta\psi) = \delta(\nabla^\mu\psi)$. Because of
the Euler-Lagrange field equation, Eq. (3.197),
the first two terms vanish and the variation
of the Lagrangian density becomes

$$\delta\mathcal{L} = \nabla^\mu\left(\frac{\partial\mathcal{L}}{\partial\nabla^\mu\psi_i}\,\delta\psi_i\right). \qquad (4.20)$$

This equation is convenient for deriving conservation laws by selecting suitable variations of the
field operators ψ_i. We consider as an example a
non-Hermitian field ψ. The Lagrangian density
then is a function of ψ, $\psi^\dagger$, $\nabla^\mu\psi$, and $\nabla^\mu\psi^\dagger$, and
the sum over i in Eq. (4.20) consists of two terms.
We consider the variation of the field ψ due to a
gauge transformation $\exp(-i\alpha)$, with α very small
and constant, and get

$$\delta\psi = \psi' - \psi = -i\alpha\psi, \quad \delta\psi^\dagger = +i\alpha\psi^\dagger \quad , \quad \alpha << 1 \quad . \quad (4.21)$$

Introducing these variations into Eq. (4.20) gives

$$\delta\mathcal{L} = i\alpha\nabla^\mu \left\{ \frac{\partial\mathcal{L}}{\partial\nabla^\mu\psi} \psi - \frac{\partial\mathcal{L}}{\partial\nabla^\mu\psi^\dagger} \psi^\dagger \right\} . \qquad (4.22)$$

If $\mathcal{L}$ is invariant under the gauge transformation we have

$$\delta\mathcal{L} = 0 \quad . \qquad (4.23)$$

We can thus define a current

$$j^\mu = -ic \left\{ \frac{\partial\mathcal{L}}{\partial\nabla_\mu\psi} \psi - \frac{\partial\mathcal{L}}{\partial\nabla_\mu\psi^\dagger} \psi^\dagger \right\} , \qquad (4.24)$$

where c is a real constant, and the i makes j^μ Hermitian. By virtue of Eqs. (4.22) and (4.23) j^μ satisfies the continuity equation

$$\nabla_\mu j^\mu = 0 \quad . \qquad (4.25)$$

With the proper choice of Lagrangian (Table 3.2), the current, Eq. (4.24), reduces to the previously discussed expressions Eq. (3.108) for the Klein-Gordon, and Eq. (3.168) for the Dirac field. We also note that the current vanishes for Hermitian fields. By integrating j^0 over all space and

assuming that the current is restricted to a finite
region, we find that the quantity

$$N = \int d^3x \; j^0(x)$$

is conserved. To establish the relation between
the treatment given so far and an additive conservation law, we consider for instance the Dirac
field, follow the discussion after Eq. (3.192), and
find easily that

$$N = c \sum_{p,h} \left(N_{ph}^a - N_{ph}^b \right).$$

With $c = 1$, N can be identified with the electric
charge number (if particles have positive charge).

We have pointed out in Subsection A that a
number of conserved additive quantum numbers exist.
Presumably, the Lagrangian describing the relevant
system then is invariant under a corresponding
number of separate gauge transformations. However,
since we do not know the complete Lagrangian
(and don't even know whether a Lagrangian description will ultimately be successful) the various
conserved generalized charges must be found experimentally. In the following subsections, we
discuss the evidence for the presently known additive quantum numbers.

E.   Conservation of the Electric Charge

In classical electrodynamics, conservation
of the electric charge is expressed by a continuity
equation of the form of Eq. (4.25), where the
current density and the charge density are macroscopic quantities. It is therefore possible to
test the conservation of the electric charge by
careful macroscopic experiments. However, for
theories involving elementary particles microscopic
tests are essential.

Consider a process such as $e^- p \rightarrow \nu n$. Charge
conservation does not mean that the magnitude of
the proton charge must be equal to that of the
electron. A very small charge could be carried
by the neutrino or the neutron and the total
charge in the initial and the final state could
still be the same. Thus there are two questions
to be solved experimentally: Is the electric
charge conserved? Do particles carry exact multiples of the elementary quantum e or do very
small deviations exist?

Charge conservation can be tested by studying
processes that are forbidden by charge conservation,
but allowed by all other known conservation laws.
Two decays that satisfy these conditions are

$$e \rightarrow \nu \gamma$$

$$e \rightarrow \nu \bar{\nu} \nu \; .$$

The lifetime of electrons is a test of charge con-

servation.   Goldhaber and der Mateosian have
looked for decays of electrons bound in atoms.[13]
If an atomic electron decays, it leaves a hole in
the shell.   The subsequent emission of an X ray
could be detected with a scintillation counter.
Goldhaber and der Mateosian set a lifetime limit
of about 10^{19} years; Moe and Reines improved the
limit by about two orders of magnitude.[14]   At the
present time there is no evidence for violation
of the conservation of the electric charge.

If particles carry exact multiples of the
elementary quantum e, the neutron charge should be
exactly zero and the charges of the electron and
the proton should be equal and opposite.   Shull,
Billman, and Wedgwood have studied the deflection
of a monoenergetic neutron beam by an electrostatic
field.[15]   This direct determination of the neutron
charge yields a value of $(1.9 \pm 3.7) \times 10^{-18}$e.
In indirect measurements, the deflection of neutral
atoms and molecules is investigated.   Such studies[16] indicate that the proton-electron charge
difference is less than 3×10^{-17} e and that the
neutron charge is less than 3×10^{-20} e.

We summarize the experimental results by

13.   G. Feinberg and M. Goldhaber, Proc. Nat. Acad.
      Sci. US 45, 1301 (1959).
14.   M. K. Moe and F. Reines, Phys. Rev. 140B, 992
      (1965).
15.   C. G. Shull, K. W. Billman, and F. A. Wedg-
      wood, Phys. Rev. 153, 1415 (1967).
16.   J. C. Zorn, G. E. Chamberlain, and V. W.
      Hughes, Phys. Rev. 129, 2566 (1963).   J. G.
      King, quoted in ref. 15.

saying that the electric charge q is conserved and
that a charge quantum number N = q/e can be assigned
to all particles. This quantum number obeys an
additive conservation law in all interactions,
hadronic, electromagnetic, and weak, as indicated
by the following examples:

Hadronic

$$p \quad p \rightarrow p \quad p \quad p \quad \bar{p}$$
$$1 + 1 = 1 + 1 + 1 - 1 \quad ;$$

$$p + (A,Z) \rightarrow {}^{7}_{3}Li + (A-6,Z-2),$$
$$1 + \quad Z \quad = \quad 3 \quad + \quad (Z-2) \quad ;$$

Electromagnetic

$$\gamma \quad d \rightarrow p \quad n \quad ,$$
$$0 + 1 = 1 + 0 \quad ;$$

Weak

$$n \rightarrow p \quad e^{-} \quad \bar{\nu}$$
$$0 = 1 - 1 + 0 \quad .$$

F. The Baryon Number (Heavy Charge)

In a few places, for instance in Eqs. (1.3)
and (3.80), we have described transformations that
involve protons and neutrons. If we look at these
we note that the total number of nucleons is conserved. A free neutron can decay into a proton,
an electron, and a neutrino -- but before and after
the decay one nucleon is present. Nucleons are
either stable or very long lived; to characterize
their stability, we give limits on the lifetime.
The age of the universe gives an absolute lower
limit. A better one is obtained by considering
the heat flow from the earth: subtracting the
amount due to radioactivity and ascribing the rest

to the decay of nucleons into lighter particles results in a limit of about 10^{20} years. Still better lower limits are obtained by looking for possible decay processes in a large amount of matter deep underground, well shielded from cosmic rays.[17,18] The established limits depend somewhat on the assumed decay mode; some values are quoted in Table 4.1.

Table 4.1 Lower limits on the nucleon lifetime[18]

Assumed decay mode	Lifetime limits
$p \rightarrow \mu \rightarrow e$	2×10^{30} y
$p \rightarrow \pi^+ \nu$	1.7×10^{29} y
$p \rightarrow e^+ \gamma$	3.0×10^{29} y
$n \rightarrow \pi^+ e^-$	3.6×10^{29} y

Conservation of the total number of nucleons as a fundamental law of nature was first expressed by Stueckelberg and later by Wigner.[19] To formulate the law, we assign a baryon number A = +1 to

17. F. Reines, C. L. Cowan, and M. Goldhaber, Phys. Rev. 96, 1157 (1954).
18. H. S. Gurr, W. R. Kropp, F. Reines, and B. Meyer, Phys. Rev. 158, 1321 (1967). F. Reines and M. F. Crouch, Phys. Rev. Letters 32, 493 (1974).
19. E. C. G. Stueckelberg, Helv. Phys. Acta 11, 225, 299 (1938).

nucleons, $A = -1$ to antinucleons, and state the
conservation law[20]

$$\sum_i A_i = \text{const.} \qquad (4.26)$$

Leptons, mesons, and photons have $A = 0$. This
assignment is consistent with reactions such as

$$p \; p \rightarrow p \; p \; p \; \bar{p} \quad ,$$

$$p \; n \rightarrow p \; p \; \pi^- \quad .$$

The nucleon lifetime values given in Table 4.1
indicate that the validity of the conservation law
(4.26) is established to a very high degree for
nucleons.

When strange particles were discovered, decays
such as

$$\Lambda \rightarrow p \; \pi^-$$

were found, and the law of conservation of nucleons
had to be generalized to include heavier fermions,
the so-called hyperons. It turns out that there
are a considerable number of heavy fermions that
have $A = \pm 1$; including the nucleons these are
called baryons and the assignment given for nucleons

---

20.   The baryon number is often denoted by B. We
      prefer A here because the baryon number of a
      nucleus is its mass number, traditionally
      designated by A.

is generalized to

$$A = 1 \text{ for } p \ n \ \Lambda \ \Sigma \ \Xi \ \Omega$$

and A = -1 for the corresponding antiparticles.
<u>Nuclei</u> can also be characterized by their baryon
number. Ordinary nuclei are built up from protons
and neutrons. The total charge Z of such a system
is equal to the number of protons, the total baryon
number A is equal to the total number of nucleons
and we designate such a nucleus by (A,Z). <u>Hyper-</u>
<u>nuclei</u> are similar to nuclei, but with one or more
nucleons replaced by hyperons. They can be characterized by giving the baryon number A.

In all reactions involving particles, nuclei,
and hypernuclei, no violation of Eq. (4.26) has
ever been seen.

So far, we have introduced baryon conservation
as an expression for an observed experimental fact.
We know from the discussion in Subsection D that
an additive conservation law is the consequence
of a gauge transformation. For the baryon field
ψ we then write this gauge transformation as

$$\psi(x) \rightarrow \psi(x) e^{-iA\Lambda(x)} . \qquad (4.27)$$

The observed conservation of A suggests that the
Lagrangian governing the baryons and their interactions should be invariant under such a gauge
transformation, with $\Lambda(x) = \lambda = $ const. The unsolved question, however, is whether a generalized

charge is connected with the baryon number. If
$\Lambda(x)$ is indeed a function of x, i.e., if the
relative phase of the baryon field between any
two space-time points is not a measurable quantity, then this suggests that a corresponding
gauge field exists: In Subsection C we saw that
the space-time variation of the gauge function
$\Lambda(x)$ had to be counteracted by the space-time
variation of the electromagnetic potential A(x).
The corresponding electromagnetic field is
coupled with the same strength to all particles of
the same charge. The coupling strength is ± e to
a lepton with no strong interactions, and also ± e
to a hadron of this charge. Similarly the space-
time variation of the baryon gauge function $\Lambda(x)$
should be counterbalanced by the variation of some
vector field universally coupled to all baryons.
If this hypothetical field were massless, like
the electromagnetic field, it would give rise to
a long-range force that could be observed in
Eötvös-type experiments; existing data indicate
that the universal coupling constant then would be
extremely weak.[4] However, such a field does
not have to be massless and it could have escaped
detection so far.[5,21] The question as to whether
a baryonic charge is connected with the baryonic
number is thus not solved.

---

21. J. Schwinger, in Theoretical Physics (International Atomic Energy Agency, Vienna, 1963).

Similar problems exist in connection with the other additive quantum numbers that we will discuss in the following subsections: What are the corresponding gauge transformations? Do corresponding interaction constants, i.e., generalized charges exist? At present, there is no evidence for them.

G.   The Lepton Number (Lepton Charge)[22]

We have listed in Table 1.1 four leptons, e^-, μ^-, ν_e, ν_μ; the four corresponding antileptons also exist. In order to explain the absence of certain decay modes allowed by all then known conservation laws, Konopinski and Mahmoud[23] introduced the idea of lepton conservation: Assign lepton number L = +1 to all leptons, L = -1 to all antileptons, L = 0 to all other particles. Lepton conservation then reads

$$\sum_i L_i = \text{const.} \tag{4.28}$$

Leptons, like baryons, can be destroyed or created only in particle-antiparticle pairs. In the neutron decay, for instance, an antineutrino appears with the electron; in a decay with positron emission, a neutrino is created together with the positron.

---

22.   W. Pauli, Nuovo Cimento 6, 204 (1957). T. D. Lee and C. S. Wu, Ann. Rev. Nucl. Sci. 15, 381 (1965).
23.   E. J. Konopinski and H. M. Mahmoud, Phys. Rev. 92, 1045 (1953).

Lepton conservation is intimately tied up with the
question of neutrino and antineutrino identity.
Electron and positron are easily distinguished by
their charge and lepton number can be assigned
correspondingly. However, it is much harder to
see how the neutrino and the antineutrino can be
different because they are neutral, probably have
zero mass and no magnetic moment. If neutrino
and antineutrino are identical, the lepton number
assignments given above are wrong. The experiments
that we are discussing in this subsection are aimed
at proving or disproving this identity.

Evidence for lepton conservation comes from
neutrino reactions and from double-beta decay
studies. Consider first the neutrino reactions:

$$\bar{\nu}\ p \rightarrow e^+\ n \tag{4.29}$$

$$\bar{\nu}\ ^{37}Cl \rightarrow e^-\ ^{37}Ar\ . \tag{4.30}$$

The first of these reactions is allowed by lepton
conservation, the second forbidden. Both reactions
have been studied with antineutrinos from reactors.
(Fission produces neutron-rich nuclides; neutron-
rich nuclides decay with the emission of antineutrinos since the excess neutrons are transformed
into protons via the reaction $n \rightarrow p\ e^-\ \bar{\nu}$). Reaction
(4.29) was indeed observed by Reines, Cowan, and

co-workers.[24)]   However, the cross section for the
forbidden reaction (4.30), studied by Davis with a
radiochemical method, is at least 20 times smaller
than predicted if lepton conservation were not
valid.[25)]

We describe here one of the experiments partly
because the results are important and partly because
it set a trend for elementary particle physics in
recent years. It used to be that a particle counter
could be carried around by one man. Counters for
neutrino experiments, however, are monsters and they
grow every year. The small cross sections and the
impossibility of focusing neutrinos are responsible
for this development.

The arrangement used in the anti-neutrino-
capture experiment is shown in Fig. 4.1. A beam
of antineutrinos from the Savannah River fission
reactor, of intensity 1.3×10^{13} $\bar{\nu}/cm^2$ sec, passed
through a large liquid scintillator, containing
1400 liters of scintillating liquid and 1.8 g/liter
cadmium (in the form of cadmium octoate). The
scintillator was viewed by 110 photomultipliers.

---

24. F. Reines and C. L. Cowan, Phys. Rev. 113,
     273 (1959); F. Reines, C. L. Cowan, F. B.
     Harrison, A. D. McGuire, and H. W. Kruse,
     Phys. Rev. 117, 159 (1960).
25. R. Davis, Phys. Rev. 97, 766 (1955); J. N.
     Bacall, in High Energy Physics and Nuclear
     Structure (North Holland Publishing Co.,
     Amsterdam, 1967). R. Davis, Jr., D. S.
     Harmer, and K. C. Hoffman, Phys. Rev. Letters
     20, 1205 (1968).

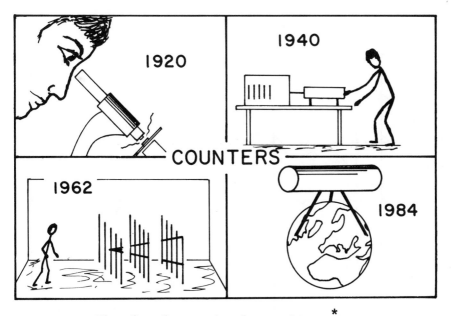

The development of counters.[*]

The entire detection system was shielded against cosmic rays and radiation other than neutrinos from the reactor. The identification of an antineutrino capture event was based on the following facts: If an antineutrino is captured by a proton, a neutron of a few keV and a positron in the range from about 0 to 8 MeV energy (depending on the energy of the incident antineutrino) are produced. The positron loses energy quickly and annihilates after about 3×10^{-10} sec, producing two 0.51 MeV gamma rays. The neutron loses its energy more

---

*Since this figure was first drawn, economic realities have somewhat delayed realization of the ultimate counter in 1984.

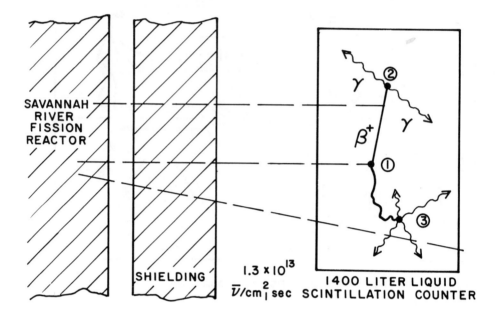

SAVANNAH
RIVER
FISSION
REACTOR

SHIELDING

1.3×10^{13}
$\overline{\nu}/cm^2 \, sec$

1400 LITER LIQUID
SCINTILLATION COUNTER

Fig. 4.1.   Antineutrino capture. An incident
            antineutrino is captured at 1, creating
            a positron and a neutron. The positron
            slows down quickly and annihilates at
            2, producing two 0.51 MeV gamma rays.
            The neutron loses energy more slowly,
            but it is finally captured by a Cd
            nucleus at 3, producing about 9.1 MeV
            of capture gamma rays.

slowly, in times of the order of μsec. It is
finally captured by a Cd nucleus which has a large
capture cross section. Capture gamma rays, of
total energy 9.1 MeV, are emitted. In the experiment this sequence of events took place in the

scintillator. The output of the photomultiplier
was displayed on oscilloscopes and the time sequence
of events could thus be followed. Only if the
correct pulse heights were observed within the
proper time intervals was an event registered.
Finally, to find the background, the entire equipment was operated with the nuclear reactor shut
down. The answer was unambiguous: Counts were
observed that could only be due to antineutrinos.
When the cross section was computed from the estimated antineutrino flux and the observed counting
rate, it was found to be

$$\sigma(\text{meas}) = (11 \pm 2.6) \times 10^{-44} \text{ cm}^2 \quad . \quad (4.31)$$

This cross section agrees with crude estimates and
is in excellent agreement with the value obtained
from weak interaction theory. However, the reader
should realize how small this cross section is and
how much the success of the capture experiment
depended on major advances in nuclear physics. To
visualize the smallness of the cross section, it
is sufficient to estimate the mean free path of
such neutrinos: A lead absorber of more than a
light year thickness is required to reduce in half
the intensity of the reactor antineutrino beam.
Indeed, when Bethe and Peierls discussed the
neutrino in 1934[26], they stated: "It is therefore

_____

26. H. Bethe and R. Peierls, Nature 133, 532,
     689 (1934).

absolutely impossible to observe processes of this
kind with the neutrinos created in nuclear transformations." Only the development of powerful
neutrino sources (namely nuclear reactors) and of
new types of detectors (scintillation counters)
made it possible to disprove this statement.

We turn to the second reaction, Eq. (4.30),
forbidden if lepton conservation holds. To search
for this reaction, Davis irradiated a large volume
of CCl_4 with antineutrinos from a reactor, then
swept out the product ^{37}Ar with helium gas, separated the argon from the helium, and counted the
^{37}Ar activity in a low-background counter. The
latest runs give a cross section limit
$\sigma \lesssim 0.2 \times 10^{-45}$ cm^2, as compared to the prediction
$\sigma_{id} = 4.3 \times 10^{-45}$ cm^2 if neutrinos and antineutrinos
are not distinguished by the lepton number.

A better limit on lepton conservation is
reached from studies of double beta decay.[27] We
consider here the specific case, given in Fig. 4.2,
of ^{48}Ca decaying into ^{48}Ti. The beta decay
$^{48}Ca \rightarrow ^{48}Sc$ is extremely slow because the decay
energy is very small. However, there is more than
enough energy available for the process

$$^{48}Ca \rightarrow {}^{48}Ti + 2\ e^- + 2\ \bar{\nu} \ . \qquad (4.32)$$

---

27.   H. Primakoff and S. P. Rosen, Alpha-, Beta-,
      and Gamma-Ray Spectroscopy, K. Siegbahn, ed.,
      (North Holland, Amsterdam, 1965) Vol. II,
      p. 1499. V. R. Lazarenko, Soviet Phys.
      Uspekhi 9, 860 (1967).

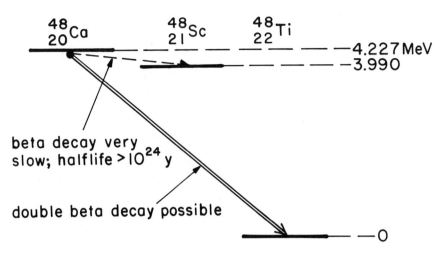

Fig. 4.2.   Double beta decay in the isobars with
            A = 48.

If neutrinos and antineutrinos are identical, the
following virtual decay can occur

$$^{48}Ca \rightarrow \, ^{48}Sc + e^- + \nu \quad \text{(virtual)} \quad .$$

The virtual neutrino is never emitted, but it
induces the second step,

$$\text{virtual} \quad \nu + \, ^{48}Sc \rightarrow \, ^{48}Ti + e^- \quad .$$

The following reaction thus should be possible:

$$^{48}Ca \rightarrow \, ^{48}Ti + 2\, e^- \quad . \tag{4.33}$$

In this case, only two particles are emitted and
the lifetime is much shorter than for the decay,

Eq. (4.32). However, lepton conservation would be violated! Experimentally, the second decay mode, Eq. (4.33), would be easier to see: Since only two leptons are emitted and the recoil energy of the nucleus can be neglected, the two electrons together share the entire decay energy. The energies of the two electrons thus add up to the decay energy and this fact can be used to search for a "sum line." In the decay (4.32), four leptons are emitted and the sum of the electron energies varies from 0 to the decay energy.

The search for double beta decay has been performed with counters and with radiochemical tools. Counter experiments are difficult because of the background from radioactive contaminants. The best limit has been obtained by searching for coincidences in a small two-gap discharge chamber, a dwarf relative of the streamer chambers used in high-energy physics. To reduce the background further, the apparatus was set up in a salt mine 600m below ground level.[28a] The limits obtained are given in Table 4.2. In radiochemical studies, one tries to detect the accumulation of decay products during geological time periods ($\sim 10^9$ y) in minerals that are rich in suspected double-beta decay active nuclides. Two nuclides have yielded positive results[28b,c] and

28.   a.   R. K. Bardin, P. J. Gollon, J. D. Ullman, and C. S. Wu, Phys. Letters 26B, 112 (1967). b. T. Kirsten, W. Gentner, and O. A. Schaeffer, Z. Physik 202, 273 (1967). c. T. Kirsten, O. A. Schaeffer, E. Norton, and R. W. Stoenner, Phys. Rev. Letters 20, 1300 (1968).

these are also shown in Table 4.2.

Table 4.2. Halflife for double beta decay.

Decay	Theoretical halflife (y)		Exp. halflife (y)	Ref.
	Neutrino-less	Two neutrinos		
$^{48}Ca \xrightarrow{\text{e}^-\text{e}^-} {}^{48}Ti$	$\approx 10^{15}$	$\approx 10^{21}$	$\{ \begin{matrix} >10^{21*} \\ >10^{19.5\#} \end{matrix}$	28a
$^{82}Se \xrightarrow{\text{e}^-\text{e}^-} {}^{82}Kr$	$10^{16\pm2}$	$10^{22\pm2.5}$	$10^{19.8\pm0.3}$	28b
$^{130}Te \xrightarrow{\text{e}^-\text{e}^-} {}^{130}Xe$	$10^{16.3\pm2}$	$10^{22.5\pm2.5}$	$10^{21.3\pm0.1}$	28c

*assuming a neutrinoless decay.

#assuming a two-neutrino decay.

The experiments discussed in this subsection show that the lepton number is conserved. However, a small contribution of the lepton-nonconserving neutrinoless decay mode cannot be ruled out. The results summarized in Table 4.2 indicate that the amplitude ratio of a lepton non-conserving inter-action to a lepton-conserving one is at most of the order $(10^{16}$ y/10^{21} y$)^{1/2}$, or $\approx 3 \times 10^{-3}$.

H.   The Muon Number

The radiative decay of the muon

$$\mu \rightarrow e\gamma \qquad\qquad (4.34)$$

is allowed by all the conservation laws discussed so far. Despite valiant efforts by many groups, such a decay has never been seen and a limit

$$\frac{R(\mu \rightarrow e\gamma)}{R(\mu \rightarrow e\bar{\nu}\nu)} < 2 \times 10^{-8} \qquad\qquad (4.35)$$

has been obtained.[29] Does the absence of the radiative muon decay point to some law that we have not yet discussed or can we understand it in terms of already explained facts?

The first question that comes to mind concerns the possibility of an electromagnetic transition.[30] How fast would an electromagnetic transition be? The neutral pion decays into two gamma rays with a lifetime of about 10^{-16} sec; if the process (4.34) were allowed to go by the electromagnetic interaction, it would swamp the normal muon decay.

If we assume that muon and electron satisfy the Dirac equation and that the electromagnetic

---

29.   S. Parker, H. L. Anderson, and C. Rey, Phys. Rev. 133, B768 (1964).
30.   G. Feinberg, P. Kabir, and S. Weinberg, Phys. Rev. Letters 3, 527 (1959).

interaction is minimal, it is trivial* to see that
the electromagnetic transition $\mu \rightarrow e\gamma$ is forbidden.
Inserting the minimal electromagnetic interaction
Eq. (4.17) into the Dirac Lagrangian (Table 3.1)
gives a Lagrangian density

$$\mathcal{L} = \bar{\psi}(i\slashed{\partial} - q\slashed{A} - m)\psi \quad . \tag{4.36}$$

The second term describes the interaction with an
electromagnetic field. For the radiative muon
decay, we write the interaction term as

$$\mathcal{L}_{int} = -q\, \bar{\psi}_e\, \slashed{A}\, \psi_\mu + H.c.$$

$$\tag{4.37}$$

$$\equiv -q\, \bar{\psi}_e\, \gamma^\nu\, \psi_\mu\, A_\nu + H.c. \quad .$$

In second quantized language, $\mathcal{L}_{int}$ is an operator
that describes the radiative muon decay: ψ_μ
destroys a muon, $\bar{\psi}_e$ creates an electron, and A_ν
creates (or destroys) a photon. The corresponding
Feynman diagram is given in Fig. 4.3.

---

*"Trivial," as used by physicists, is a technical
 term. In one case, B. A. Jacobsohn questioned
 C. N. Yang about a certain "trivial" point.
 After a short interchange, they retired to an
 office and re-emerged after a day, agreeing
 that the point was trivial.

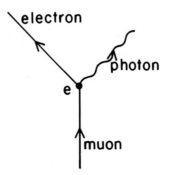

Fig. 4.3.   Feynman diagram for the radiative
            muon decay.

The term $\bar{\psi}\gamma^\nu\psi$ is just the current operator for
Dirac particles; thus $\mathcal{L}_{int}$ has the familiar form

$$\mathcal{L}_{int} = -q\, j^\nu A_\nu + \text{H.c.} \quad , \tag{4.38}$$

where j^ν and A^ν are field operators. We will now
show that the phenomenological Lagrangian density
(4.38) is not consistent with current conservation.
Using Eq. (4.25) we write

$$0 = \nabla^\nu j_\nu = \nabla^\nu (\bar{\psi}_e \gamma_\nu \psi_\mu) = (\bar{\psi}_e \overleftarrow{\slashed{\nabla}}) \psi_\mu + \bar{\psi}_e \slashed{\nabla} \psi_\mu \quad .$$

The Dirac equation (3.167) gives

$$i\slashed{\nabla}\psi_\mu = m_\mu \psi_\mu \quad , \quad i\bar{\psi}_e \slashed{\nabla} = -m_e \bar{\psi}_e \quad ,$$

and current conservation thus yields

$$(m_\mu - m_e) \overline{\psi}_e \psi_\mu = 0 \quad . \tag{4.39}$$

Since $m_\mu \neq m_e$, we get a contradiction and the
radiative decay of the muon cannot go by a minimal
electromagnetic interaction.

The weak interaction can give rise to the
radiative muon decay through electromagnetic corrections to the ordinary decay. Two typical
Feynman diagrams are shown in Fig. 4.4. In the
first, the muon decays virtually, the electron
radiates a photon, and then reabsorbs the neutrino
and the antineutrino. In the second, the muon

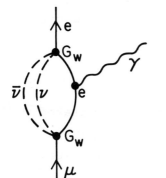
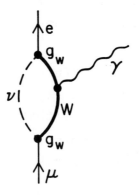

Fig. 4.4. Diagrams for the radiative decay of
           the muon. The first diagram is second
           order in G_w, the second one first
           order (Eq. (1.13)).

decays into a neutrino and a W particle. (We have

mentioned the hypothetical W particle in Section
1.3.) The W radiates a photon and then transforms
into an electron with absorption of the virtual
neutrino. Computations[31] for the branching ratio
$(\mu \to e\gamma)/(\mu \to e\nu\bar{\nu})$ with diagrams of the type shown
in Fig. 4.4 give values of the order of 10^{-4} or
10^{-5} - much larger than the observed limit.

Explanations for the absence of the radiative
decay were put forward by many authors;[32] common
to these is the occurrence of a difference between
the neutrinos accompanying electrons and the
neutral particles associated with muons. A way
to introduce such a difference is to postulate a
new additive quantum number, the muon number L_μ,
and assign L_μ = +1 to the negative, L_μ = -1 to
the positive muon, respectively. The reaction

$$\pi^+ \to \mu^+ \nu_\mu$$

then indicates that the neutrinos associated with
muons also have $|L_\mu|$ = 1. All other particles
would have L_μ = 0. The absence of the decay (4.34)

---

31. L. Di Lella, Lecture Notes on Weak Inter-
     actions, C. Fronsdal, ed. (W. A. Benjamin,
     Reading, Mass., 1963), p. 122.
32. K. Nishijima, Phys. Rev. 108, 907 (1957);
     J. Schwinger, Ann. Phys. 2, 407 (1957); N.
     Cabibbo and R. Gatto, Phys. Rev. Letters 5,
     114 (1960); G. Feinberg and S. Weinberg,
     Phys. Rev. Letters 6, 381 (1961).

is then "trivial."

The lepton and muon number assignments to all leptons are

$$
\begin{array}{ccccccccc}
 & e^- & e^+ & \nu_e & \bar{\nu}_e & \mu^- & \mu^+ & \nu_\mu & \bar{\nu}_\mu & \\
L & 1 & -1 & 1 & -1 & 1 & -1 & 1 & -1 & \quad (4.40) \\
L_\mu & 0 & 0 & 0 & 0 & 1 & -1 & 1 & -1 &
\end{array}
$$

Unless such an assignment leads to new predictions, it is meaningless. However, a reaction with muon neutrinos should lead only to muons, as in

$$\nu_\mu \; n \rightarrow \mu^- \; p$$

but the corresponding electron production,

$$\nu_\mu \; n \rightarrow e^- \; p$$

should be absent. To test this prediction, extensive experiments were performed by a Columbia group at Brookhaven and by a CERN group.[33] The essential idea is simple: A neutrino "beam" is generated by decay in flight of pions (Eq. 4.40).

33. G. Danby, J. M. Gaillard, K. Goulianos, L. M. Lederman, N. Mistry, M. Schwartz, and J. Steinberger, Phys. Rev. Letters 9, 36 (1962). J. K. Bienlein, Böhm, Von Dardel, Faissner, Ferrero, Gaillard, Gerber, Hahn, Kaftanov, Krinen, Reinharz, Salmeron, Seiler, Staude, Stein, Steiner, Phys. Letters 13, 80 (1964).

The pions are produced by high-energy protons
striking a target. All particles except neutrinos
are then removed from the beam by an iron shield,
about 14m thick. The neutrino interactions are
observed in a giant spark chamber and (at CERN)
also in a bubble chamber.[34) The basic arrangement is sketched in Fig. 4.5.

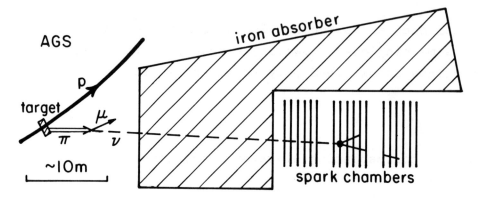

Fig. 4.5.   Basic set-up for the detection of
            muon neutrinos.

While the idea of the experiment is simple, the
execution produces a large number of headaches.
The spark chambers must be shielded on all sides
against neutrons; cosmic ray particles must be
recognized; electrons and muons must be distin-

34.   M. M. Block, Burmeister, Cundy, Eiben,
      Franzinetti, Keren, Møllerud, Myatt,
      Nikolic, Orkin-Lecourtois, Paty, Perkins,
      Ramm, Schultze, Sletten, Soop, Stump, Venus,
      and Yoshiki, Phys. Letters 12, 281 (1964).

guished in the chambers.   Some remarks concerning
experimental details may shed additional light on
the difficulties:

a)   The accelerator yields about 1 pulse per sec.
     The radiation from the target occurs during
     individual bursts 20 nsec wide, spaced about
     220 nsec.   To minimize the background due to
     other radiations, the chambers are made
     sensitive only during the time neutrinos are
     expected.

b)   To be certain that the events seen in the
     spark chamber are really due to decaying pions
     (and kaons), the first 1.2m of iron are removed
     and a similar quantity of lead is set as
     closely to the target as possible.   The pions
     and kaons then interact before decaying and
     the neutrino beam is "switched off," while the
     background is essentially unchanged.

     The result of the experiments is unambiguous:
the muon neutrino can be distinguished from the
electron neutrino; the introduction of an additional
quantum number is justified.   How well is the muon
number conserved?   By using a pure neutrino beam,
and by measuring the charge ratio of the muons
produced in neutrino interactions, a limit of
6×10^{-2} has been obtained for the amplitude of
a possible violation.[35]

---

35.   K. Borer, B. Hahn, H. Hofer, H. Kaspar, F.
      Krienen, and P.-G. Seiler, Phys. Letters
      29B, 614 (1969).

To explain the absence of the decay $\mu \rightarrow e\gamma$, we have introduced an additive quantum number here, but a multiplicative one will also explain all present experiments. A multiplicative conservation law is less restrictive and future experiments hopefully will distinguish between the two possibilities.[36]

I. The Hypercharge (Strangeness)

Denn eben, wo Begriffe fehlen,
Da stellt ein Wort zur rechten Zeit sich ein.

MEPHISTOPHELES in GOETHE's FAUST

In 1947, a new class of particles was discovered.[37] They were neutral and upon decay gave rise to two tracks that formed a V. Because of these tracks, they were called V particles. A sketch from a bubble chamber photograph is shown in Fig. 4.6; the tracks are designated with the current names of the particles.

---

36. R. Gatto, in Proceedings of the International School of Physics "Enrico Fermi." Selected Topics on Elementary Particle Physics (Academic Press, New York, 1963), Course 26, p. 254. N. Van Hieu and B. Pontecorvo, JETP Letters 7, 105 (1968). J. J. Amato, P. Crane, V. W. Hughes, J. E. Rothberg, and P. A. Thompson, Phys. Rev. Letters 21, 1709 (1968). C. Y. Chang, Phys. Rev. Letters 24, 79 (1970).
37. G. D. Rochester and C. C. Butler, Nature 160, 855 (1947).

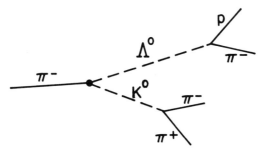

Fig. 4.6. V particles in a hydrogen bubble
chamber.

By about 1952, a sizable number of V particles
had been observed and these particles seemed to be
strange indeed: They were produced copiously,
but decayed very slowly. With our present knowledge
and with the current names, we can describe the
puzzle in the following way: The total cross section near threshold for the reaction shown in Fig.
4.6,

$$\pi^- p \rightarrow K^\circ \Lambda^\circ \quad , \qquad (4.41)$$

is sketched in Fig. 4.7 as a function of the pion
kinetic energy. The production cross section is
of the order of a mb and hence comparable to that
of pions by pions on nucleons. The production is
clearly an hadronic process. The decays of the
lambda and the kaon, however, are much slower:

$$\Lambda^\circ \rightarrow p\pi^-, \ \tau = 2.5 \times 10^{-10} \ \text{sec},$$
$$K^\circ \rightarrow \pi^+\pi^-, \ \tau = 0.87 \times 10^{-10} \ \text{sec}.$$

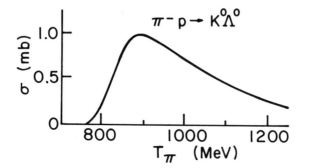

Fig. 4.7. Total cross section for the production of strange particles (V particles) by pions. T_π is the pion laboratory kinetic energy.

These mean lifes are about 12 orders of magnitude longer than expected for hadronic interactions, but about right for the weak interactions. Kaons and hyperons are produced strongly, but decay weakly! This discrepancy was a major concern around 1952. The first step to the presently accepted solution was advanced by Pais[38] who suggested the hypothesis of associated production: "Strange particles are always produced in pairs." Gell-Mann and Nishijima independently analyzed the situation further and introduced a new quantum number, strangeness[9,39,40]. Today, with our detailed

38. A. Pais, Phys. Rev. 86, 663 (1952).
39. M. Gell-Mann, Phys. Rev. 92, 833 (1953).
40. T. Nakano and K. Nishijima, Progr. Theoret. Phys. 10, 581 (1953); K. Nishijima, Fortschr. Phys. 4, 519 (1956); Progr. Theoret. Phys. 13, 285 (1955).

knowledge of reactions involving strange particles,
it is easy to justify the assignments that we will
give below; in 1953, however, the experimental
data were meager and the introduction of strangeness
was truly imaginative.

We assign the strangeness quantum number by
using a number of reactions that have been observed
to lead to copious production of strange particles.
We begin with

$$p\ p \rightarrow p\ \Lambda^0\ K^+\ ,\qquad\qquad (4.42)$$

and assign <u>strangeness S</u> by <u>definition</u>,

$$S = +1 \text{ for the } K^+$$
$$S = 0 \text{ for all normal particles}\qquad (4.43)$$
$$\text{(pions and nucleons)}.$$

For particles that do not interact hadronically,
strangeness is not defined. We now assume that
strangeness is an additive quantum number, conserved in all hadronic and electromagnetic interactions, but violated by the weak interaction. We
can then explain the discrepancy between hadronic
production and weak decay of kaons and hyperons.
As long as we use nonstrange particles in a reaction, strange particles can only be produced in
pairs and the associated production is "explained."
In reaction (4.42), the lambda particle would be
assigned $S = -1$ and strangeness would be conserved.
In the decays of the hyperons and the kaons,

strangeness is violated and the decays must proceed via a weak interaction. (Actually strange particles can also be produced singly through a weak interaction. The cross section for such processes is very small and it will require "meson factories" to produce them in this way.)

From observed hadronic interactions and with $S(K^+) = 1$, $S(\Lambda^0) = -1$, further strangeness assignments can be made:

$$\pi^- p \to K^+ \Sigma^- \qquad \to \quad S(\Sigma^-) = -1$$
$$p\,p \to n\,\Sigma^+ K^+ \qquad \to \quad S(\Sigma^+) = -1 \qquad (4.44)$$
$$p\,p \to p\,\Sigma^0 K^+ \qquad \to \quad S(\Sigma^0) = -1.$$

Since the three Σ particles also have about the same masses (Table 1.1), we assume that they belong together and form an isospin triplet. To get the strangeness assignment for the cascade particle, we note that it is produced hadronically in the reaction

$$\pi^- p \to K^+ K^0 \Xi^- \quad ;$$

with Eq. (4.41), we get $S(K^0) = +1$ and hence $S(\Xi^-) = -2$. Similarly, the strangeness of the neutral cascade particle can be determined and it is also found to be -2.

Reactions involving the K^- bring a surprise;

$$\pi^- p \to K^+ K^- n$$

indicates that $S(K^-) = -S(K^+) = -1$. The K^+ and the K^- have different strangeness, unlike the case of the three Σ hyperons in Eq. (4.44). What does this difference mean? We remember that S is an additive quantum number and we expect thus that

$$S(\text{particle}) = - S(\text{antiparticle}) \qquad . \qquad (4.45)$$

The K^+ could be the antiparticle of the K^-, but how then do the three sigmas relate to each other? The Σ^+ cannot be the antiparticle of the Σ^-. If we assume that the three sigmas of Eq. (4.44) belong together and that Eq. (4.45) holds, we expect three antisigmas. These antisigmas should have $S = +1$ and $A = -1$. Such antisigmas have been found.

We now return to the kaons. If the K^- is the antiparticle of the K^+, then the K° should also have an antiparticle with $S = -1$; this $\overline{K^\circ}$ must differ from the K°. The $\overline{K^\circ}$ can be produced, for instance by the hadronic reaction.

$$\pi^+ p \rightarrow \overline{K^\circ} K^+ p \qquad .$$

It thus appears that the K^+ and the K° belong together and that the K^- and the $\overline{K^\circ}$ are their antiparticles. Nature has produced two sets of kaons. While the Σ° and the $\overline{\Sigma^\circ}$ differ in their strangeness and their baryon number, the neutral kaons only differ in their strangeness. When Gell-Mann proposed the existence of the two neutral kaons, Fermi

challenged him to devise an experiment that would
show an observable difference between the K^0 and
the $\overline{K^0}$. We will see in Chapter 5 that Gell-Mann
and Pais answered this challenge brilliantly.

We can now state when a particle can be identical to its antiparticle: Only a completely
neutral particle, for which all additive quantum
numbers are zero, can be identical to its antiparticle. The neutral pion, for instance, has
$A = q = S = 0$ and hence the π^0 and the $\overline{\pi^0}$ can be
identical. They are identical and the antiparticle
of the π^+ is the π^-.

For many discussions it is more useful to
introduce the hypercharge instead of the strangeness.[41]   Since A and S are conserved in hadronic
and electromagnetic interactions, we can use a
linear combination. The hypercharge Y is defined
by

$$Y = A + S \quad .$$ (4.46)

In most of the following discussions we will use
the hypercharge rather than strangeness.

In Table 4.3 we list the values of strangeness,
hypercharge, and baryon number for a number of
hadrons. In the last column, we give the average
charge of the particles listed in the relevant row.
We will later make use of the fact that for each
multiplet $\frac{1}{2}Y = <Q>/e$.

---

41. B. d'Espagnat and J. Prentki, Nucl. Phys. $\underline{1}$,
    33 (1956); $\underline{3}$, 446 (1957).

J. Hypercharge Selection Rules

If a quantity A is absolutely conserved, the corresponding selection rule is $\Delta A = 0$ in all interactions. When a quantum number, such as hypercharge, is conserved in the strong, but is violated in the weak interaction, is there a limit to the violation? Do there exist selection rules? To answer these questions, we look at the decay scheme shown in Fig. 4.8.

Table 4.3.   Baryon number A, strangeness S, hyper-
             charge Y, and average charge <Q/e> for
             some hadrons and antihadrons.

Particle		A	S	Y	<Q/e>
Pion	π^+, π^o, π^-	0	0	0	0
Kaon	K^+, K^o	0	1	1	1/2
	$K^-, \overline{K^o}$	0	-1	-1	-1/2
Nucleon	p, n	1	0	1	1/2
	$\overline{p}, \overline{n}$	-1	0	-1	-1/2
Lambda	Λ^o	1	-1	0	0
	$\overline{\Lambda^o}$	-1	1	0	0
Sigma	$\Sigma^+, \Sigma^o, \Sigma^-$	+1	-1	0	0
	$\overline{\Sigma}^-, \overline{\Sigma}^o \, \overline{\Sigma}^+$	-1	1	0	0
Cascade	Ξ^-, Ξ^o	1	-2	-1	-1/2
	$\overline{\Xi}^+, \overline{\Xi}^o$	-1	2	1	1/2

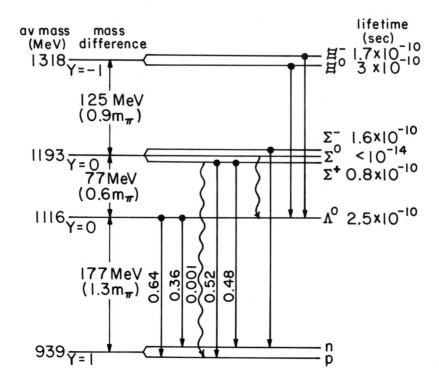

Fig. 4.8. Decay scheme of hyperons. Straight arrows represent pions, wavy lines represent photons. (Relative intensities are given.) Beta decays are not shown.

As pointed out before, all lifetimes (except that of the Σ^0) are very long, of the order of 10^{-10} sec. If strong interactions were responsible for the decays, the lifetime with pion emission would be of the order of 10^{-22} sec. In the one case where decay with $\Delta Y = 0$ can occur, namely from the Σ^0 to the Λ^0, the energy difference is not large enough to create a pion; the decay proceeds electromagnetically, with a lifetime that is shorter than 10^{-14} sec. Such a lifetime is consistent with no inhibition. Hence we conclude

$$\Delta Y = 0 \quad \text{in hadronic and electromagnetic}$$
$$\text{interactions.} \qquad (4.47)$$

We now turn to the weak interaction. The decay scheme Fig. 4.8 shows that particles differing by zero or one unit of hypercharge are connected by weak decays. (One should not be confused by the fact that the pion occurs both in strong and in weak decay modes. The pion has strong, electromagnetic and weak interactions. The weak interaction of the pion can only be observed when strong and electromagnetic ones are forbidden.) Since no decay of the Ξ^- or Ξ^0 to the nucleon is observed, we conclude that $\Delta Y = 2$ transitions are forbidden or strongly reduced so that

$$\Delta Y = 0, \pm 1 \quad \text{for the weak interaction.} (4.48)$$

The two selection rules (4.47) and (4.48) still appear to leave one question unanswered: The decay $\Sigma^+ \to p\gamma$ occurs with a partial lifetime of about 10^{-8} sec. For no inhibition of the electromagnetic decay we would expect a lifetime of about 10^{-20} sec. (The decay energy is about twice the energy of the π^0 and only one photon is emitted.) With the selection rule (4.47), we would not

expect a gamma decay at all. How is the gamma ray
emitted in the decay of the Σ^+? A Feynman diagram
that can explain the long lifetime is shown in
Fig. 4.9.[42] The sigma decays weakly into a neutron and a virtual charged pion. The pion emits

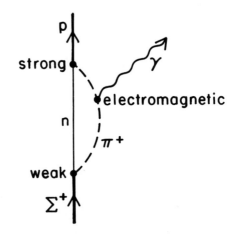

Fig. 4.9.   Feynman diagram for the decay of
            $\Sigma^+ \rightarrow$ pγ by a combination of interactions.

a photon and is then reabsorbed strongly by the
neutron. The hypercharge change occurs at the
weak vertex. Other Feynman diagrams contribute
to the same decay, but the idea should be clear
from Fig. 4.9. The slow decay of the Σ^+ is a
radiative correction to the weak decay and is
additional evidence for the selection rule $\Delta Y = 0$
for pure electromagnetic interactions.

---

42. R. E. Behrends, Phys. Rev. 111, 1691 (1958).

4.3   ROTATIONS IN ORDINARY SPACE

In Section 4.1 we mentioned that invariance
under rotations in ordinary space leads to conservation of angular momentum; in the present section
we derive this connection and outline the formal
apparatus that we will use in later chapters.

A.   Rotations

We describe a physical system by the scalar
wave function $\psi(\vec{x})$ as sketched in Fig. 4.10a. By
rotating the physical system around the z axis by
a positive angle α we get the system $\psi^{\mathcal{R}}(\vec{x}^{\mathcal{R}})$ as
shown in Fig. 4.10b. The same rotation can be
described, in a mathematically equivalent form, by
rotating the coordinate axes by the angle $-\alpha$.[*] We
denote the operator that rotates the <u>physical
system</u> by the angle α around the z axis by $R_z(\alpha)$;
the rotation from Fig. 4.10 a to b is given by

$$\vec{x}^{\mathcal{R}} = \mathcal{R}_z(\alpha)\,\vec{x} \quad . \tag{4.49}$$

$\mathcal{R}$ is not a unitary operator, but denotes a rotation
in ordinary space. Comparing Figs. 4.10a and 4.10b
we notice that

$$\psi^{\mathcal{R}}(\vec{x}^{\mathcal{R}}) = \psi(\vec{x}) \quad . \tag{4.50}$$

---

[*]Rotation of the physical system is often called
an <u>active</u> rotation, in contrast to the <u>passive</u>
rotation of the coordinate axes; we will use
active rotations.

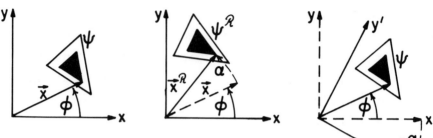

Fig. 4.10.   Rotation in space. The rotation of
the system $\psi(\vec{x})$ can either be described
by a rotation of the physical system
by the angle α around the z axis (Fig.
b), or by a rotation of the coordinate
axes by an angle $-\alpha$ (Fig. c). The
angle ϕ defines the position of $\vec{x}$; it
does not denote a rotation.

This relation is really trivial. It states that
the rotated scalar wave function at the rotated
point $\vec{x}^{\mathcal{R}}$ is the same as the original wave function
at the original point $\vec{x}$. Of course this argument
assumes invariance under rotations. With
$\vec{x} = \mathcal{R}^{-1} \vec{x}^{\mathcal{R}}$, where $\mathcal{R}^{-1}$ denotes the rotation inverse to $\mathcal{R}$, Eq. (4.50) can be rewritten as
$\psi^{\mathcal{R}}(\vec{x}^{\mathcal{R}}) = \psi(\mathcal{R}^{-1}\vec{x}^{\mathcal{R}})$. Since $\vec{x}$ is an arbitrary
vector, we can drop the superscript $\mathcal{R}$ and write

$$\psi^{\mathcal{R}}(\vec{x}) = \psi(\mathcal{R}^{-1}\vec{x}) \qquad . \qquad (4.51)$$

The validity of this relation is also obvious from

Fig. 4.10.   Consider now an infinitesimal rotation
about the z axis by the angle $\alpha \rightarrow \varepsilon$, with $\varepsilon << 1$.
Expanding the right-hand side of Eq. (4.51) and
keeping only terms that are linear in ε gives

$$\psi^{\mathcal{R}}(\vec{x}) = \psi(\mathcal{R}_z^{-1}\vec{x}) = \psi(\vec{x}) + \frac{\partial \psi}{\partial \phi}(-\varepsilon)$$

$$= (1 - \varepsilon \frac{\partial}{\partial \phi})\psi(\vec{x})$$

or

$$\psi^{\mathcal{R}}(\vec{x}) = (1 - i\varepsilon L_z)\psi(\vec{x}) \quad . \qquad (4.52)$$

Since ψ and $\psi^{\mathcal{R}}$ describe the same physical situa-
tion, the corresponding state vectors $|\psi\rangle$ and
$|\psi^{\mathcal{R}}\rangle$ must be connected by a unitary transformation,
Eq. (4.3b). Because the transformation is contin-
uous, U is of the form of Eq. (4.9),

$$|\psi^{\mathcal{R}}\rangle = U(\varepsilon)|\psi\rangle = (1 + i\varepsilon F)|\psi\rangle \quad ,$$

and

$$\psi^{\mathcal{R}}(\vec{x}) = \langle\vec{x}|\psi^{\mathcal{R}}\rangle = \langle\vec{x}|(1 + i\varepsilon F)|\psi\rangle$$

$$= \psi(\vec{x}) + i\varepsilon \langle\vec{x}|F|\psi\rangle \quad .$$

With the closure relation (3.30) we write

$$\psi^{\mathcal{R}}(x) = \psi(\vec{x}) + i\varepsilon \sum_{\vec{x}'} \langle\vec{x}|F|\vec{x}'\rangle \langle\vec{x}'|\psi\rangle$$

$$= \psi(\vec{x}) + i\varepsilon \sum_{\vec{x}'} \langle\vec{x}|F|\vec{x}'\rangle \psi(\vec{x}') \quad .$$

This expression agrees with Eq. (4.52) if we set $F = -L_z$; the unitary transformation $U_z(\alpha)$ for finite angles α is indeed given by Eq. (4.11). The generalization to a rotation by an angle δ around an arbitrary direction $\hat{n}$ is straightforward and gives

$$U_{\hat{n}}(\delta) = e^{-i\delta\hat{n}\cdot\vec{L}} \quad . \tag{4.53}$$

If the system that is rotated has a spin $\vec{S}$, U still has the form of Eq. (4.53), but $\vec{L}$ is replaced by the total angular momentum $\vec{J} = \vec{L} + \vec{S}$.

If the system is invariant under any rotation about $\hat{n}$, the total Hamiltonian H will commute with $U_{\hat{n}}$, and hence also with $\hat{n}\cdot\vec{J}$:

$$[H,U_{\hat{n}}] = 0 \rightarrow [H,\hat{n}\cdot\vec{J}] = 0 \quad . \tag{4.54}$$

The component of the angular momentum along $\hat{n}$ is conserved. If there were a preferred axis $\hat{z}$ in space, invariance would occur only for rotations about this axis. Isotropy means that we can choose $\hat{n}$ to be any axis, say along $\hat{x}$, $\hat{y}$, or $\hat{z}$.

Of course, whether the angular momentum $\vec{J}$ is a constant of the motion can only be determined by experiment. No evidence for even a very small violation of angular momentum conservation has ever been found.

B. Angular Momentum

Rotations do not commute; from Eq. (4.53) follows the generalization of Eq. (3.12),

$$[J_x, J_y] = iJ_z \quad \text{and cyclic,} \tag{4.55}$$

which holds for any angular momentum, spin, orbital, and total. Equation (4.54) gives

$$[H, J_x] = [H, J_y] = [H, J_z] = [H, J^2] = 0, \tag{4.56}$$

and with Eq. (4.55) also

$$[J^2, \hat{n} \cdot \vec{J}] = 0 \quad . \tag{4.57}$$

J^2 and $\hat{n} \cdot \vec{J}$ can be diagonalized simultaneously; $\hat{n}$ is usually chosen along the z axis. The eigenfunctions of J^2 and J_z are given by generalizations of Eq. (3.13),

$$J^2 | j\ m\ a\rangle = j(j + 1) | j\ m\ a\rangle \tag{4.58}$$

$$J_z | j\ m\ a\rangle = m | j\ m\ a\rangle \tag{4.59}$$

where a denotes additional quantum numbers used to
characterize the state. It is often useful to
define two non-Hermitian operators, $J_\pm$

$$J_\pm = J_x \pm iJ_y \quad , \quad J_+ = J_-^\dagger \quad , \tag{4.60}$$

for which the following commutation rules can be
readily established:

$$[J_\pm , J^2] = 0 \quad , \quad [J_z , J_\pm] = \pm J_\pm ,$$

$$[J_+ , J_-] = 2J_z \quad . \tag{4.61}$$

From these commutation relations and $|m| \leq j$ it
follows that

$$J_\pm |j\ m\ a> = +\left[(j \mp m)(j \pm m + 1)\right]^{\frac{1}{2}} |j,m\pm1,a> \quad .$$

$$\tag{4.62}$$

The + sign is chosen as the conventional arbitrary
phase. Because J_+ increases m by one unit, it is
called a raising operator; J_- lowers m by one unit
and is called a lowering operator.

C. Rotation Matrices[*]

We select a representation $|j\ m\ a>$ where J^2 and J_z are diagonal and all operators that specify a are unaffected by rotations. Equation (4.3b) then reads, for a rotation by an angle δ around the axis $\hat{n}$,

$$|j\ m\ a>^{\mathcal{R}} = U_{\hat{n}}(\delta)|j\ m\ a> \quad .$$

where the superscript $\mathcal{R}$ indicates that the ket $|\ >^{\mathcal{R}}$ describes the rotated physical system. To find explicit expressions for the transformation, we use Eq. (3.30) and write

$$|j\ m\ a>^{\mathcal{R}} = \sum_{j'\ m'} |j'\ m'\ a><j'\ m'\ a|U_{\hat{n}}(\delta)|j\ m\ a>.$$

---

[*]Various definitions of the rotation matrices (D functions) are in use; this fact often leads to confusion and even experts spend hours to get their minus signs right. We wish we could claim that our treatment avoids all the pitfalls but we know better. We only point out that the best way to clear up uncertainties is to select a simple example and carry out the calculations explicitly. We also mention again that we rotate the physical system and not the coordinate axes. This "active" definition is commonly used in particle physics; in nuclear physics, passive rotations are preferred. (See e.g. Bohr and Mottelson.)

Since $[U_{\hat{n}}, J^2] = [U_{\hat{n}}, H] = 0$, the rotation cannot induce transitions between different values of j; therefore we have j' = j and the sum extends only over m'. For a rotation around the z axis, the matrix element $<j\ m'a|U_z(\alpha)|j\ m\ a>$ is diagonal; with Eq. (4.8) defining the operation of the exponential operator Eq. (4.53), the matrix element is

$$<j\ m'\ a|U_z(\alpha)|j\ m\ a> = e^{-im\alpha}\ \delta_{m'm} \qquad . \quad (4.63)$$

For a general rotation, the matrix element can no longer be evaluated so simply. However, because the rotation operator is used frequently, the matrix elements have been tabulated (See Appendix B); they are denoted by

$$<j\ m'\ a|U_{\hat{n}}(\delta)|j\ m\ a> = D^j_{m'm}(\mathcal{R}) \qquad . \qquad (4.64)$$

With this abbreviation, the transformation properties of a state $|j\ m\ a>$ under the rotation $\mathcal{R}_{\hat{n}}(\delta)$ are given by

$$|j\ m\ a>^{\mathcal{R}} = \sum_{m'} D^j_{m'm}(\mathcal{R})\ |j\ m'\ a> \qquad . \qquad (4.65)$$

Sometimes it is convenient to have the transformation properties of a wave function rather than a state. To get such an expression, we multiply Eq. (4.65) by $<\vec{x}|$ and find

$$\Psi_{jm}^{\mathcal{R}}(\vec{x}) = \sum_{m'} \Psi_{jm'}(\vec{x}) \, D_{m'm}^{j}(\mathcal{R}) \quad . \qquad (4.66)$$

This equation relates the wave function of the original and the rotated state at the same point in space. We will give an example of the use of Eq. (4.66) in Subsection D. Finally we note that in all rotations, the coordinate axes remain fixed in space and all magnetic quantum numbers therefore refer to one and the same z axis.

So far we have described a general rotation by giving the rotation axis $\hat{n}$ and the corresponding rotation angle δ. Since two angles are required to describe $\hat{n}$, three angles completely specify a rotation. For the explicit evaluation of the rotation matrices, a different set of angles is more convenient, namely the three Euler angles α, β, and γ. To define these, we distinguish two sets of axes, the coordinate axes, (x,y,z), that are fixed in space and the systems axes, (X,Y,Z), that are fixed in the physical system. Before the rotation, the two sets coincide. The Euler angles are defined by the three rotations of the physical system shown in Fig. 4.11:

1. Positive rotation through α about z to X_1,Y_1,z; $(0 \leq \alpha < 2\pi)$
2. Positive rotation through β about Y_1 to X_2,Y_1,z'; $(0 \leq \beta < \pi)$

3.   Positive rotation through γ about Z' to
     X',Y',Z'; (0 ≤ γ ≤ 2π).

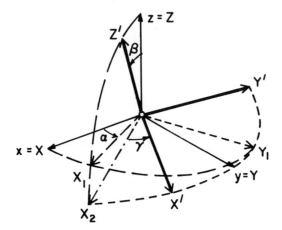

Fig. 4.11.   Rotation of a physical system,
             described by the Euler angles
             α, β, and γ. The coordinate axes
             (x,y,z) remain fixed in space. The
             system axes (X,Y,Z) are fixed in
             the physical system and rotate
             with it.

Overall, the physical system characterized by
(X,Y,Z) is transformed to a position characterized
by (X',Y',Z'). The unitary operator $U(\alpha,\beta,\gamma)$ is
then given by

$$U(\alpha,\beta,\gamma) = U_{Z'}(\gamma)\, U_{Y_1}(\beta)\, U_z(\alpha) , \qquad (4.67)$$

where the order of operation is from the right to

the left.   For explicit computations, this form is
not useful because rotations around the axes Y_1
and Z' are involved, and these move with the system
that is rotated.   However, Fig. 4.11 shows that the
same final orientation can be obtained by inverting
the order of the three rotations and carrying them
out in the fixed coordinate system.   To show this
fact formally, we first note that the operator
$U_{Y_1}(\beta)$ rotates the physical system from the direction z to Z'.   With $\mathscr{O} = U_z(\gamma)$, we get from Eq.
(4.3a)

$$U_{Z'}(\gamma) = U_{Y_1}(\beta)\, U_z(\gamma)\, U_{Y_1}^{\dagger}(\beta) \quad .$$

Inserting this expression into Eq. (4.67) and
using $U_{Y_1}^{\dagger}(\beta)\, U_{Y_1}(\beta) = 1$ gives

$$U(\alpha,\beta,\gamma) = U_{Y_1}(\beta)\, U_z(\gamma)\, U_z(\alpha)$$

$$= U_{Y_1}(\beta)\, U_z(\alpha)\, U_z(\gamma) \quad .$$

Similarly we eliminate $U_{Y_1}(\beta)$ by using

$$U_{Y_1}(\beta) = U_z(\alpha)\, U_y(\beta)\, U_z^{\dagger}(\alpha)$$

and find

$$U(\alpha,\beta,\gamma) = U_z(\alpha)\, U_y(\beta)\, U_z(\gamma) \quad . \qquad (4.68)$$

With the expression (4.53) for U we finally get

$$U(\alpha,\beta,\gamma) = e^{-i\alpha J_z}\, e^{-i\beta J_y}\, e^{-i\gamma J_z} \quad , (4.69)$$

and we can evaluate the matrix element $D^j_{m'm}$, Eq. (4.64):

$$D^j_{m'm}(\alpha,\beta,\gamma) = e^{-i\alpha m'}\, d^j_{m'm}(\beta)\, e^{-i\gamma m} \quad ,(4.70)$$

$$d^j_{m'm}(\beta) = <jm'a|e^{-i\beta J_y}|jma> \quad . \quad (4.71)$$

The properties of the rotation matrices D and d can be obtained from those of the rotation operator U. Some of the more important ones, and some Tables are given in Appendix B.

D.   The Symmetric Top[*]

As an application of the previous considerations we derive the wave function of the symmetric
top. Some nuclei have permanent deformations and
can be idealized as shown in Fig. 4.12. We now
distinguish two sets of axes: x, y, and z are
fixed in space and are not rotated. The principal

---

[*]For a discussion of the phase conventions see K.
Alder and A. Winther, Phys. Letters 34B, 357
(1971).

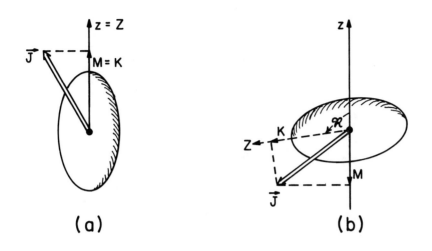

(a)   **(b)**

Fig. 4.12.   Symmetric top.   The axes X, Y, and Z
are fixed in the top, the axes x, y,
and z are fixed in space.   The rotation
$\mathcal{R}$ carries the top from the special
position, where the z and the Z axes
coincide, to the general position
shown at right. $\vec{J}$ is the angular
momentum of the top.

axes X, Y, and Z are fixed in the body and rotate
with it.   Classically, the Hamiltonian of the top
is

$$H_{rot} = \frac{J_X^2}{2\mathcal{I}_X} + \frac{J_Y^2}{2\mathcal{I}_Y} + \frac{J_Z^2}{2\mathcal{I}_Z} \tag{4.72}$$

where the $\mathcal{I}_i$ are the moments of inertia about the principal axes. Nuclei generally are axially symmetric; with Z as the symmetry axis and $\mathcal{J}_X = \mathcal{J}_Y \equiv \mathcal{J}$, the Hamiltonian becomes

$$
\begin{aligned}
H_{rot} &= \frac{1}{2\mathcal{J}} (J_X^2 + J_Y^2) + \frac{1}{2\mathcal{J}_Z} J_Z^2 \\
&= \frac{1}{2\mathcal{J}} J^2 + \frac{1}{2} \left(\frac{1}{\mathcal{J}_Z} - \frac{1}{\mathcal{J}} \right) J_Z^2 \quad .
\end{aligned}
\tag{4.73}
$$

The energy eigenvalues then are

$$
E_{rot} = \frac{1}{2\mathcal{J}} j(j+1) + \frac{1}{2} \left(\frac{1}{\mathcal{J}_Z} - \frac{1}{\mathcal{J}} \right) K^2 \quad , \tag{4.74}
$$

where K is the eigenvalue of J_Z , as shown in Fig. 4.12.

Next we consider the angular momentum eigenfunctions of the top. Such eigenfunctions are easiest to find if the principal axis Z coincides with the quantization axis z, as shown in Fig. 4.12 a, and we denote these functions by $\psi_{jm}^O(\hat{x})$. They become particularly simple if the top (or nucleus) is axially symmetric. In this case, the angular eigenfunctions do not depend upon rotation around the z axis and are therefore constants. Moreover, since the component of $\vec{J}$ along $\hat{z} = \hat{Z}$ is conserved and has the value m = K, only the one function ψ_{jK}^O is different from zero:

$$
\psi_{jm}^O (\text{symmetric top}) = \delta_{mk} \quad . \tag{4.75}
$$

The angular eigenfunctions of the symmetric
top in the general position shown in Fig. 4.12b
are now easy to find. The operator $\mathcal{R}$ rotates
the top and with it the angular momentum $\vec{J}$ from
the preferred position shown in Fig. 4.12a to the
general position in Fig. 4.12b. The component of
$\vec{J}$ along the body-fixed axis Z is denoted by K; it
is unchanged by the rotation. The component of J
along the quantization axis naturally changes
under the rotation; in the general position it is
denoted by M. The angular eigenfunctions in the
special and the general case are connected by the
relation Eq. (4.66); with Eq. (4.75) we thus get
for the function $\psi^{\mathcal{R}}$:

$$\Psi_{jM}^{\mathcal{R}} \equiv \Psi_{jMK}^{\mathcal{R}} = \sqrt{\frac{2j+1}{8\pi^2}}\ D_{KM}^{j}(\mathcal{R}) \quad . \tag{4.76}$$

The normalization constant has been computed with
Eq. (B 6); the indices M and K indicate that the
eigenvalues of J_z and J_Z are M and K, respectively.
     The nuclear spectrum is further restricted by
symmetry of the nucleus about the XY plane.   To
see the effect of this symmetry on the wave
function, we first consider the nucleus to be in
the special position of Fig. 4.12a.   Rotation of
the nucleus about the y axis by the angle π changes
the wave function Eq. (4.75) into

$$(\psi_{jm}^{o})^{\mathcal{R}} = D_{KM}^{j}(0,\pi,0) = d_{KM}^{j}(\pi) = (-1)^{j+K}\ \delta_{K,-m} \ ,$$

where we used Eqs. (4.66) and (B.15). With
$\psi^0 = \delta_{mK} + (-1)^{j+K}\delta_{m,-K}$ the rotational nuclear
wave function for the general position becomes,
after proper normalization,

$$\psi^{\mathcal{R}}_{jMK} = \left(\frac{2j + 1}{16\pi^2(1+\delta_{K0})}\right)^{1/2}$$

$$\times \left\{D^j_{KM} + (-1)^{j+K}D^j_{-KM}\right\}. \tag{4.77}$$

The simplest situation arises for $K = 0$, where
Eq. (4.77) shows that j must be even. Energy
eigenvalues and wave function are then given by

$$E_{rot} = \frac{1}{2\mathcal{J}}j(j+1) \quad , \ j \text{ even}$$

$$\psi_{rot} = \sqrt{\frac{1}{2\pi}}(-1)^M Y^{M*}_j(\beta,\gamma) \quad . \tag{4.78}$$

E. Scalar, Vector, and Tensor Operators

A scalar operator, S, is invariant under rotation; with Eq. (4.53) the invariance condition is

$$[\vec{J},S] = 0.$$

A state $S|j',m',a'\rangle$ hence is an eigenstate of J^2
and J_z with eigenvalues $j'(j' + 1)$ and m' and is
thus orthogonal to any state $|j,m,a\rangle$ unless

$j' = j$, $m' = m$; the matrix element
$<j',m',a'|S|j,m,a>$ vanishes if $j' \neq j$ or $m' \neq m$.
Moreover, J_+ commutes with S, and for $j' = j$ and
$m' = m$ we thus get with Eq. (4.62)

$$<j \; m \; a'|S|j \; m \; a>$$

$$= [(j-m+1)(j+m)]^{-1/2} <j \; m \; a'|SJ_+|j,m-1,a>$$

$$= [(j-m+1)(j+m)]^{-1/2} <j \; m \; a'|J_+S|j,m-1,a>$$

$$= <j,m-1,a'|S|j,m-1,a> \quad .$$

The matrix element of a scalar operator hence is
independent of m and vanishes unless $j = j'$ and
$m = m'$:

$$<j'm'a'|S|jma> = \delta_{jj'} \; \delta_{mm'} <ja'|S|ja> \quad ,(4.79)$$

or

$$<j'm'a'|S|jma>$$

$$= \frac{1}{\sqrt{2j+1}} <jm00|j'm'><ja'||S||ja> \quad . \quad (4.79a)$$

These equations express the fact that the matrix
element of a scalar operator between angular
momentum eigenstates can be written as a product
of two factors. The first factor gives the dependence on the orientation in space, the second is
independent of m and contains the dynamics of the

problem. We will see below that the second equation is a special case of the Wigner-Eckart theorem
in its customary form; the m-independent factor
<ja'|| S || ja> is called the <u>reduced</u> <u>matrix</u> <u>element</u>
and is defined through Eq. (4.79a).

We call $\vec{V}$ a vector operator if its expectation
value transforms like the vector $\vec{x}$, Eq. (4.49).
The expectation value before rotation is $<a|\vec{V}|a>$;
rotation changes the operator $\vec{V}$ to $U \vec{V} U^\dagger$ and the
expectation value to $<a|U \vec{V} U^\dagger|a>$. $\vec{V}$ then is a
vector operator if

$$<a|U \vec{V} U^\dagger|a> = \mathcal{R}<a|\vec{V}|a> \quad ,$$

or if

$$U \vec{V} U^\dagger = \mathcal{R} \vec{V} \quad .$$

With this relation, Eq. (4.53) and using infinitesimal rotations about the three coordinate axes,
we find the commutation relations

$$[J_x,V_x] = 0, \quad [J_x,V_y] = iV_z, \quad [J_x,V_z] = -iV_y ,$$

$$+ \text{ cycl.} \qquad (4.80)$$

The angular momentum commutation rules are a
special case of Eq. (4.80). Any operator satisfying the commutation relations (4.80) with $\vec{J}$ is
called a vector operator.

As the next step, we could evaluate the matrix

elements of a vector operator. Instead of doing it, we turn directly to irreducible tensor operators[43] and quote a number of results without proof. Detailed discussions can be found in the books listed at the end of this chapter.

In Equation (4.65) we have given the expression for a rotated state $|j\ m\ a>^{\mathcal{R}}$ in terms of the original states $|j\ m'\ a>$ and the rotation matrix $D^j_{m'm}$. In analogy, an irreducible tensor operator $T^{(k)}$, of order (or rank) k, is defined as consisting of a set of 2k + 1 operators $T^{(k)}_q$, q = k, k-1, ... , -k, that transform under rotation among themselves according to

$$\left(T^{(k)}_q\right)^{\mathcal{R}} = U\ T^{(k)}_q\ U^{\dagger} = \sum_{q'=-k}^{k} T^{(k)}_{q'}\ D^{(k)}_{q'q} \quad .(4.81)$$

By considering infinitesimal rotations, the commutation rules with the angular momentum operator $\vec{J}$ are obtained:

$$\left[\vec{J}, T^{(k)}_q\right] = \sum_{q'} T^{(k)}_{q'}\ <kq'|\vec{J}|kq> \quad , \qquad (4.82)$$

43. G. Racah, Phys. Rev. underline{61}, 186 (1942); underline{62}, 438 (1942); underline{63}, 367 (1943). E. P. Wigner, On the Matrices Which Reduce the Kronecker Products of Representations of Simply Reducible Groups (unpublished); Am. J. Math. underline{63}, 57 (1941). These papers are reprinted in Quantum Theory of Angular Momentum, L. C. Biedenharn and H. Van Dam, eds. (Academic Press, New York, 1965).

or

$$\left[J_z, T_q^{(k)} \right] = q \, T_q^{(k)}$$

(4.82a)

$$\left[J_\pm, T_q^{(k)} \right] = [k(k+1) - q(q\pm1)]^{1/2} \, T_{q\pm1}^{(k)} \quad .$$

Scalars are irreducible tensor operators of order zero. Vector operators are irreducible tensor operators of order 1. The Cartesian components of vector operators satisfy the commutation rules (4.80). By comparing this equation with Eq. (4.82) it is straightforward to find the "spherical components" of a vector operator; they are given by

$$T_1^{(1)} = \frac{-1}{\sqrt{2}} \, (V_x + iV_y) ,$$

$$T_0^{(1)} = V_z ,$$

(4.83)

$$T_{-1}^{(1)} = \frac{1}{\sqrt{2}} \, (V_x - iV_y) \quad .$$

We have already encountered an example of such components in the circular polarization vectors $\hat{\varepsilon}_\pm$ which were defined in Eq. (3.18) and which become operators in field theory. More generally, the 2k+1 spherical harmonics y_k^q, considered as operators, are an example of an irreducible tensor operator of order k.

The most important property of irreducible

tensor operators is expressed by the Wigner-Eckart theorem:

$$<j'm'a'|T_q^{(k)}|jma>$$

(4.84)

$$= \frac{1}{\sqrt{2j'+1}} <jmkq|j'm'><j'a'||T^{(k)}||ja> \quad .$$

The theorem states that the matrix element of a component, $T_q^{(k)}$, of an irreducible tensor operator between angular momentum eigenstates factors into a Clebsch-Gordan coefficient, $<jmkq|j'm'>$, and a reduced matrix element, $<j'a'||T^{(k)}||ja>$. The reduced matrix element, denoted by the double bars, is the same for all components of $T^{(k)}$ and is independent of the orientation in space. It contains the dynamics. The "geometrical" information is contained in the Clebsch-Gordan coefficient.

4.4   ISOSPIN

Elementary particles and nuclear states are distinguished by their quantum numbers, such as mass, charge, spin, baryon number, and parity. Since over a hundred elementary particles and thousands of nuclear states are known, it is desirable to establish unifying relations. One such relation is the particle-antiparticle connection. There also exist groups of particles and of nuclear levels that appear to differ solely in their electromagnetic and weak interaction processes, but have the same space-time properties and strong interactions. Examples are (n,p), (π^+, π^o, π^-), $(^3\text{He}, ^3\text{H})$, $(^{14}\text{C}, ^{14}\text{N}^*, ^{14}\text{O})$. How do we describe such "multiplets"? What determines the number of members of a multiplet?

An examination of the systematics of nuclear isobars suggests the existence of a basic symmetry between protons and neutrons. The hadronic interactions between proton and neutron, neutron and neutron, and proton and proton are very closely the same in the same space and spin states. Such symmetries exist also among mesons and hyperons. In order to describe the symmetry, an additional conserved quantum-mechanical property, isospin, is introduced. Isospin has many properties similar to ordinary spin. The number of particles in a given "isospin multiplet" is $2I+1$, where I is the isospin quantum number. Particles belonging to one isospin multiplet have the same strong interactions and the same space-time properties; this fact can be expressed as "invariance under rotation in isospin space" and the formal considerations treated in the previous section can be taken over.

In the following subsections we first discuss evidence that led to the formulation of the principle of charge independence of the strong forces, then introduce the isospin formalism, and finally present some applications.

A. Charge Symmetry

In 1932, Heisenberg postulated neutrons and protons to be the nuclear constituents.[44] In order to explain the observed fact that the mass of most nuclides, in units of m_p, is about twice their charge, in units of e, he assumed that the attraction between neutron and proton is stronger than that between two neutrons and that only the Coulomb repulsion acts between protons. Somewhat later it was conjectured that the p-p hadronic force should be equal to the n-n hadronic force because otherwise the maximum of nuclear stability would not occur for equal numbers of neutrons and protons. This principle of charge symmetry can be expressed more generally: Apart from electro-magnetic and weak effects, a system of mesons and baryons has properties identical to its charge symmetric counterpart, provided the particles are in the same space-spin states.

Evidence for charge symmetry in nuclear physics is abundant. Mirror nuclei, for instance ^{3}H and ^{3}He, have very closely the same binding energy after Coulomb effects have been subtracted. The level structure in mirror nuclei is very similar. In Fig. 4.13, we show the first few excited states in ^{7}Li and ^{7}Be; the similarity is obvious.

---

44. W. Heisenberg, Z. Physik 77, 1 (1932); Z. Physik 78, 156 (1956). Translated in D. M. Brink, Nuclear Forces (Pergamon Press, Ltd., London, 1965).

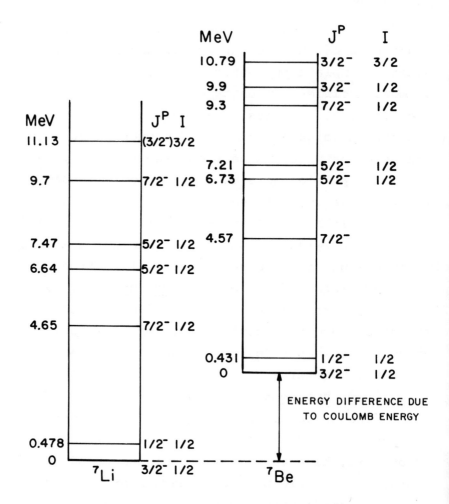

Fig. 4.13.   Level structure in mirror nuclides
             (isobars). Spins, parities, and
             isospins of the levels are shown.

    Considerable evidence for charge symmetry
exists also in particle physics. After correcting
for Coulomb scattering, the scattering of positive

and negative pions from deuterons shows the same cross section. The systems $p\pi^-$ and $n\pi^+$ behave very similarly, as do the systems $n\pi^-$ and $p\pi^+$.

B.   Charge Independence of Nuclear Forces

In 1936, evidence[45] accumulated that the strength of the np and pp interactions were approximately equal if identical ranges were assumed. Breit, Condon, and Present proposed that the two interactions are exactly equal if the Coulomb force is neglected.[46]   This assumption, together with charge symmetry, leads to the hypothesis of charge independence, expressed by Breit and Feenberg:[47]   The forces between any two nucleons in the same orbital angular momentum and spin states are the same, apart from electromagnetic effects. As an example, the pp, nn, and np forces in the $^1S_0$ state should be the same.

The experimental evidence for charge independence is strong.   After correction for Coulomb effects, the np and pp scattering in the same space-time states are characterized by similar parameters.   Similar states (isobaric states) exist in nuclei, for instance in ^6He, $^6Li^*$, ^6Be, in ^{14}C, $^{14}N^*$, ^{14}O; and in ^{91}Zr, $^{91}Nb^*$.   (For

45.   M. A. Tuve, N. Heydenberg, and L. R. Hafstad, Phys. Rev. 50, 806 (1936).
46.   G. Breit, E. U. Condon, and R. D. Present, Phys. Rev. 50, 825 (1936).
47.   G. Breit and E. Feenberg, Phys. Rev. 50, 850 (1936).

nuclei, stars indicate excited states.) We will
treat such states below. The principle of charge
independence also applies to meson-nucleon scattering and to hyperons. One of the more accurate
experimental verifications of charge independence
involves the pion-nucleon and nucleon-nucleon
forces. The experiment compares the production of
neutral and charged pions from the collision of
protons with deuterons. The experimental ratio

$$\sigma(pd \rightarrow \pi^+ \, ^3H) \big/ \sigma(pd \rightarrow \pi^0 \, ^3He)$$

was found to be 2.13 ± 0.06 at 590 MeV.[48] The
theoretical ratio is 2 if electromagnetic effects
are neglected and 2.20 ± 0.07 if these effects
are taken into account.[49]

We cannot expect charge independence to be an
exact symmetry, even after subtracting Coulomb
effects, since there are other effects of electromagnetic origin. We have stated earlier that
nuclear forces are mediated by the exchange of
mesons. The neutral and the charged mesons have
slightly different masses (presumably also because
of electromagnetic effects) and we must anticipate
small deviations from charge independence, even
after Coulomb effects have been removed.[50]

48. D. Harting, J. C. Kluyver, A. Kusumegi, R.
     Rigopoulos, A. M. Sachs, G. Tibell, G. Van-
     derhaeghe, and G. Weber, Phys. Rev. 119, 1716
     (1960).
49. H. S. Köhler, Phys. Rev. 118, 1345 (1960).
50. E. M. Henley, in Isobaric Spin in Nuclear
     Physics, J. D. Fox and D. Robson, eds.
     (Academic Press, Inc., New York, 1966).

C. The Isospin for Nucleons

The similarity between neutrons and protons led Heisenberg to treat them as two states of one particle, now called the nucleon N.[44] Without the electric charge, the two states would presumably have the same mass. In the presence of the electric interaction, their masses are slightly different,

$$(m_n - m_p) / \frac{1}{2}(m_n + m_p) = 1.38 \times 10^{-3} \quad .$$

To describe the two states of the nucleon, we introduce an isospin space (charge space) and make the analogy

	spin $\frac{1}{2}$ particle in ordinary space	nucleon in charge space
orientations	$\begin{cases} \text{up} \\ \text{down} \end{cases}$	$\begin{cases} \text{up (proton)} \\ \text{down (neutron)} \end{cases}$

In the spin case, we are used to treating the spin-up and spin-down particles as two different

states of the same particle. Similarly we now
consider neutron and proton as two states of the
nucleon. An analogy to the Pauli spin formalism,
given in Subsection 3.6A, we introduce two iso-
spinors[*]

$$|p> \equiv \begin{pmatrix} 1 \\ 0 \end{pmatrix} \quad , \text{ proton} \quad ,$$

$$|n> \equiv \begin{pmatrix} 0 \\ 1 \end{pmatrix} \quad , \text{ neutron} \quad .$$

We will, however, use the matrix formalism only
rarely and usually characterize a state with isospin I and third component I_3 by the state vector
$|I,I_3>$.[**] Proton and neutron then are given by

$$|p> \equiv |\tfrac{1}{2},\tfrac{1}{2}> \quad ,$$

$$|n> \equiv |\tfrac{1}{2},-\tfrac{1}{2}> \quad .$$

The charge operator for the neutron and proton can
be written in the form

$$Q = e(I_3 + \tfrac{1}{2}) \quad . \tag{4.85}$$

---

[*]Nuclear physicists often use the opposite defini-
tion, $|n> = \begin{pmatrix} 1 \\ 0 \end{pmatrix}$, $|p> = \begin{pmatrix} 0 \\ 1 \end{pmatrix}$, because most nuclei
contain more neutrons than protons. Instead of I,
isospin is also often denoted by T; sometimes it
is called isobaric spin.

[**]We designate operators and quantum numbers with
I and I_3; where confusion can arise we add the
subscript "op."

and the eigenvalue of Q for the state |p> is e,
for the state |n> is 0.

Treating the proton and the neutron as two
states of one particle, the nucleon N, implies
that "switching off" the electromagnetic and weak
interactions would result in equal proton and
neutron masses. Is this assumption correct?
Estimates give the correct order of magnitude for
the mass splitting, but satisfactory calculations
have yet to be seen.[51]

D.   Isospin and Charge Independence

To get a better understanding of the isospin
concept, we compare it with the mechanical spin.
We consider first an atom with spin $\vec{J}$ and magnetic
moment μ. In the absence of a magnetic field, we
can turn the spin in any direction and we can select
the quantization axis z arbitrarily without changing
the energy (mass) of the system. The energy is
invariant under rotation and we express this fact
by

$$[H_o, \vec{J}] = 0 \quad,$$
(4.86)

where H_o is the energy operator in the absence of
a magnetic field. The energy level of the atom
is (2j+1) fold degenerate, as indicated in Fig.
4.14.

51.   S. D. Drell, Comments on Nuclear and Particle
      Physics 1, 94 (1967). A. Zee, Physics Report
      3C, 129 (1972).

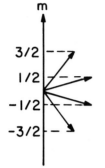

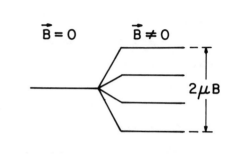

orientation of the spin
j = 3/2 in space

energy levels
without field $\vec{B}$ with field $\vec{B}$

Fig. 4.14.   The energy levels of a spin j = 3/2
system without and with external
magnetic field $\vec{B}$ (along z axis).   The
magnetic moment of the system is
denoted by μ.

The degeneracy of the state is lifted if we apply
a magnetic field $\vec{B}$; it is then natural to choose
the direction of $\vec{B}$ as quantization axis.   The
state splits into (2j+1) magnetic sublevels, characterized by magnetic quantum numbers m.   It is
said that the magnetic interaction "breaks" the
symmetry; the total Hamiltonian H = H_o + M_{mag} no
longer commutes with $\vec{J}$:

$$[H_o + H_{mag}, \vec{J}] \neq 0 \quad . \tag{4.87}$$

The component of the angular momentum along the

magnetic field still remains conserved,

$$[H_o + H_{mag}, J_z] = 0 \quad . \tag{4.88}$$

Analogous considerations apply to isospin. Charge independence states that the hadronic forces do not distinguish between neutron and proton. The two states have the same mass and the isospin vector $\vec{I}$ can point in any direction[*] as long as only the hadronic interaction H_h is present:

$$[H_h, \vec{I}] = 0 \quad . \tag{4.89}$$

With only H_h present, we would have $2I + 1 = 2$ particles of the same mass, the proton and the neutron. The electromagnetic interaction destroys the isotropy of charge space; proton and neutron have different energies and we hence have

$$[H_h + H_\gamma, \vec{I}] \neq 0 \quad . \tag{4.90}$$

However, the electric charge is conserved, even in the presence of H_γ:

$$[H_h + H_\gamma, Q] = 0 \quad . \tag{4.91}$$

With Eq. (4.85) we then get

$$[H_h + H_\gamma, I_3] = 0 \quad . \tag{4.92}$$

---

[*] We stress again that the "direction" in isospin space has nothing to do with directions in ordinary space.

The third component of the isospin is a good quantum
number even in the presence of the electromagnetic
interaction.

We pointed out in subsection B that charge
independence of the hadronic forces holds also
for hadrons other than the nucleons. We tentatively
generalize Eq. (4.89) to all hadrons; I then can
have values other than $\frac{1}{2}$. We will give examples
later, but treat here the general consequences of
Eq. (4.89), following the arguments of Section 4.3.

In analogy to Eq. (4.53), a rotation in iso-
spin space by the angle ω about the direction $\hat{\alpha}$ can
be written as

$$U_{\hat{\alpha}}(\omega) = e^{-i\omega\hat{\alpha}\cdot\vec{I}} \quad . \tag{4.93}$$

Invariance of the Hamiltonian under such a rotation
for any axis $\hat{\alpha}$ leads to conservation of
$\vec{I} \equiv (I_1, I_2, I_3)$. The argument leading from Eq.
(4.53) to Eq. (4.55) yields the commutation relations

$$[I_1, I_2] = i\, I_3, \quad + \text{cyclic} \quad . \tag{4.94}$$

In analogy to Eqs. (4.58) and (4.59), we find the
eigenvalue equations

$$I_{op}^2 |I, I_3\rangle = I(I+1)|I, I_3\rangle$$

$$\tag{4.95}$$

$$I_{3,op} |I, I_3\rangle = I_3 |I, I_3\rangle \quad .$$

Although there are close similarities between spin and isospin, there are also differences.[52] In particular, the direction of the symmetry-breaking magnetic field $\vec{B}$ can be chosen arbitrarily whereas the "direction" of H_γ is dictated by nature through the electric charge.

E.   The Isospin for Nuclei[53]

We consider a nucleus with A nucleons, Z protons and N = A-Z neutrons. The total charge of the nucleus is Ze. We assume that isospins can be added just as angular momenta and get for the nucleus

$$\vec{I} = \sum_{i=1}^{A} \vec{I}_i$$

$$\tag{4.96}$$

$$I_3 = \sum_{i=1}^{A} I_{3,i}$$

$$Q = \sum_{i=1}^{A} Q_i = e(I_3 + \tfrac{1}{2}A) \quad . \tag{4.97}$$

52.   L. L. Foldy, Phys. Rev. 93, 1395 (1954).
53.   B. Cassen and E. U. Condon, Phys. Rev. 50, 846 (1936).
      E. Wigner, Phys. Rev. 51, 106 (1937); 56, 519 (1939).
      D. Robson, Ann. Rev. Nuclear Science 16, 119 (1966).
      E. P. Wigner, "Isotopic Spin-A Quantum Number for Nuclei." Proceedings of the Robert A. Welch Foundation Conferences on Chemical Research, W. O. Milligan, ed. (Rice University, Houston, 1958), Vol. I.

For a given nuclide, Z and A are constant and the
eigenvalue of the operator Q is Ze. All states
of this nuclide have the same value of I_3, given
by Eq. (4.97) as

$$I_3 = Z - \frac{1}{2}A = \frac{1}{2}(Z-N) \quad .\tag{4.98}$$

Consider now an assembly of A nucleons and
assume $H_\gamma = H_w = 0$. Equation (4.89) implies that
each state of this assembly can be characterized
by a value of I and is (2I+1) fold degenerate. In
reality, the electromagnetic interaction cannot
be switched off and it breaks the degeneracy, as
indicated in Fig. 4.15. Each sublevel appears in
a different isobar. Since the electromagnetic

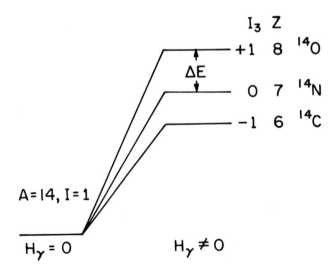

Fig. 4.15.   Splitting of an isospin triplet by
             the electromagnetic interaction.

interaction is weak ($Ze^2 \ll 1$), we expect to find I to be approximately conserved and hope to be able to label real nuclear states by I.

How do we find the members of an isospin multiplet and how do we assign I? We first note that I is limited to the range

$$\tfrac{1}{2}A \geq I \geq |Z - \tfrac{1}{2}A| . \qquad (4.99)$$

By Equation (4.89), a rotation in isospin space does not affect H_h; all members of an isospin multiplet must have the same hadronic quantum numbers and should be similar in all properties that are dominated by the hadronic interaction. We can therefore start with a given energy level in a nuclide (A,Z) and look for levels with similar properties in neighbouring isobars. We can predict where we should find the other members of the multiplet: H_γ gives rise to the Coulomb energy and the neutron-proton mass difference. The energy difference between members of an isospin multiplet in isobars (A,Z+1) and (A,Z) then is

$$\Delta E = E(A,Z+1) - E(A,Z)$$

$$\approx \Delta E_{Coul} - (m_n - m_p) , \qquad (4.100)$$

where $(m_n - m_p) = 1.293$ MeV. To estimate the Coulomb energy difference, E_{Coul}, between members of an isomultiplet, we assume the charge Ze to be uniformly distributed through a sphere of radius R;

the classical electrostatic energy is then given
by

$$E_{Coul} = \frac{3}{5} \frac{(Ze)^2}{4 \pi R} \, .$$ (4.101)

The Coulomb energy difference between isobars
(A,Z+1) and (A,Z) is

$$\Delta E_{Coul} \underset{\sim}{\sim} \frac{6}{5} \frac{e^2}{4\pi R} Z \, ,$$ (4.102)

if we assume equal radii. For better results, the
actual charge distribution in nuclei has to be
taken into account and the calculation must be
performed on the basis of quantum mechanics, including exchange (antisymmetrization) effects.[54]
    Starting from a given level with energy E_o
in a nuclide (A,Z) we can now search for a level
with the same hadronic quantum numbers in the
isobar (A,Z+1) at the energy E_o + ΔE (Fig. 4.16)
or in (A,Z-1) at E_o-ΔE. The search can be continued
in the neighbouring isobars till all (2I+1) components of the isomultiplet have been found; I is
then determined. Unfortunately, it is often
difficult to reach the extreme members of a multiplet; the number of components then only gives a
lower limit on I.

---

54. J. A. Nolen, Jr., and J. P. Schiffer, Ann.
     Rev. Nuclear Science 19, 471 (1969).

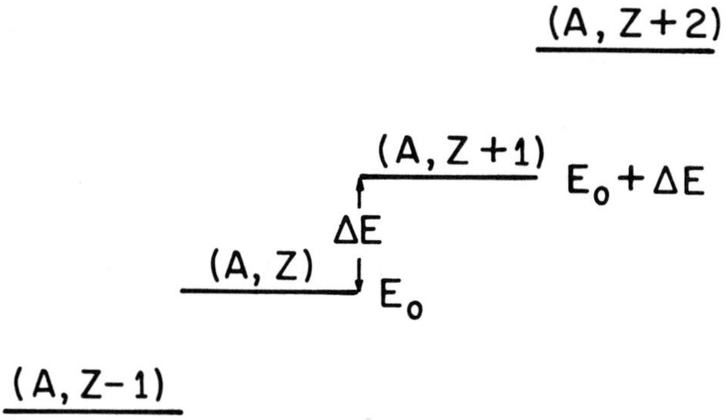

$$\underline{(A, Z+2)}$$

$$\underline{(A, Z+1)} \quad E_0 + \Delta E$$

$$\Delta E$$

$$\underline{(A, Z)} \quad E_0$$

$$\underline{(A, Z-1)}$$

Fig. 4.16.   Isospin multiplet in isobars. The
members of the multiplet have the
same hadronic quantum numbers and
are similar in all properties that
are dominated by the hadronic
interaction.

Many complete isospin multiplets have already
been found; we list a few representative examples
here. Equation (4.98) shows that singlets can
occur only in self-conjugate nuclides, for which
$N = Z$. The ground states of ^{2}H, ^{4}He, ^{6}Li, ^{8}Be, ^{12}C,
^{14}N, and ^{16}O have $I = 0$. The conclusion that these
states are not just the $I_3 = 0$ component of a
higher isospin multiplet, say $I = 1$, follows from
the fact that the $I_3 = \pm 1$ levels do not exist.
Even the ground states of the neighbouring isobars
lie considerably higher than predicted by Eq.
(4.100). Isospin doublets occur in mirror nuclei
for which $Z = (1/2)(A \pm 1)$ with A odd. Two examples

are shown in Fig. 4.13:   the ground and the first
excited states of ^{7}Li and ^{7}Be form doublets.   The
two members of each doublet have identical quantum
numbers (apart from Z and I_3) and similar lifetimes.
Triplets occur; perhaps the most famous one is
^{14}C, ^{14}N*, ^{14}O; another is ^{6}He, ^{6}Li*, ^{6}Be.   Quar-
tets (I=3/2) and quintets (I=2) have also been
found.[55]

As we have pointed out above, counting levels
to determine I is not always possible.   Fortunately
other methods are available and many assignments
are based on the use of isospin selection rules
in reactions and decays.[56,57]   As a simple example,
consider the reactions

$$d \; {}^{16}O \rightarrow \alpha \; {}^{14}N \quad ,$$

$$d \; {}^{12}C \rightarrow p \; {}^{13}C,$$

$$\alpha \; {}^6Li \rightarrow p \; {}^9Be \quad .$$

The three initial states have isospin 0, since I
for the ground states of light even-A nuclides is

55.   J. Cerny, R. H. Pehl, G. Butler, D. G. Fleming,
      C. Maples, and C. Detraz, Phys. Letters 20,
      35 (1966).   R. Leonardi, P. Loncke, and J.
      Pradal, Phys. Rev. 146, 615 (1966).   J. Cerny,
      Annual Review Nuclear Science 18, 27 (1968).
56.   R. K. Adair, Phys. Rev. 87, 1041 (1952).
57.   W. M. MacDonald, in Nuclear Spectroscopy, F.
      Ajzenberg-Selove, ed. (Academic Press, Inc.,
      New York, 1960), Vol. B., p. 937.

zero. In the final state, the α also has I=0, the proton I=1/2. If isospin is conserved, the first reaction can only lead to states of isospin 0 in ^{14}N, the other two only to states with I=1/2 in ^{13}C and ^{9}Be, respectively. The observation of a strongly produced state in the reactions given above therefore determines the isospin of that state.

The last statement must be qualified. Since we do not expect isospin to be completely conserved, transitions to other isospin states cannot be ruled out. The extent to which isospin is a good quantum number can be studied by comparing the strength of isospin-forbidden and isospin allowed transitions. In particular, the reaction d ^{16}O → α ^{14}N should not lead to the first excited state of ^{14}N, because its isospin is 1. (Actually, this transition is also forbidden by charge symmetry, a weaker condition). The forbidden reaction has nevertheless been observed, but with a relative intensity of less than a few percent of isospin-allowed reactions.[58]

The ground-state isospins, determined by counting or through selection rules, all show a remarkable property: They always take on the smallest possible value, $I_o = |I_3| = (1/2)(N-Z)$. This feature can be understood on the basis of the isospin-dependence of the nuclear forces and

58. C. P. Browne, Phys. Rev. 104, 1598 (1956).
     L. Meyer-Schützmeister, D. von Ehrenstein,
     and R. G. Allas, Phys. Rev. 147, 743 (1966).

of the Pauli principle: Nucleons are best able
to take advantage of their short-range attraction
in S - states. The spin-isospin wave function
must therefore be antisymmetric and this require-
ment leads to small values of the isospin.

How meaningful is the assignment of the isospin quantum number to a nuclear level? In order
to answer this question, we first note that spin
and isospin satisfy commutation relations that
have the same form. We can therefore use all
results of Subsection 4.3.E also for isospin.
Next we remark that H_h and $\vec{I}$ commute, as expressed
by Eq. (4.89): H_h is an isoscalar. If only the
hadronic interaction were present, I would be a
good quantum number. The presence of the electromagnetic interaction, H_γ, alters the picture, as
we have stressed above. It lifts the degeneracy
of the various I_3 sublevels, but it also admixes
to a state $|I_o,a\rangle$ other states $|I,b\rangle$ which have
the same angular momentum and parity, but have
$I \neq I_o$. Since $H_\gamma \ll H_h$, perturbation theory can
be used, at least for light nuclei, and it gives
in first order

$$|\Psi\rangle = |I_o,a\rangle + \sum_{b \neq a} C_b |I,b\rangle \quad ,$$

$$(4.103)$$

$$C_b = \frac{\langle I,b|H_\gamma|I_o,a\rangle}{E_o - E_b} \quad ,$$

where the sum extends over all states $|I,b\rangle$ with the same angular momentum and parity as $|I_o,a\rangle$, but with different isospins I. The isospin impurity of $|\Psi\rangle$ is defined as the sum of the squares of the expansion coefficients in Eq. (4.103),

$$\sum_b |c_b|^2 \quad .$$

We can get more information about the matrix element in Eq. (4.103) by expressing H_γ in terms of isospin operators. If we assume H_γ to be the Coulomb interaction, Eq. (1.2), we can write it for a nucleus as

$$H_\gamma = \frac{e^2}{4\pi} \sum_{i<j} \frac{1}{r_{ij}} (I_{i3}+1/2)(I_{j3}+1/2) \quad , \quad (4.104)$$

where the sum extends over all nucleons. This form can be used to verify that H_γ and I^2 do not commute, but it is not convenient for the evaluation of matrix elements because it does not have simple transformation properties in isospin space. The transformation properties become clearer after rewriting H_γ as a sum of irreducible tensors $T_q^{(k)}$:

$$H_\gamma = T^{(0)} + T_0^{(1)} + T_0^{(2)} \quad ,$$

$$T^{(0)} = \frac{e^2}{4\pi} \sum_{i<j} \frac{1}{r_{ij}} \left(\frac{1}{4} + \frac{1}{3}\, \vec{I}_i \cdot \vec{I}_j\right) \quad ,$$

$$T_0^{(1)} = \frac{e^2}{8\pi} \sum_{i<j} \frac{1}{r_{ij}} \left(I_{i3} + I_{j3}\right) \quad ,$$

(4.105)

$$T_0^{(2)} = \frac{e^2}{4\pi} \sum_{i<j} \frac{1}{r_{ij}} \left(I_{i3}I_{j3} - \frac{1}{3}\, \vec{I}_i \cdot \vec{I}_j\right) \quad .$$

The transformation properties of the three terms can be recognized easily: $T^{(0)}$ is an isoscalar. $T_0^{(1)}$ is the third component of an isovector and, according to Eq. (4.83), it is therefore the zero-component of an irreducible tensor of order 1. Finally, with Eq. (4.82a) it is straightforward to establish that $T_0^{(2)}$ is the zero-component of an irreducible tensor of order 2.

Clearly, the isoscalar $T^{(0)}$ does not lead to isospin impurities. To see which isospins can be admixed to the state $|I_o,a\rangle$ by $T_0^{(1)}$, we use the Wigner-Eckart theorem, Eq. (4.84), to write

$$\langle II_3,b|T_0^{(1)}|I_o I_{o3},a\rangle$$

(4.106)

$$= \frac{1}{\sqrt{2I+1}} \, \langle I_o I_{o3} 10|I\ I_3\rangle \langle I,b|| T_0^{(1)} || I_o,a\rangle \quad .$$

The Clebsch-Gordan coefficient vanishes unless $I_{o3} = I_3$, and $I = I_o$, $I_o \pm 1$. The states $I = I_o$ do not introduce impurities so that the only impurities admixed by the isovector part of H_γ come from state $I = I_o \pm 1$. The discussion can easily be extended to $T_0^{(2)}$ and shows that it admixes states with $I = I_o \pm 1$, $I_o \pm 2$.

With the results obtained so far we can understand why isospin is a surprisingly good quantum number in light nuclei. Consider for instance the ground state of ^{7}Li. It has spin 3/2, negative parity, and isospin 1/2. The next level with the same spin and parity occurs at an excitation energy of 11.1 MeV. (Fig. 4.13). All other states below this energy cannot be admixed to the ground state. Now the Coulomb force between two protons at the average internucleon distance in nuclei is about 1 MeV. Even without further computation, Eq. (4.103) tells us that the admixture of the 11.1 MeV state to the ground state must be small because of the large energy denominator. Actual calculations have been performed by using various nuclear models; these indicate that the impurities are indeed very small in light nuclei.[59]

For medium-weight and heavy nuclei, the arguments used so far break down. The Coulomb interaction is no longer small as attested to by the fact that stable nuclei have

---

59. W. M. MacDonald, Phys. Rev. 100, 51 (1955); 101, 271 (1956).

$|I_3| = (1/2)(N-Z) \gg 1$. Furthermore, the spacing of energy levels is much smaller in heavy than in light nuclei. The coefficients C_b in Eq. (4.103) therefore are no longer expected to be small and for many years it was assumed that isospin was meaningless in heavy nuclei. It came therefore as a great surprise when Anderson and Wong[60] discovered what are now called "isobaric analog states" in ^{51}V and ^{51}Cr. We will explain here why the isospin is still a very good quantum number in the ground states of heavy nuclei. The ground state isospin is usually given by $I_o = (1/2)(N-Z)$ and in practice only states with I_o+1 are admixed to it. States with I_o+2 lie very high and states with I_o-1 and I_o-2 would have $I < I_3$ and thus do not exist. The admixture of states I_o+1 is reduced by two effects: (1) the constancy of the average Coulomb field inside the nucleus and (2) the dilution of the isospin impurity by the (N-Z) excess neutrons. (1) A constant electric field leads to a splitting of the different I_3 states, but does not induce transitions between the I states. In a heavy nucleus, the field over most of the nuclear volume is nearly constant and the matrix element in Eq. (4.103) is therefore considerably reduced. (2) The excess

---

60. J. D. Anderson and C. Wong, Phys. Rev. Letters 7, 250 (1961). J. D. Anderson, C. Wong and T. W. McClure, Phys. Rev. 126, 2170 (1962). J. D. Fox, C. F. Moore and D. Robson, Phys. Rev. Letters 13, 198 (1964).

neutrons have no Coulomb interaction. Since
their number, N-Z, is large, they dilute the
isospin impurity introduced by the protons. We
can get an estimate for the dilution by comparing
the diagonal and off-diagonal elements of the
matrix element Eq. (4.106). With Table A 2 we
get for the ratio of the corresponding Clebsch-
Gordan coefficients

$$\frac{\langle I_o I_o \; 1 \; 0 | I_o+1, I_o \rangle}{\langle I_o I_o \; 1 \; 0 | I_o I_o \rangle} = \frac{1}{\sqrt{I_o}} = \sqrt{\frac{2}{N-Z}} \quad .$$

Other things being equal, the isospin-admixing
off-diagonal element is reduced by a factor
$[(N-Z)/2]^{-1/2}$; for heavy elements, the corresponding impurity reduction can be considerable. Calculations of the isospin impurity in medium-
weight and heavy nuclei, based on specific models,
indicate that the impurity is indeed smaller than
a few percent.[61]

61. A. M. Lane and J. M. Soper, Nuclear Physics
     37, 663 (1962). L. A. Sliv and Yu. I.
     Kharitonov, Physics Letters 16, 176 (1965).

F. Isospin Multiplets in Particle Physics[9]

The success of isospin in nuclear physics
suggests that it should be tried on elementary
particles. Historically the first candidate
granted an isospin degree was the <u>pion</u>. It occurs
in three different charge states, 1, 0, -1, and
the most likely assignment is I = 1. Actually,
long before the experimental discovery of pions,
Kemmer suggested the existence of a meson
triplet.[62] He based his suggestion on the charge
independence of nuclear forces and their explanation in terms of Yukawa's meson exchange.

To justify the assignment of I = 1 to the
pion, we consider the basic Yukawa reaction

$$N \rightarrow N' + \pi$$

$$I: \quad \frac{1}{2} \quad \frac{1}{2} \; + \; ?$$

If we assume conservation of isospin, the isospin
of the pion can only be 0 or 1. The former is
excluded by the existence of more than one pion.
Hence we assume that the pions form an isospin
triplet, also called an <u>isovector</u>:

62. N. Kemmer, Proc. Camb. Phil. Soc. <u>34</u>, 354
     (1938).

$$\text{Pion} \quad I = 1 \quad \begin{cases} I_3 = 1 & \pi^+ \\ I_3 = 0 & \pi^o \\ I_3 = -1 & \pi^- \end{cases}$$

The charge operator Q is connected to I_3 by

$$Q = e\, I_3 \quad , \tag{4.107}$$

which agrees with Eq. (4.97).

Again, as in the case of nucleons and nuclei, we assume that all members of the pion triplet would have the same mass if we could switch off H_γ and H_w. Actually, their masses are different by

$$m_{\pi^\pm} - m_{\pi^o} = 4.60 \text{ MeV}, \quad \Delta m / \bar{m} = 0.033 \quad .$$

Calculations give the correct sign, and a magnitude of the mass splitting of about 5 MeV.[51,63]

The isospin assignment to the kaons is not as straightforward as for pions. Since three

---

63. I. S. Gerstein, B. W. Lee, H. T. Nieh, and H. J. Schnitzer, Phys. Rev. Letters 19, 1064 (1967). T. Das, G. S. Guralnik, V. S. Mathur, F. E. Low, and J. E. Young, Phys. Rev. Letters 18, 759 (1967). F. E. Low, Comments on Nuclear and Particle Physics, 2, 111 (1968). See, however, S. Weinberg, Phys. Rev. Letters 29, 388 (1972).

charge states appear, we are tempted to use I = 1 also. However, the evidence against such an assignment is very strong. The K^- and K^+ behave differently in their strong interactions with matter and they have opposite values of strangeness. Since we assume that rotation in isospin space leaves hadronic quantum numbers unchanged, they cannot belong to the same multiplet. Since we have seen in Subsection 4.2.I that two different neutral kaons exist, we assume that K^+ and K^o form an isospin doublet and that K^- and $\overline{K^o}$ form the corresponding antiparticle doublet. This assignment is in agreement with all experimental facts.

The lambda occurs singly, no other hyperon is close in mass. We assume that the lambda is an isospin singlet.

The sigma occurs in three charge states and we assume that it is a triplet. However, while the pion triplet is identical to its antitriplet (which is possible because the pion has A = Y = 0), the sigma triplet and antitriplet are different.

The cascade particle (Ξ) occurs in two states, the Ξ^- and Ξ^o. The situation here is similar to the kaon and we assume that the Ξ is a doublet, and that the antiparticle doublet is $\overline{\Xi^-}$ and $\overline{\Xi^o}$.

The assignments given here can all be tested by selection rules and predictions of cross section ratios.

G. Displaced Charge Multiplets. Gell-Mann-Nishijima Relation

Each hadron encountered so far is characterized by a set of quantum numbers (Q,A,Y,I_3) that are conserved by the hadronic and the electromagnetic interaction. In Eq. (4.97) we have found a connection between the charge and I_3 for nuclei and for pions. Can we generalize this relation to strange particles? This question was answered by Gell-Mann[9,64] and by Nishijima.[65]

Following their argument we first note that for nucleons, nuclei, and pions the electric charge operator is linearly related to I_3:

$$Q/e = I_3 + C.$$

Since I_3 ranges from $-I$ to I, we get for the average charge of a multiplet

$$\langle Q/e \rangle = C \quad .$$

We can determine C for hadrons other than pions or nuclei by locating the center of charge $\langle Q/e \rangle$. Table 4.3 shows that only the nonstrange mesons have their center of charge at 0, all

---

64. M. Gell-Mann, Phys. Rev. 92, 833 (1953).
65. T. Nakano and K. Nishijima, Progr. Theoret. Phys. 10, 581 (1953); K. Nishijima, Fortschritte Phys. 4, 519 (1956); Progr. Theoret. Phys. 13, 285 (1955).

other hadrons have their center displaced.   With
Table 4.3, we get

$$C = \langle Q/e \rangle = \frac{1}{2} Y \quad ,$$

and write the Gell-Mann-Nishijima relation

$$Q/e = I_3 + Y/2 = I_3 + (S + A)/2. \qquad (4.108)$$

This relation can be used to define strangeness.
Since Q, I_3, and A are conserved in the hadronic
and the electromagnetic interactions, S must be
conserved too.

Equation (4.108) shows that the operator of
the electromagnetic charge is composed of an isoscalar and the third component of an isovector.
The current $\vec{j}$ and therefore the entire four-current
j^μ also have these transformation properties in
isospin space. The interaction of a charged
particle with the electromagnetic field can thus
be said to satisfy the isospin selection rule.
Fig. 1.1 shows that the force between two electric
charges involves the interaction of the electromagnetic field twice and the selection rule should
therefore be

$$\Delta I = 0, \pm 1, \pm 2 \quad .$$

We have already derived this rule in a more formal
way after Eq. (4.106).

In 1956, nucleons, lambda, Σ^+ and Σ^-, and the charged cascade particle were known, even if not all were completely certified. Gell-Mann then asked the question[9]: "What possibilities are there for more particles without introducing multiple charges?" He predicted the Σ° and Ξ° and stated in addition: "Among baryons we have already a doublet with S = 0 (n,p), a singlet and a triplet with S = -1(Λ and Σ), and a doublet with S = -2(Ξ). The only other possibilities are evidently a singlet with S = -3, which we may call Ω^-, and a singlet with S = +1, which we call z^+...".

In 1956, Gell-Mann could only give limits on the mass of the Ω^-. Later, with the eightfold way, he could also predict its mass. The first Ω^- was found after a difficult hunt by a large team at Brookhaven.[65] In the meantime, more Ω^- have been found, but the world supply is still very limited!

The second particle predicted by Gell-Mann, the z^+, has not yet been certified beyond doubts.

H. Isospin State Functions, Pauli Principle

We turn to the isospin state functions for combined systems and begin with two $I = \frac{1}{2}$ hadrons, for instance two nucleons N_1 and N_2. The individual particles have the four isospin states $|p_1>$, $|p_2>$, $|n_1>$, and $|n_2>$, corresponding to

---

65. V. E. Barnes, et al., Phys. Rev. Letters 12, 204 (1964).

$I = \frac{1}{2}$, $I_3 = \pm \frac{1}{2}$. For the interacting two-particle system, the total isospin is a constant of motion if electromagnetic effects are neglected. Just as for two spin $\frac{1}{2}$ particles, we can form a symmetric and an antisymmetric state, corresponding to I = 1 and I = 0, respectively. These isospin state functions can be obtained in several ways, for instance using Eq. (A2); the states $|I,I_3\rangle$ become

$$I = 1 \quad |1,1\rangle = |p_1\rangle\, |p_2\rangle$$

$$|1,0\rangle = \sqrt{1/2}\{|p_1\rangle\, |n_2\rangle + |n_1\rangle\, |p_2\rangle\}$$

$$|1,-1\rangle = |n_1\rangle\, |n_2\rangle \qquad\qquad (4.110)$$

$$I = 0 \quad |0,0\rangle = \sqrt{1/2}\{|p_1\rangle\, |n_2\rangle - |n_1\rangle\, |p_2\rangle\} \quad .$$

Equation (4.110) describes the isospin part of the state function of two $I = \frac{1}{2}$ hadrons. The total wave function also contains a spin and a space part,

$$|total\rangle = |space,\ spin,\ isospin\rangle \quad .$$

Since for hadronic forces neutron and proton are now treated as two states of the same particle, the Pauli principle must be reformulated to include the isospin quantum numbers. The state function of a system of identical fermions must be <u>antisymmetric</u> under the exchange of any two particles, the state function of a system of identicle bosons must be <u>symmetric</u> under exchange of any two particles. The interchange of two particles now means not only $\vec{x}_1 \rightleftarrows \vec{x}_2$, $\vec{\sigma}_1 \rightleftarrows \vec{\sigma}_2$, $\vec{p}_1 \rightleftarrows \vec{p}_2$, but also $\vec{I}_1 \rightleftarrows \vec{I}_2$. The experimental foundation and the theoretical basis of the symmetrization postulates have been studied by Messiah and co-workers.[66]

For the two-nucleon system, the Pauli principle allows only certain combinations of space-spin-isospin state functions. For relative angular momenta $L \leq 3$, the states are listed in Table 4.4.[*]

We turn for a moment to the physical interpretation of isospin invariance for the two-nucleon system. Isospin invariance states that the hadronic forces in the three $I = 1$ states are the same, but it does not make any statement connecting the $I = 0$ and $I = 1$ states. From the point-of-view of hadronic interaction theory, the classification of the two-nucleon system in terms of I is fundamental. From the point of view

---

66. A. M. L. Messiah,and O. W. Greenberg, Phys. Rev. <u>136</u>, B248 (1964). J. Bros, A. M. L. Messiah, and D. N. Williams, Phys. Rev. <u>149</u>, 1008 (1966).

  [*] We denote a state with total angular momentum J, spin S, and orbital angular momentum L with $^{2S+1}L_J$. Low values of L are replaced by letters: S, P, D, F,...denote $L = 0, 1, 2, 3,...$ in that order.

Table 4.4.   Allowed states for the two-nucleon
             system, for angular momenta L $\leq$ 3.
             S,P,D,F denote states with L=0,1,2,3,
             respectively; the superscript indicates
             singlet or triplet spin states.

Symmetry			Notation		System
space	spin	isospin			
s	s	a	$^3S_1$	$^3D_{1,2,3}$	np
s	a	s	$^1S_0$	$^1D_2$	
					} pp,nn,np
a	s	s	$^3P_{0,1,2}$	$^3F_{2,3,4}$	
a	a	a	$^1P_1$	$^1F_3$	np

of the experimental physicist, however, the system
|pn> is more fundamental than the state |1,0>
because the states that can be prepared in the
laboratory are the |n> and the |p>. The state
$\sqrt{1/2}\{|p_1>|n_2> + |n_1>|p_2>\}$, for instance, cannot be
realized directly, but only through a reaction.

     Similar remarks apply to the pion-nucleon
system. In the laboratory, pions of well defined
charge can be scattered off protons or neutrons.
For the theoretical analysis, however, it is more
convenient to use states $|I,I_3>$. Since the pion

has isospin 1 and the nucleon isospin $\frac{1}{2}$, the isospin of the combined system can be $\frac{1}{2}$ or $\frac{3}{2}$. The
isospin state functions for the pion-nucleon
system are easily obtained with Eq. (A.2); they
are given in Table 4.5.

Table 4.5.   Isospin state functions for the
             pion-nucleon system.

Charge q/e	I	State	Expansion
2	3/2	$\lvert 3/2,3/2\rangle$ =	$\lvert \pi^+ p\rangle$
1		$\lvert 3/2,1/2\rangle$ =	$\sqrt{1/3}\{\lvert \pi^+ n\rangle + \sqrt{2}\,\lvert \pi^\circ p\rangle\}$
0		$\lvert 3/2,-1/2\rangle$ =	$\sqrt{1/3}\{\sqrt{2}\,\lvert \pi^\circ n\rangle + \lvert \pi^- p\rangle\}$
-1		$\lvert 3/2,-3/2\rangle$ =	$\lvert \pi^- n\rangle$
1	1/2	$\lvert 1/2,1/2\rangle$ =	$\sqrt{1/3}\{\sqrt{2}\,\lvert \pi^+ n\rangle - \lvert \pi^\circ p\rangle\}$
0		$\lvert 1/2,-1/2\rangle$ =	$\sqrt{1/3}\{\lvert \pi^\circ n\rangle - \sqrt{2}\,\lvert \pi^- p\rangle\}$

   One word of warning concerns the phases.*
The order of the pions and nucleons in the kets of
Table 4.5 is not important for the final result,
but writing $|N\pi>$ instead of $|\pi N>$ introduces minus-
signs in some amplitudes. In the final result, a
cross section or transition probability, the sign
ambiguity drops out, but it can lead to confusion
in intermediate steps. It is best to stick to a
given order of the particles in a state, for instance writing the particle with the higher isospin
first.

   As an example, we consider the reaction

$$K^+p \to N^*K^* \quad K^* \quad \text{mass 891 MeV}$$
$$I = 1/2 \ (K^{*+}, K^{*0})$$
$$N^* \quad \text{mass 1238 MeV}$$
$$I = 3/2 \ (N^{*++} \ . \ . \ N^{*-}) \ .$$

Charge conservation limits the possible final states
for the reaction to $N^{*++}K^{*0}$ and $N^{*+}K^{*+}$. (Actually,
we do not have to restrict the discussion to these

---

*Lipkin describes the problem in appendix D "Phases,
a Perennial Headache" in his book <u>Lie Groups for
Pedestrians</u> (North Holland Publishing Co., Amsterdam, 1965) in the following words: "However, once
a convention is chosen, confusion and errors are
avoided by using the same convention throughout
a particular calculation." And then he adds a
footnote: "Do not believe this sentence. There
is always confusion and errors. You have to live
with them."

two states.   Charge conservation is built into
isospin conservation; the Clebsch-Gordan coefficients leading to charge non-conserving states
will be zero.)   To determine the relative amounts
going into the various final states, we expand
the initial state in terms of the possible final
states; we consider only the isospin parts, which
we write as $|I(N^*)I_3(N^*), I(K^*)I_3(K^*)>$

$$|K^+,p> = |1\ 1> = \sqrt{3/4}\ |3/2\ 3/2,1/2 - 1/2>$$

$$- \sqrt{1/4}\ |3/2\ 1/2,1/2\ 1/2>$$

$$= \sqrt{3/4}\ |N^{*++}\ K^{*0}>$$

$$- \sqrt{1/4}\ |N^{*+}\ K^{*+}> \quad .$$

The probability of finding the final state
$|I_f, I_{3f}>$ is equal to the square of the corresponding expansion coefficient, and we expect the
intensity ratio

$$\frac{K^+p \rightarrow N^{*++}\ K^{*0}}{K^+p \rightarrow N^{*+}\ K^{*+}} = \frac{(\sqrt{3/4})^2}{(\sqrt{1/4})^2} = 3 \quad ,$$

in good agreement with experiment.

I. The Pion Isospin in Field Theory

In Section 3.5, we introduced separate and unrelated field operators for the neutral and the charged pions. We know now that neutral and charged pions are members of one isospin triplet; their field operators must therefore be related. We present a unified description of the field operators in the present subsection. Rather than develop the description from the one given in Section 3.5, we state the final results and then show how to regain the earlier expressions.

In order to describe the three pion states together, we introduce the isovector operator $\vec{\Phi} = (\Phi_1, \Phi_2, \Phi_3)$ through the expansion

$$\vec{\Phi} = \sum_p \frac{1}{\sqrt{2E}} \left\{ \vec{a}_p\, e^{-ip \cdot x} + \vec{a}_p^{\dagger}\, e^{ip \cdot x} \right\} \quad . \quad (4.111)$$

For the annihilation and creation operators $\vec{a} \equiv (a_1, a_2, a_3)$ and $\vec{a}^{\dagger} \equiv (a_1^{\dagger}, a_2^{\dagger}, a_3^{\dagger})$ we postulate the commutation relations[*]

$$\left[a_{p,i} \quad , \quad a_{p',j}^{\dagger} \right] = \delta_{p,p'}\, \delta_{i,j}, i,j=1,2,3.$$

$$(4.112)$$

All other commutators = 0.

---

[*] Note that each operator $a_{p,i}$ now carries two indices; the first indicates which momentum it annihilates and the second is the component in isospin space.

We further define an operator $\vec{I}$ by

$$\vec{I} = -i \sum_{p} \vec{a}_{p}^{\dagger} \times \vec{a}_{p} \quad .$$

(4.113)

With the commutation relations (4.112) it is straightforward to see that $\vec{I}$ satisfies the angular momentum commutation relations Eq. (4.55); $\vec{I}$ can therefore be interpreted as the isospin operator. Moreover, $\vec{I}$ and $\vec{\Phi}$ satisfy the relation

$$\left[I_1, \Phi_1 \right] = 0 \quad , \quad \left[I_1, \Phi_2 \right] = i\Phi_3 \quad ,$$

$$\left[I_1, \Phi_3 \right] = -i\Phi_2, \, + \text{cycl.} \quad .$$

(4.114)

According to Subsection 4.3.E., $\vec{\Phi}$ thus is a vector operator in isospin space.

To connect the isovector $\vec{\Phi}$ to the observed particles, we consider the charge operator,

$$Q = e \, I_3 = -i \, e \sum_{p} \left(a_{p,1}^{\dagger} \, a_{p,2} - a_{p,2}^{\dagger} \, a_{p,1} \right) \quad .$$

(4.115)

Neutral pions must satisfy

$$Q|\pi^o\rangle = 0|\pi^o\rangle \quad .$$

(4.116)

Since a_3 does not appear in Q, Eq. (4.116) is

satisfied if we identify the field operator for neutral pions, Φ_o, with Φ_3. We then have

$$\Phi_3|0> = |\pi^o> \quad , \tag{4.117}$$

where $|0>$ is the vacuum state. To find the operators creating and destroying charged pions, we introduce in analogy to Eq. (4.60)

$$I_\pm = I_1 \pm iI_2 \quad . \tag{4.118}$$

I_+ increases I_3 and hence the charge by one unit, I_- decreases I_3 and charge by one unit. Eq. (4.62) gives

$$I_\pm|\pi^o> = \sqrt{2} \; |\pi^\pm> \quad . \tag{4.119}$$

With Eq. (4.114), we find

$$I_\pm\Phi_3 - \Phi_3 I_\pm = \mp \; (\Phi_1 \pm i\Phi_2) \quad . \tag{4.120}$$

If we apply the operator (4.120) to the vacuum state, the second term on the left-hand side gives zero and we have with Eq. (4.116)

$$I_\pm\Phi_3|0> = I_\pm|\pi^o> = \mp(\Phi_1 \pm i\Phi_2) \; |0> \quad .$$

Comparing with Eq. (4.119) yields

$$|\pi^\pm> = \mp \frac{1}{\sqrt{2}} \; (\Phi_1 \pm i\Phi_2) \; |0> \quad . \tag{4.121}$$

We therefore write

$$\Phi_+ = \frac{-1}{\sqrt{2}} (\Phi_1 + i\Phi_2)$$

$$\Phi_o = \Phi_3 \qquad\qquad\qquad\qquad (4.122)$$

$$\Phi_- = \frac{1}{\sqrt{2}} (\Phi_1 - i\Phi_2)$$

and identify Φ_+ with the operator that creates positive and destroys negative pions, and Φ_- with the operator that creates negative and destroys positive pions.

4.5 PARITY

Courtesy Elizabeth M. Watson, Seattle, Washington

Our macroscopic world is not symmetric under
the interchange left-right. Our body shows asymmetries; DNA is not left-right symmetric, and the
earth rotates in a preferred direction.[67]
Nevertheless physicists for many years believed
that nature at the microscopic level is symmetric
and that no microscopic experiment can distinguish
between our world and its mirrored (space reflected)
image. When this belief was rudely shattered in
1956 by the discovery of parity nonconservation,
physicists in airplanes and trains could be recog-

---

67. In this context, the reader should consult
     the amusing contribution by P. Jordan and R.
     Kronig, Nature, December 3, 1927.

nized by the fact that they were reading Weyl's
beautiful book on symmetries in art and nature.[68]

    As we will show in Subsection A, invariance
of a quantum mechanical system under space reflection leads to conservation of parity, a quantum
number that has no classical analogue. This
quantum number was discovered empirically before
the advent of wave mechanics: Laporte found that
two types of atomic energy levels appeared in
iron, "gestrichene" (barred) and "ungestrichene"
(unbarred), and he established the selection rules
for transitions between the two types.[69] However,
no convincing physical explanation for the existence of the two types of levels could be found
within the framework of the "old" quantum theory.
Wigner then showed that the two types of levels
and the selection rules followed from the invariance
of the Schrödinger equation under space reflection.[70] Since invariance under space reflection
is intuitively so appealing (why should a left-
and a right-handed system be different?), conservation of parity quickly became a sacred cow. It
was therefore a major upheaval when Lee and Yang
in 1956 showed that no evidence existed for parity
conservation in weak interactions[71], and when it

---

68. H. Weyl, Symmetry (Princeton University Press,
    Princeton, 1952).
69. O. Laporte, Z. Physik 23, 135 (1924).
70. E. P. Wigner, Z. Physik 43, 624 (1927).
71. T. D. Lee and C. N. Yang, Phys. Rev. 104,
    254 (1956).

was indeed found that parity was not conserved in
these interactions. Since then, physicists have
become much more careful in adopting appealing
symmetries without careful experimental checks.

A.   Parity in Quantum Mechanics

The space reflection or parity operation P
reverses the sign of true (polar) vectors;

$$\vec{x} \rightarrow -\vec{x}, \qquad \vec{p} \rightarrow -\vec{p} \quad . \tag{4.123}$$

The corresponding operators transform as

$$P \, \vec{x}_{op} \, P^{-1} = -\vec{x}_{op}, $$

$$P \, \vec{p}_{op} \, P^{-1} = -\vec{p}_{op} \quad . \tag{4.124}$$

Axial vectors do not change under P ; for example
the orbital angular momentum operator $\vec{L} = \vec{r} \times \vec{p}$
transforms as

$$P \, \vec{L}_{op} \, P^{-1} = \vec{L}_{op} \quad . $$

This equation also follows from the fact that the
parity operator commutes with the operator of
infinitesimal rotations. It can be generalized to
apply to any angular momentum, spin or orbital:

$$P \, \vec{J} \, P^{-1} = \vec{J} \quad \text{or} \quad [P, \vec{J}] = 0 \quad . \tag{4.125}$$

Under the parity operation, states change as

$$P|\vec{x}> = \eta_P |-\vec{x}> ,$$

(4.126)

$$P|\vec{p},\vec{s}> = \eta_P |-\vec{p},\vec{s}> ,$$

where $\vec{s}$ is the spin and η_P is a phase factor. P is now chosen to be unitary so that it preserves the normalization and orthogonality of states. The operator P^2 leads back to the original state and with proper phase choice it is possible to set $P^2 = 1$. (For fermions, P^2 is actually equal to -1; the problems that arise from this fact will be discussed in Subsection C.) With $P^2 = 1$ and $P^\dagger P = 1$, P is Hermitian and it is thus an observable with eigenvalues $+1$ and -1.

No invariance arguments are involved in Eq. (4.126); the parity operator can be applied to any state. Invariance arguments come into play if the Hamiltonian H of a system is reflection invariant:

$$[H,P] = 0 .$$

(4.127)

In this case, if Ψ is a solution of the Schrödinger equation $i\dot{\Psi} = H\Psi$, so is $P\Psi$.

In terms of the Hamiltonian density (3.199) the condition for reflection invariance is written as

$$P \mathcal{H}(t,\vec{x}) = \mathcal{H}(t,-\vec{x}) P .$$

(4.128)

Note that $\mathcal{H}(t,-\vec{x})$, and not $\mathcal{H}(t,\vec{x})$, must appear
on the right-hand side; otherwise, the operators
$\mathcal{H}P$ and $P\mathcal{H}$, acting on a state, would not give
the same result.

For systems obeying Eq. (4.127) simultaneous
eigenfunctions of H and P can be found; they
satisfy

$$P\Psi_P = \eta_P \, \Psi_P \quad ,$$

$$\eta_P = \pm 1 \quad . \tag{4.129}$$

Thus, the phase η_P is the observable associated
with the operator P and is called the <u>parity</u> of
the state ψ_P . As an example, we discuss the
angular part of the wave function of a particle
with orbital angular momentum ℓ, given by the
spherical harmonics $Y_\ell^m(\theta,\phi)$. Under space reflection we get $\theta \to \pi - \theta$, $\phi \to \phi + \pi$; and the
spherical harmonics change as

$$Y_\ell^m(\pi-\theta, \ \phi+\pi) = (-1)^\ell \ Y_\ell^m(\theta,\phi) \quad . \tag{4.130}$$

The state function $|\ell,m\rangle$ therefore has the parity

$$P|\ell,m\rangle = (-1)^\ell \ |\ell,m\rangle \quad . \tag{4.131}$$

Next we consider the behavior of matrix ele-
ments and start with an operator $\mathcal{O}_e$ that is even
under space reflection, $P \, \mathcal{O}_e \, P^{-1} = \mathcal{O}_e$. The matrix
element of $\mathcal{O}_e$ between states $|\alpha\rangle$ and $|\beta\rangle$ with

parities η_α and η_β then satisfies*

$$<\beta|\sigma_e|\alpha> = <\beta|P^{-1}P\,\sigma_e\,P^{-1}P|\alpha>$$

$$\quad\quad (4.132)$$

$$= \eta_\alpha\,\eta_\beta\,<\beta|\sigma_e|\alpha> \quad ;$$

it is different from zero only if $\eta_\alpha\,\eta_\beta = +1$. Even operators only connect states of equal parity. Similarly we have for odd operators

$$<\beta|\sigma_o|\alpha> = -\eta_\alpha\,\eta_\beta\,<\beta|\sigma_o|\alpha> \quad ; \quad\quad (4.133)$$

such operators only connect states of opposite parities.

Finally we note that Eq. (4.125) shows that the parity and the momentum operator do not commute. A momentum eigenstate does not have a definite parity and a parity eigenstate does not have a definite momentum.[72] This remark will be important in the discussion of the intrinsic parity of massless particles.

---

*Note that the unitarity of P is crucial here, so that $P^{-1} = P^\dagger$.

72. The relevance of these facts for the classification of particles is discussed by H. J. Lipkin in Preludes in Theoretical Physics, A. DeShalit, H. Feshbach, and L. Van Hove, eds. (North-Holland Publishing Co., Amsterdam, 1966).

B.   Intrinsic Parity and Superselection Rules[73,74)]

Does the parity of a state depend only on the
relative motion of its constituents or do the con-
stituent particles have "intrinsic parities" that
have to be taken into account? Space inversion
is easily imagined for a microscopic system, but
much harder to visualize if we consider the structure of elementary particles. However, it will
become clear in the present and the following
subsections that the concept of an intrinsic
parity must be introduced. This notion is important in discussing interactions. Consider for
instance the reaction

$$\pi^- d \rightarrow nn \quad . \tag{4.134}$$

If the relative orbital angular momentum of the
two incident particles is given by L, that of the
two outgoing particles by L', and if we denote the
intrinsic parities of the particles with η_i,
parity conservation and Eq. (4.131) give

$$\eta_\pi \eta_d \ (-1)^L = \eta_n \eta_n \ (-1)^{L'} = (-1)^{L'} \quad . \tag{4.135}$$

If the values of L and L' can be determined exper-

---

73.   G. C. Wick, A. S. Wightman, and E. P. Wigner,
     Phys. Rev. 88, 101 (1952).
74.   P. T. Matthews, Nuovo Cimento 6, 642 (1957).

imentally, the <u>relative</u> intrinsic parities of π^-
and d can be found. The example raises a question:
Can the <u>absolute</u> intrinsic parity of particles be
determined? By absolute intrinsic parity, we mean
the parity relative to that of the vacuum state
where the parity of the vacuum state is taken to
be positive,

$$P|0> = + |0> .$$

(4.136)

We will show here that the absolute intrinsic
parity can only be determined for particles or
systems that have the same additive quantum numbers
as the vacuum, i.e., $Q = A = Y = 0$.
   We define the intrinsic parity η_X of a particle
at rest by

$$P|X> = \eta_X |X> , \quad \eta_X = \pm 1 .$$

(4.137)

Assume that X is specified, in part, by the conserved additive quantum number N_X. The interactions
of X then are invariant under the gauge transformation Eq. (4.15), and we introduce the new
transformation operator

$$P' = P\, e^{iN_X\lambda} .$$

(4.138)

P' satisfies Eq. (4.124) and it provides a definition of the intrinsic parity that is physically
indistinguishable from P; in particular we get

$$P'|X> = e^{iN_x \lambda} \eta_x |X> \quad .$$

If we select the arbitrary parameter λ equal to
π/N_x, the intrinsic parities of particle X as
given by P' and P are opposite! The gauge transformation leaves unchanged only the relative
parities of those particles that have the same
additive quantum numbers Q, A, and Y.

At this point, the situation looks bleak.
Experimentally, only very few intrinsic parities
can be determined absolutely, and relative parities
are restricted to small groups of particles. Two
theoretical arguments, however, improve the situation. We noted in Section 4.5, particularly in
Eq. (4.89), that all members of an isospin multiplet should have the same hadronic quantum
numbers. Since the intrinsic parity of hadrons
is presumably determined by the strong interactions,
we will assume that all members of an isospin
multiplet have the same intrinsic parity. The
second theoretical argument states that bosons and
antibosons have the same parity, while fermions
and antifermions have opposite intrinsic parities.
We will give a proof of this assertion in Subsection C. The identical parity of bosons and
antibosons is in agreement with the fact that
some bosons are identical to their antibosons,
for instance the photon and the neutral pion.
The opposite parity of a fermion and its antiparticle has been observed experimentally by

determining the polarization of the gamma rays in
two-photon annihilation of an electron-positron
pair.[75]

We now return to the question of when the
relative parity of two particles can be determined.
The question leads to the concept of superselection
rules, which was introduced by Wick, Wightman, and
Wigner in a famous paper usually quoted as WWW[73],
and was further explored by Foldy.[52]   Consider
two states |1> and |2>. The absolute phase of |1>
or |2> cannot be measured. In general, however, it
is possible to determine the relative phase between
the state vectors |1> and |2>; it often takes the
form exp $i(E_1-E_2)t$. Sometimes, however, no spon-
taneous transitions between the states |1> and |2>
occur and no measurable quantities with finite
matrix elements between the states |1> and |2>
exist. In this case, even the determination of
the relative phase is impossible and it is said
that a superselection rule forbids the measurement
of the relative phase and hence also the determination of the relative parity.

At the present time, only experiments can tell
when a superselection rule operates. No transitions
between two states can occur if the states belong
to different eigenvalues of an absolutely conserved
additive quantum number, such as the electric charge
Q or the baryonic charge A. The hypercharge Y is

75.   C. S. Wu and I. Shaknov, Phys. Rev. 77, 136
        (1950).

not absolutely conserved as shown by Eq. (4.48);
the weak interaction can connect states that differ
by one unit of Y. However, we will see in Subsection F that the weak interactions do not conserve parity and hence cannot be used to find
relative parities: the relative parity of particles
that differ by one unit of Y also cannot be determined. Finally we note that no transitions occur
between states of integral and half-integral spins
(see Subsection 4.6B). We will give examples of
the determination of intrinsic parities in Subsections D and E.

In our discussions we assume validity of the
superselection principle. However, Aharonov and
Susskind[76] have reinvestigated superselection
rules and suggest that it may be possible to
observe interference between states of different
charge. If this suggestion can be verified, some
of the previous arguments will have to be re-
examined.

76.   Y. Aharonov and L. Susskind, Phys. Rev. 155,
      1428 (1967); see also W. B. Rolnick, Phys.
      Rev. Letters 19, 717 (1967); R. Mirman, Phys.
      Rev. 186, 1380 (1969).

C. Parity in Field Theory and Dirac Theory

Some of the finer points involving the parity
operation are most easily treated in the language
of field theory. If Φ_o denotes the operator that
describes a free neutral spinless boson (Subsection
3.5.D), we assume that the parity operator acts on
Φ as

$$P\Phi_o(t,\vec{x})P^{-1} = \eta_B \, \Phi_o(t,-\vec{x}),$$

$$\eta_B = \pm 1 \; ;$$

(4.139)

particles with $\eta_B = +1$ are called scalar, whereas
those of $\eta_B = -1$ are called pseudoscalar bosons.
In order to discuss the parity of a state, we first
study how the creation and annihilation operators
behave under P. With the expansion (3.105) and
noting that P is an operator in Hilbert space and
does not act on the exponential factor, we find

$$P\Phi_o(t,\vec{x})P^{-1} = \sum_p \frac{1}{\sqrt{2E\,V}} \left\{ Pa_p P^{-1} \, e^{-ip\cdot x} \right.$$

$$\left. + Pa_p^\dagger P^{-1} \, e^{ip\cdot x} \right\} \quad .$$

The operator $\eta_B\Phi_o(t,-\vec{x})$ has the expansion

$$\eta_B \Phi_0 (t, -\vec{x}) = \eta_B \sum_p \frac{1}{\sqrt{2E\ V'}}$$

$$\times \left\{ a_p e^{-i(Et + \vec{p} \cdot \vec{x})} + a_p^\dagger e^{i(Et + \vec{p} \cdot \vec{x})} \right\}$$

$$= \eta_B \sum_p \frac{1}{\sqrt{2E\ V'}} \left\{ a_{-p} e^{-ip \cdot x} + a_{-p}^\dagger e^{ip \cdot x} \right\} \quad ,$$

where the subscript -p is defined to mean that the three-vector $\vec{p}$ has been inverted. With Eq. (4.139), the transformation properties of the creation and destruction operators follow as

$$P a_p P^{-1} = \eta_B\ a_{-p} \quad ,$$

$$\text{(4.140)}$$

$$P a_p^\dagger P^{-1} = \eta_B\ a_{-p}^\dagger \quad .$$

For a one-boson state of momentum $\vec{p}$, $|\vec{p}> = a_p^\dagger |0>$, we get

$$P|\vec{p}\rangle = Pa_p^\dagger P^{-1} P|0\rangle$$

$$(4.141)$$

$$= \eta_B a_{-p}^\dagger |0\rangle = \eta_B |-\vec{p}\rangle \quad,$$

where the parity of the vacuum state is given by Eq. (4.136). Equation (4.141) states that the momentum $\vec{p}$ of a boson changes sign under the parity operation; it is just Eq. (4.123) in different form. It also shows that the single free boson state of definite momentum does not have a fixed parity, a fact that is expected because the linear momentum and parity operators do not commute.[*] However in the rest system of the particle we get

$$P|\vec{p} = 0\rangle = \eta_B |\vec{p} = 0\rangle \quad ;$$

the particle does have a well-defined intrinsic parity, $\eta_B = +1$ for a scalar and $\eta_B = -1$ for a pseudoscalar boson. This relation agrees with

---

[*]In Chapter 3 we treated the expansion of field operators in terms of plane wave states. The operator Φ can also be expanded in spherical harmonics in which case the various terms correspond to states of well-defined orbital angular momentum. These states are eigenstates of the parity operator with parity $(-1)^\ell \eta_B$. See E. M. Henley and W. Thirring, Elementary Quantum Field Theory (McGraw-Hill, New York, 1962), p. 38.

Eq. (4.137).

To discuss the relative parity of bosons and antibosons, we repeat the arguments leading to Eq. (4.140) with the expansion Eq. (3.115) and get

$$Pa_p P^{-1} = \eta_B a_{-p} \quad , \quad Pa_{\bar{p}} P^{-1} = \eta_B a_{-\bar{p}}$$

$$\text{(4.140a)}$$

$$Pa_p^\dagger P^{-1} = \eta_B a_{-p}^\dagger \quad , \quad Pa_{\bar{p}}^\dagger P^{-1} = \eta_B a_{-\bar{p}}^\dagger \quad .$$

From these relations and Eq. (4.141) it follows immediately that bosons and antibosons have the same intrinsic parity.

For _fermions_, we first consider the ordinary (first quantized) theory and write for the action of the parity operator on the four-component wave function $\Psi(t,\vec{x})$

$$P\Psi(t,\vec{x}) = \Psi'(t',\vec{x}') = \Psi'(t,-\vec{x}) \quad .$$

To establish a connection between Ψ' and Ψ, we assume that they are related by

$$\Psi' = U\Psi$$

where U is a unitary operator (four-by-four matrix). We further assume that Ψ satisfies the Dirac equation, Eq. (3.167),

$$\left(i\gamma^0 \frac{\partial}{\partial t} + i\vec{\gamma}\cdot\vec{\nabla} - m \right) \Psi = 0 \quad .$$

Invariance under the parity operation demands that Ψ' satisfy formally the same equation in the primed coordinate system:

$$\left(i\gamma^0 \frac{\partial}{\partial t'} + i\vec{\gamma}\cdot\vec{\nabla}' - m \right) \Psi' = 0 \quad .$$

After expressing this equation in terms of $t = t'$, $\vec{x} = -\vec{x}'$, and $\Psi = U^{-1}\Psi'$,

$$\left(i\gamma^0 \frac{\partial}{\partial t} - i\vec{\gamma}\cdot\vec{\nabla} - m \right) U^{-1}\Psi = 0 \quad ,$$

multiplying from the left with U and comparing with the original form, we get

$$U\gamma^0 U^{-1} = \gamma^0$$

$$U\vec{\gamma}U^{-1} = -\vec{\gamma} \quad .$$

These conditions are satisfied with $U = \eta_F \gamma^0$, $|\eta_F|^2 = 1$; the parity operator in the Dirac theory can hence be written as

$$P\Psi(t,\vec{x}) = \eta_F \gamma^0 \Psi(t,-\vec{x}) \quad . \tag{4.142}$$

This relation corresponds to Eq. (4.126) for the Schrödinger case. There we have assumed that P^2 leads back to the original state, $P^2 = 1$. However,

it is well known that the situation for fermions
is more complicated; a rotation by 2π results in
a multiplication of the wave function by -1, as
can be seen with Eq. (4.53). In order to return
to the original situation, a rotation by 4π is
necessary.[*] We assume that the rotation by 4π
corresponds to four parity operations and thus set

$$P^4 = 1 \quad , \quad \eta_F = \pm 1, \pm i \quad . \tag{4.143}$$

We are faced here with the existence of four
different phase choices[77] and the identification
of η_F with the intrinsic parity is not clear.
Fortunately, superselection rules save the day:
Fermions do not connect to the vacuum and it is
therefore impossible to measure their absolute
intrinsic parity.

One case of special interest is the relative
parity of a fermion and the corresponding anti-
fermion. The Dirac equations for particles and
antiparticles at rest are given by Eq. (3.181);
with Eq. (4.142) we get

$$Pu(\vec{p}=0) = \eta_F \, \gamma^0 \, u(\vec{p}=0) = \eta_F \, u(\vec{p}=0) \quad ,$$
$$\tag{4.144}$$
$$Pv(\vec{p}=0) = \eta_F \, \gamma^0 \, v(\vec{p}=0) = -\eta_F \, v(\vec{p}=0) \quad .$$

---

[*] A demonstration object that behaves under
rotation like a fermion wave function, said
to have been introduced by Dirac, is sketched
by H. J. Bernstein in Scientific Research,
18, August 1969, p. 33.

77. C. N. Yang and J. Tiomno, Phys. Rev. **79**, 495
(1950). G. Morpurgo and B. F. Touschek,
Nuovo Cimento **1**, 1159 (1955).

The intrinsic parity of a Dirac particle is opposite to that of its antiparticle.

In field theory, $\Psi(t,\vec{x})$ is reinterpreted as an operator and the space reflection for a fermion field is defined by

$$P\Psi(t,\vec{x})P^{-1} = \eta_F \gamma^0 \, \Psi(t,-\vec{x}) \quad ,$$

$$P\overline{\Psi}(t,\vec{x})P^{-1} = \eta_F^* \, \overline{\Psi}(t,-\vec{x}) \, \gamma^0 \quad . \tag{4.145}$$

With the expansion (3.190) it is then straightforward to see that the intrinsic parity of a fermion-antifermion pair (which can be determined absolutely) is negative.

D. The Intrinsic Parity of Particles with A=Q=Y=0[78,79)]

Some particles and systems have the same additive quantum numbers as the vacuum state $|0>$; the absolute intrinsic parities of these particles can be determined. We discuss here two such particles, the photon and the neutral pion.

Since the photon is described by a vector field, we expect its intrinsic parity to be odd.[*]

---

78. C. N. Yang, Phys. Rev. 77, 242 (1950).
79. P. T. Matthews, Repts. Progr. Phys. 18, 453 (1955).
*Note that a vector has negative parity, a pseudovector positive parity.

Indeed, the source of the electromagnetic field
is a distribution of charges and currents so that

$$A^\mu(t,\vec{x}) = \int\int d^3x' d^3t' \ \frac{j^\mu(t',\vec{x}')}{|\vec{x} - \vec{x}'|} \ \delta(t'+|\vec{x}-\vec{x}'|-t).$$

(4.146)

The current operator j^μ transforms under parity
as

$$P \ j^\mu(t,\vec{x}) P^{-1} = P(\rho,\vec{j}) P^{-1}$$

(4.147)

$$= (\rho,-\vec{j}) \equiv j_\mu(t,-\vec{x}) \quad,$$

because the charge is a scalar and j is proportional to the momentum. With Eq. (4.146), the
transformation property of the field operator A^μ
becomes

$$P A_0(\vec{x}) P^{-1} = +A_0(-\vec{x})$$

$$P \vec{A}(\vec{x}) P^{-1} = -\vec{A}(-\vec{x})$$

(4.148)

or $\quad P A^\mu(t,\vec{x}) P^{-1} = A_\mu(t,-\vec{x}).$

The Hamiltonian for the interaction of the electromagnetic field with a current qj^μ is given by

$$H_\gamma = q \ j_\mu A^\mu \ ;$$

(4.149)

with Eqs. (4.147) and (4.148) it is seen that this

interaction is invariant under the parity operation:

$$[P,H_\gamma] = 0 \quad .$$

(4.150)

To exhibit the negative intrinsic parity of the free photon explicitly, we first note that with Eqs. (3.54) and (4.148) and with arguments similar to those leading to Eq. (4.140) we obtain

$$Pa_{k\lambda} P^{-1} = -a_{-k-\lambda}$$

$$Pa_{k\lambda}^\dagger P^{-1} = -a_{-k-\lambda}^\dagger \quad .$$

(4.151)

Not only the momentum, but also the helicity λ changes sign under the parity operation. The one-photon state, Eq. (3.47), then transforms as

$$P|\vec{k},\lambda> = Pa_{k\lambda}^\dagger P^{-1}P|0> = a_{-k-\lambda}^\dagger|0> = -|-\vec{k},-\lambda> \quad .$$

At this point we run into a difficulty: $|\vec{k},\lambda>$ is an eigenstate of P only if $\vec{k} = 0$, but photons are always relativistic and cannot be brought to rest! To circumvent this problem we return to the heavy boson and note that its intrinsic parity η_B is given the phase factor appearing in Eq. (4.141). The corresponding phase factor for the photon in the previous equation is negative and we identify it with the intrinsic photon parity.

In interactions, the photon is emitted or absorbed in certain angular momentum states. It turns out that it is impossible to separate

orbital and spin angular momentum of photons in an
unambiguous way; the only meaningful quantity is
the total angular momentum. The parity of a photon
hence cannot be written as a product of an intrinsic and an orbital factor, as can be done for
massive particles. Rather, for each value of the
total angular momentum J, the photon can carry
positive or negative parity. The two possible
states are denoted as magnetic and electric radi-
ations and the parities η_γ for a photon of total
angular momentum j are

$$\text{Electric} \quad 2^j \text{ pole} \quad \eta_\gamma = (-1)^j \quad ,$$

$$\text{Magnetic} \quad 2^j \text{ pole} \quad \eta_\gamma = (-1)^{j+1} \quad .$$

(4.152)

The two modes differ in their polarization.

The neutral pion decays predominantly into
two gamma rays with a lifetime of about 10^{-16} sec.
Consider the decay in the rest system of the
pion, as shown in Fig. 4.17. The three measurable
quantities are the momentum $\vec{k}$ of one of the gamma
rays and the polarizations $\hat{\varepsilon}_1$ and $\hat{\varepsilon}_2$ of the two
gamma rays. Each quantity can only be measured
once; the momentum $\vec{k}$ of one photon determines the
momentum $-\vec{k}$ of the other one completely. The
final wave function must contain the three vectors
$\hat{\varepsilon}_1$, $\hat{\varepsilon}_2$, and $\vec{k}$ and it must satisfy three conditions:

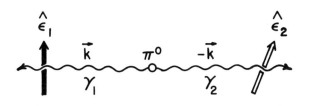

Fig. 4.17. Decay of the neutral pion in its rest system.

1. It must be linear in $\hat{\epsilon}_1$ and $\hat{\epsilon}_2$ since each photon occurs once;

2. it must be symmetric under the simultaneous operations

$$\vec{k} \rightleftarrows -\vec{k} \quad , \qquad \hat{\epsilon}_1 \rightleftarrows \hat{\epsilon}_2 \quad ,$$

   since the two photons are identical particles;

3. because of the transverse nature of the electromagnetic field, the photons must satisfy Eq. (3.16),

$$\hat{\epsilon}_1 \cdot \vec{k} = \hat{\epsilon}_2 \cdot \vec{k} = 0 \quad .$$

   We assume first that the spin of the neutral pion is 1. The final state, consisting of the two photons, then must also be a spin 1 state, i.e., it must be a vector or a pseudovector. From $\hat{\epsilon}_1$, $\hat{\epsilon}_2$, and $\vec{k}$, only three independent spin 1 quantities can be formed that satisfy condition 1:

$$\hat{\epsilon}_1 \times \hat{\epsilon}_2$$

$$(\hat{\epsilon}_1 \cdot \hat{\epsilon}_2)\vec{k} \qquad\qquad (4.153)$$

$$\vec{k} \times (\hat{\epsilon}_1 \times \hat{\epsilon}_2) = \hat{\epsilon}_1(\vec{k}\cdot\hat{\epsilon}_2) - \hat{\epsilon}_2(\vec{k}\cdot\hat{\epsilon}_1) \quad .$$

The first two combinations are excluded because they are not symmetric, and the third one because of transversality. The result is general: <u>A spin one particle cannot decay into two photons.</u> Hence, the neutral pion does not have spin 1.

Before we turn to the spin 0 case, we insert some remarks concerning the choice of the forms in Eq. (4.153). Experience shows that simple invariance arguments are difficult to understand at first. We therefore justify the forms in somewhat more detail. The pion decay is described by the Feynman diagram in Fig. 4.18. To write down the correct matrix element, we would have to

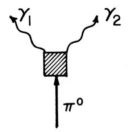

Fig. 4.18. Feynmann diagram for the two-gamma decay of the neutral pion.

know the interaction Lagrangian. Since we don't
know it, we write in symbolic form for the black
box

$$\mathcal{M} = \langle \gamma_1 \gamma_2 | \ f(\vec{A}_1 \vec{A}_2 \Phi_o) | \pi^o \rangle \quad , \qquad (4.154)$$

where Φ_o destroys the neutral pion, $\vec{A}_1$ and $\vec{A}_2$
create the two photons, and f is an unknown
function. One property of f is clear: It must
be a scalar operator formed from $\vec{A}_1, \vec{A}_2$ and Φ_o
because $\mathcal{M}$ is a scalar. In addition, it must be
linear in $\vec{A}_1, \vec{A}_2$ and Φ_o; otherwise the matrix
element vanishes. Equation (3.69) then shows that
$\hat{\epsilon}_1$ and $\hat{\epsilon}_2$ indeed appear linearly. The momentum $\vec{k}$
appears if f prescribes differentiation.

　　We return to the invariance arguments. We
now know that the neutral pion spin cannot be 1.
Spin 2 or higher is unlikely: The decay into 2
photons would be slower than 10^{-16} sec and the
spin of the charged pions turns out to be zero.
If we maintain that the three pions form an
isospin triplet, the neutral member must also have
spin 0.

　　We can form two spin zero expressions with
the three vectors at our disposal; these two
combinations and their properties are shown in
Table 4.6. To determine the parity of the neutral
pion, we look at the polarizations of these
combinations. The decay rate of the neutral pion
is proportional to the square of the matrix element.
If ϕ denotes the angle between the polarization

Table 4.6.   Independent spin 0 combinations of
$\hat{\varepsilon}_1$, $\hat{\varepsilon}_2$, and $\vec{k}$.

Combination	Transformation property	Spin	Parity
$\hat{\varepsilon}_1 \cdot \hat{\varepsilon}_2$	Scalar	0	+
$(\hat{\varepsilon}_1 \times \hat{\varepsilon}_2) \cdot \vec{k}$	Pseudoscalar	0	−

vectors $\hat{\varepsilon}_1$ and $\hat{\varepsilon}_2$, the two cases in Table 4.6 thus
have the following angular dependences:

Scalar $\qquad |\mathcal{m}|^2 \propto |\hat{\varepsilon}_1 \cdot \hat{\varepsilon}_2|^2 \propto \cos^2\phi$,

Pseudoscalar $\quad |\mathcal{m}|^2 \propto |\hat{\varepsilon}_1 \times \hat{\varepsilon}_2|^2 \propto \sin^2\phi$

If the pion is a scalar particle, the two decay
photons will have predominantly parallel linear
polarizations; if the pion is a pseudoscalar
particle, the two photons will be emitted predominantly with perpendicular linear polarizations.
The polarization of high-energy gamma rays is
difficult to measure. However, in about one
decay out of 30,000, the pion decays by emitting
two electron-positron pairs,

$$\pi^o \rightarrow e^+e^-e^+e^- .$$

In a bubble chamber, such a "double Dalitz pair" looks like a bow tie, as sketched in Fig. 4.19. The planes of the pairs remember the polarization

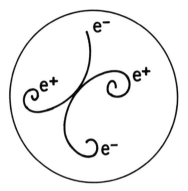

Fig. 4.19.   Double Dalitz pair (bow tie) in a
                   bubble chamber with magnetic field.

of the (virtual) photons. For the pseudoscalar case, the planes of the two internally converted pairs should be more often perpendicular.[80] Plano et al. have looked at about 700,000 pictures taken with a 30 cm hydrogen bubble chamber.[81] The determination of the planes of about 100 pairs is unambiguous: the neutral pion is a pseudoscalar.

80.   N. Kroll and W. Wade, Phys. Rev. 98, 1355
       (1955).
81.   R. Plano, A. Prodell, N. Samios, M. Schwartz,
       and J. Steinberger, Phys. Rev. Letters 3,
       525 (1959).

E.   The Intrinsic Parity of Particles and Nuclei

We begin the parity assignment to particles
and nuclei with the nucleons. Proton and neutron
have the same baryon number and hypercharge, but
differ in their electric charge. No spontaneous
transitions between the two particles have ever
been observed and no known observable connects
them. There is therefore no way to measure their
relative parity. However, because the proton and
the neutron are members of an isospin doublet we
assign even parity to both. Once we have assigned
the parities of the neutron and the proton, we can,
at least in principle, measure the parities of all
nuclides. Consider for instance the deuteron. We
can study the radiative neutron-proton capture

$$n\ p \rightarrow d\ \gamma \qquad\qquad\qquad (4.155)$$

or the photodisintegration of the deuteron

$$\gamma\ d \rightarrow n\ p \quad . \qquad\qquad\qquad (4.156)$$

The parity of the deuteron can be determined from
the capture reaction (4.155) by measuring the
parity of the emitted photon and the relative
angular momentum of the neutron and the proton in
the initial state. In principle, the parity of
all other nuclides can be found by the same technique. The parities of excited states can also
be determined by measuring the parity of gamma
transitions leading to lower states.

We now turn to a point that often leads to questions. We have stated that no transitions exist between neutron and proton. However the neutron can decay by the process

$$n \rightarrow p \; e^- \; \bar{\nu} \quad ,$$

and virtual Yukawa reactions

$$n \rightleftarrows p \; \pi^- \qquad p \rightleftarrows n \; \pi^+ \qquad\qquad (4.157)$$

do occur. Are these not transitions between neutron and proton? The point here is that these processes indeed involve neutron and proton, but that they must include other particles in order to satisfy all conservation laws. It is therefore possible to determine the relative parity of the neutron with respect to the proton-pion system, but not with respect to the proton alone.

Next come the intrinsic parities of other par-ticles. We first ask: How many parities do we have to assign by definition in order to satisfy all superselection rules? We require two (n,p) because of the electric charge and a third, usually taken to be that of the Λ, because of Y. Using reactions involving particles and antiparticles, we can then reach all hyperons. Moreover, because two fermions together form an integer-spin system, we can also reach all bosons. We give some examples to indi-cate how the various parities have been determined, assuming

$$\eta_p = \eta_n = \eta_\Lambda = +1 \quad . \qquad\qquad (4.158)$$

The parity of the charged pion can, in principle, be found from the Yukawa reaction, Eq. (4.157). In practice reaction (4.134), $\pi^- d \to nn$, is used. When a deuterium target is bombarded by negative pions, the pions slow down and are predominantly captured in an atomic orbit, forming a pionic atom.[82] The pion usually reaches an S state before nuclear capture takes place[83] and in about 2/3 of all cases, the reaction $\pi^- d \to nn$ occurs. Since the deuteron has spin 1, the pion spin 0, and the capture occurs from an S state, the angular momentum in the final state must be 1. The two neutrons must be in a totally antisymmetric state; the only such state with total angular momentum 1, $^3P_1$, has an orbital angular momentum $\ell = 1$ and hence odd parity. Since the deuteron has even parity, the negative pion must have odd parity. By Eq. (4.141), the positive pion then should also have odd parity. This assignment agrees with the prediction of Eq. (4.89) that all members of an isospin multiplet have the same intrinsic parity.

The parity of the negative kaon has been

82. A. S. Wightman, Phys. Rev. 77, 521 (1950).
83. T. B. Day, G. A. Snow, and J. Sucher, Phys. Rev. Letters 3, 61 (1960). M. Leon and H. A. Bethe, Phys. Rev. 127, 636 (1962).

determined from the reactions[84]

$$K^- \; {}^4He \rightarrow \pi^- \; {}^4_\Lambda He$$

(4.159)

$$K^- \; {}^4He \rightarrow \pi^o \; {}^4_\Lambda H \quad .$$

Here, ${}^4_\Lambda He$ and ${}^4_\Lambda H$ are hyperfragments, nuclides in
which one neutron is replaced by a lambda. With
the assumption that the spins of all particles in
the reactions (4.159) are zero[84,85], it follows
that the orbital angular momentum must be the same
in the initial and final states, and the intrinsic
parities of ${}^4He$, ${}^4_\Lambda He$, and ${}^4_\Lambda H$ must be even. The
fact that the reactions (4.159) occur shows that
the intrinsic parity of the K^- is odd.

As an example of the determination of hyperon
parities, consider the Σ^o. As shown in Fig. 4.8,
the Σ^o decays to the Λ^o with the emission of a
gamma ray. Initial and final state of this transition have spin $\frac{1}{2}$; the transition must occur by
dipole radiation. The relative parity $\eta_\Lambda - \eta_\Sigma$ is

84. M. M. Block, E. B. Brucker, I. S. Hughes,
    T. Kikuchi, C. Meltzer, F. Anderson, A.
    Pevsner, E. M. Harth, J. Leitner, and H. O.
    Cohn, Phys. Rev. Letters 3, 291 (1959).
    M. M. Block, L. Lendinara, and L. Monari,
    Proc. Intern. Conf. High Energy Phys. CERN
    Geneva, 1962, 371.
85. R. G. Ammar, R. Levi-Setti, W. E. Slater,
    S. Limentani, P. E. Schlein, and P. H.
    Steinberg, Nuovo Cimento 19, 20 (1961).

fixed if the electric or magnetic character of
the dipole radiation is measured (Eq. (4.152)).
A direct determination of the polarization of the
gamma ray is difficult. However, about 5×10^{-3}
of the decays occur by emission of a Dalitz pair,

$$\Sigma^{0} \rightarrow \Lambda^{0}\ e^{-}\ e^{+}\quad .$$

Such Dalitz pairs have been investigated in a
hydrogen bubble chamber and it was found that
the decay goes by magnetic dipole radiation. Σ^{0}
and Λ^{0} hence have the same parity.[86]

F.   The Breakdown of Parity

> "The disturbing possibility remains
> that charge conjugation and parity
> are both only approximate."
>
> (G. C. Wick, A. S. Wightman, and
> E. P. Wigner, Phys. Rev. 88, 101
> (1952).

So far we have tacitly assumed that parity is
a good quantum number. This assumption appears to
be reasonable because invariance under space reflection is aesthetically pleasing. Moreover,

---

86.   H. Courant, H. Filthuth, P. Franzini, R. G.
      Glasser, A. Minguzzi-Ranzi, A. Segar, W.
      Willis, R. A. Burnstein, T. B. Day, B. Kehoe,
      A. J. Herz, M. Sakitt, B. Sechi-Zorn, N.
      Seeman, and G. A. Snow, Phys. Rev. Letters
      10, 409 (1963).

parity as a quantum number was introduced on the
basis of experimental evidence, namely the validity
of Laporte's rule.[71] While some doubts were expressed by Purcell and Ramsey[87] and by Wick,
Wightman, and Wigner[73], the possibility of a
parity violation was not taken seriously. However,
in 1956 the roof fell in. As is usual, the catas-
trophe did not come unannounced, but the warning
signs were largely overlooked. The problem began
with an attempt to determine the parity of kaons
from their decay,[88] and resulted in a paradox
that we will describe below. One way to solve the
paradox was proposed by Lee and Yang who suggested
that parity could be violated in the weak interactions.[72] The suggestion was rejected by most
physicists,* but was dramatically verified in a
series of beautiful experiments.[89-91]

The problem that led to the discovery of
parity non-conservation was the apparent existence
of two particles, called θ and τ , decaying weakly

87. E. M. Purcell and N. F. Ramsey, Phys. Rev.
    78, 807 (1950).
88. R. H. Dalitz, Phil. Mag. 44, 1068 (1953);
    Phys. Rev. 94, 1048 (1954); Rept. Progr.
    Phys. 20, 163 (1957).
89. C. S. Wu, E. Ambler, R. W. Hayward, D. D.
    Hoppes, and R. P. Hudson, Phys. Rev. 105,
    1413 (1957).
90. R. L. Garwin, L. M. Lederman, and M. Weinrich,
    Phys. Rev. 105, 1415 (1957).
91. J. L. Friedman and V. L. Telegdi, Phys. Rev.
    105, 1681 (1957); 106, 1290 (1957).
  *R. P. Feynman, for instance, bet N. F. Ramsey
  $50 to $1 that parity is conserved. Feynman
  paid.

as follows

$$\theta \to 2\pi , \qquad \tau \to 3\pi .$$ (4.160)

Studies of the two decays, particularly by Dalitz[88], showed that the two particles had the same lifetimes (about 10^{-8} sec), the same masses (about 490 MeV), and the same production cross sections. Such repeated accidental identities are hard to accept unless τ and θ are different decay modes of one particle, but this assumption leads to difficulties in understanding the decays.

The decay properties of the θ are simple. The mode $\theta^0 \to \pi^0 \pi^0$ has been observed; Bose statistics requires the $\pi^0 - \pi^0$ system to have even parity and therefore even orbital angular momentum. The intrinsic parity of the θ must be even and its spin can be 0,2,4... . The values 2,4,... would allow the decay $\theta \to \pi \gamma$ via weak interactions (compare Subsection 4.2.H) with a rate comparable to that for the decay $\theta \to \pi\pi$.[92] Since no such radiative decays have been observed, we assume the spin of the θ to be zero. The τ has the same lifetime and the same production characteristics as the θ and we therefore assume that it also has spin zero.

We have already noted that the parity of the θ must be even. To find the parity of the τ , we denote by $\vec{\ell}$ the relative angular momentum of the

92.   R. H. Dalitz, Proc. Phys. Soc. A69, 527 (1956).

two π^+ in their c.m., and with $\vec{L}$ the angular momentum of the π^- relative to the c.m. system of the two π^+. (Fig. 4.20) Assuming the τ spin to be zero, we have $\vec{\ell} + \vec{L} = 0$ and hence $\ell = L$. The parity of the final state and therefore also the

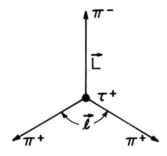

Fig. 4.20. Decay of the τ^+.

intrinsic parity of the τ is

$$\eta_\tau = (-1)^3(-1)^{\ell + L} = -(-1)^{2L} = -1 \quad . \quad (4.161)$$

Hence, either the θ and the τ are different particles despite their equal properties, or parity is not conserved in the weak decay of the τ and the θ. This, then, is the famous $\tau-\theta$ puzzle.

When Lee and Yang analyzed the puzzle, they did not simply ask "Is parity conserved?," but more specifically inquired about parity conservation in each interaction. They found good

evidence for parity conservation in the hadronic
and the electromagnetic interaction, but discovered
much to everyone's surprise that none of the numerous beta decay experiments provided evidence for
parity conservation in the weak interaction.[*72)]
They then suggested three experiments to test
parity conservation in beta and meson decays. All
three were performed and all indicated that parity
was not conserved in the weak interaction.[89-91,94)]
We treat here a fourth experiment, the detection
of the helicity of electrons, selected because we
use the information that it yields also in the
discussion of charge conjugation.

In Subsection 3.6B we have defined the helicity
operator for spin 1/2 particles as $\vec{\sigma}\cdot\hat{p}$, where $\vec{\sigma}$ is
the spin operator and $\hat{p}$ a unit vector in the direc-
tion of the electron momentum. With Eqs. (4.128)

---

93.   R. T. Cox, C. G. McIlwraith, and B. Kurrel-
      meyer, Proc. Nat. Acad. Sci. 14, 544 (1928);
      C. T. Chase, Phys. Rev. 34, 1069 (1929); 36,
      894, 1060 (1930).
94.   F. Boehm and A. H. Wapstra, Phys. Rev. 106,
      1364 (1957); H. Schopper, Phil. Mag. 2, 710
      (1957).
   *Actually it turned out that experiments
   around 1930 had given evidence for parity
   non-conservation.[93)] In these experiments,
   the helicity of electrons from radioactive
   sources was determined and found to be
   nonvanishing. The result disagreed with
   expectations; as a cure the beta decay electrons were replaced with those from a hot
   filament, the effect disappeared and every-
   body was satisfied.

and (4.129), we get

$$P \, \vec{\sigma} \cdot \hat{p} \, P^{-1} = - \, \vec{\sigma} \cdot \hat{p} \quad . \tag{4.162}$$

The helicity operator is odd under space reflection. The expectation value of the helicity operator for a state $|\alpha\rangle$ with <u>well-defined</u> parity η_α is given by Eq. (4.133) as

$$\langle \alpha | \vec{\sigma} \cdot \hat{p} | \alpha \rangle = - \, \eta_\alpha^2 \, \langle \alpha | \vec{\sigma} \cdot \hat{p} | \alpha \rangle = - \, \langle \alpha | \vec{\sigma} \cdot \hat{p} | \alpha \rangle \quad .$$

$$\tag{4.163}$$

The expectation value of the helicity operator vanishes if parity is conserved and the state $|\alpha\rangle$ is non-degenerate. If, however, $|\alpha\rangle$ is a super-position of states with even (e) and odd (o) parity,

$$|\alpha\rangle = (1 - |\mathcal{F}|^2)^{1/2} |e\rangle + \mathcal{F} |o\rangle \quad , \tag{4.164}$$

the conclusion is no longer valid. We then find with Eq. (4.133) and $\langle e | \vec{\sigma} \cdot \hat{p} | e \rangle = \langle o | \vec{\sigma} \cdot \hat{p} | o \rangle = 0$

$$\langle \alpha | \vec{\sigma} \cdot \hat{p} | \alpha \rangle = 2 (1 - |\mathcal{F}|^2)^{1/2} \, \text{Re}\{\mathcal{F} \langle e | \vec{\sigma} \cdot \hat{p} | o \rangle\} \quad .$$

$$\tag{4.165}$$

A nonvanishing expectation value of the helicity

operator indicates that $\mathcal{F} \neq 0$, i.e., the state
$|\alpha\rangle$ is not an eigenstate of the parity operator.

   The helicity of electrons can be measured in
various ways.[95]   We sketch here only the method
based on electron-electron scattering.[96,97]   If
two colliding electrons have parallel spins, the
differential cross section for scattering is
smaller than if they have antiparallel spins. For
low-energy negative electrons, this spin dependence
can be understood by a straightforward application
of the Pauli principle: If electrons are in an
even spin state, they cannot collide if they are
also in even spatial states. Such a situation is
shown in Fig. 4.21. The spin dependence is
strongest for scattering by 90° in the c.m. system,
since parallel spins require an antisymmetric
spatial wave function and the Legendre polynomial
$P_L(\theta)$ vanishes at $\theta = 90°$ for odd values of L.
Positron-electron collisions (Bhabha scattering)
have the same polarization dependence in the extreme
relativisitic limit as electron-electron collisions
(Møller scattering.)[97]

---

95.   H. Frauenfelder and A. Rossi, in Methods of
      Experimental Physics, L. C. Yuan and C. S. Wu,
      eds. (Academic Press, New York, 1963), Vol. 5B.
96.   H. Frauenfelder, A. O. Hanson, N. Levine, A.
      Rossi, and G. DePasquali, Phys. Rev. 107, 643
      (1957); J. D. Ullman, H. Frauenfelder, H. J.
      Lipkin, and A. Rossi, Phys. Rev. 122, 536
      (1961).
97.   A. M. Bincer, Phys. Rev. 107, 1434 (1957).

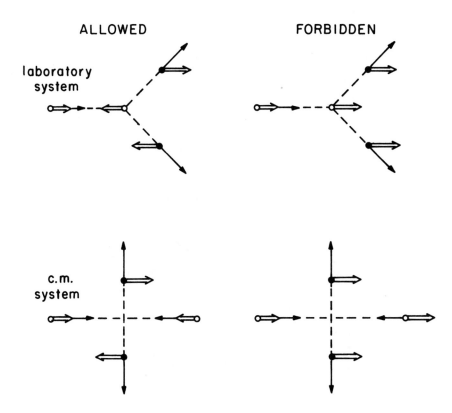

Fig. 4.21.   Møller scattering for non-relativistic
             electrons. The scattering is shown in
             the lab and the c.m. system. Before the
             scattering, the electrons are indicated
             by empty circles, after the collisions
             with solid ones. Only the totally anti-
             symmetric state (space and spin) is
             allowed by the Pauli principle. We
             indicate spins by o⇒ and momentum
             by ⟶ .

A simple arrangement to determine the helicity
of electrons by electron-electron scattering is
shown in Fig. 4.22. The momentum of the incident
electrons is selected by a monochromator. The
electrons then impinge on a thin magnetized ferro-
magnetic foil. Unfortunately, only 2 out of 26

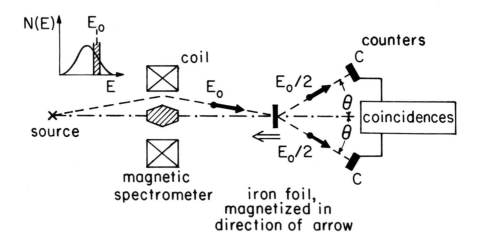

Fig. 4.22. Schematic representation of an arrange-
            ment to determine the electron helicity
            by electron-electron scattering.

electrons can be polarized in iron, but the resulting polarization is still large enough to make the
experiment feasible. We are interested in a scattering through 90° in the c.m. With stationary
target electrons, these events correspond in the
laboratory to the case where the incident and the
target electron leave the foil with equal energy
and with equal angles θ (Fig. 4.22). Since there

are about two orders of magnitude more electrons
scattered by Rutherford scattering, the few
electron-electron events must be carefully selected. Their characteristic energies and their
angular correlation make such a selection feasible:
For an incident energy E_o, each of the two
outgoing electrons has an energy $E_o/2$. The two
counters are then placed at a laboratory angle
that corresponds to 90° c.m. scattering; only
energies of about $E_o/2$ are accepted in the two
counters. Furthermore, the two counters are used
in coincidence; only if they give pulses corresponding to the energy $E_o/2$ within a resolving
time of a few nsec is an event recorded. Counting
rates are then determined for both directions of
magnetization.

As sources, beta emitters are used. The
lifetimes of these radioactive nuclides is so
long that the decay must be due to a weak interaction. For instance for the decay

$$^{32}P \rightarrow {}^{32}S \ e^- \ \bar{\nu} \quad ,$$

the half life is 14 d, the maximum energy of the
electron spectrum is 1.71 MeV. The electron helicity is determined as a function of the electron
energy. The results of all measurements on nuclear
beta decays are unambiguous: Electrons and positrons emitted in weak interactions are longitud-
inally polarized. In almost all cases, the degree
of polarization turns out to be proportional to

the velocity of the beta particles:

$$\text{Electrons} \quad \langle \vec{\sigma} \cdot \hat{p} \rangle = - v \quad ,$$
$$\text{Positrons} \quad \langle \vec{\sigma} \cdot \hat{p} \rangle = + v \quad . \qquad (4.166)$$

For high energy electrons and positrons, the expectation value of the helicity approaches -1 and 1, respectively.

The electron-electron scattering method has also been applied to measurements of the helicities of the high energy electrons and positrons emitted in the decay of the muons, $\mu^{\pm} \to e^{\pm} \bar{\nu} \nu$. The results[98] are again unambiguous; electrons and positrons are polarized and the helicities satisfy Eq. (4.166).

Equation (4.165) demonstrates that a non-vanishing expectation value of an odd observable proves that the corresponding state is not an eigenstate of the parity operator. We now connect the formalism more directly to a violation of the invariance under P. To do so, we consider the famous experiment by Wu and collaborators[89] which was performed with the decay $^{60}Co \to {}^{60}Ni\ e^{-}\ \bar{\nu}$. The cobalt nuclei were polarized so that their nuclear spins $\vec{J}$ pointed in the same direction, and the number of electrons emitted parallel and antiparallel to $\vec{J}$ was then compared. We denote the initial

---

98. J. Duclos, J. Heintze, A. DeRujula, and V. Soergel, Phys. Letters $\underline{9}$, 62 (1964) (μ^+); D. M. Schwartz, Phys. Rev. $\underline{162}$, 1306 (1967) (μ^-).

state with $|\vec{J}>$, the two different final states with $|\vec{p}>$ and $|-\vec{p}>$, and note that, according to Eqs. (4.128) and (4.129) and with suitable choice of phase we have

$$P|\vec{J}> = |\vec{J}> , \quad P|\vec{p}> = |-\vec{p}> \quad .$$

In first-order perturbation theory the number of electrons is proportional to the square of the matrix element and the measured ratio of electrons "up" to electrons "down" is given by

$$\frac{N_+}{N_-} = \frac{|<\vec{p}|H_w|\vec{J}>|^2}{|<-\vec{p}|H_w|\vec{J}>|^2} = \frac{|<\vec{p}|P^{-1}PH_wP^{-1}P|\vec{J}>|^2}{|<-\vec{p}|H_w|\vec{J}>|^2}$$

$$= \frac{|<-\vec{p}|PH_w P^{-1}|\vec{J}>|^2}{|<-\vec{p}|H_w|\vec{J}>|^2} \quad .$$

H_w is the weak Hamiltonian that is responsible for the decay of ^{60}Ni. As long as H_w is either even or odd, the ratio N_+/N_- is unity. However, if H_w consists of an even and an odd part,

$$H_w = H_e + H_o \quad , \tag{4.167}$$

we get

$$\frac{N_+}{N_-} = \frac{|<-\vec{p}|H_e-H_o|\vec{J}>|^2}{|<-\vec{p}|H_e+H_o|\vec{J}>|^2} \neq 1 \quad . \tag{4.168}$$

Experimentally, mirror symmetry is indeed violated,
N_+/N_- differs convincingly from one, and the weak
Hamiltonian therefore must contain an even and an
odd part.

G. Parity Conservation in Hadronic and Electro-
    magnetic Interactions

Before the breakdown of parity, there had
already been considerable evidence for the conservation of parity in hadronic and electromagnetic
interactions.[72]   After 1956, physicists, burned
by the experience, returned to the lab and checked
these invariances again. Before sketching some
tests, we mention the outcome: In hadronic and
electromagnetic interactions, P is conserved.
Hence we can write

$$[H_h, P] = 0 \qquad\qquad\qquad (4.169)$$

$$[H_\gamma, P] = 0 \quad . \qquad\qquad\qquad (4.170)$$

Parity conservation in the electromagnetic
interaction is tested in systems that are not
subject to hadronic forces, for instance the atomic
shell. Consider an electromagnetic transition
between atomic states of equal parity, for instance
$1^+ \to 1^+$. The parity selection rules (4.152) then
allow electric quadrupole and magnetic dipole
radiation, but forbid electric dipole radiation.
In atomic systems, the rate at which electric dipole transitions would occur can be calculated

easily. The fact that transitions between such states are not observed gives a limit of $|\mathcal{F}|^2 < 10^{-6}$ for the electromagnetic interaction. A better limit comes from a search for a circular polarization of photons emitted in a "forbidden" atomic transition. The circular polarization of photons (Eq. (3.26)) is an observable similar to the helicity of electrons and Eq. (4.165) shows that a non-vanishing expectation value indicates parity violation. The absence of an observable circular polarization of the 461.8 nm $^3P_1 \rightarrow {}^1S_0$ line in Pb I implies a limit $|\mathcal{F}|^2 < 5 \times 10^{-9}$.[99)]

In systems involving hadrons, <u>electromagnetic and hadronic forces</u> are present and tests to better than 1% involve both. One class of tests is based on transitions that are forbidden by parity selection rules. The first experiment was performed by Tanner,[100)] who searched for a parity-forbidden decay, $J^P = 1^+ \xrightarrow{\alpha} 0^+$, occurring in ^{20}Ne. We discuss here another parity-forbidden alpha decay, namely the decay of a 2^- state of ^{16}O, because it has led to a value for $|\mathcal{F}|^2$.

Consider a decay $2^- \xrightarrow{\alpha} 0^+$. The alpha particle has spin 0 and positive intrinsic parity; its parity is given by $(-1)^L$ if it is emitted with orbital angular momentum L. Conservation of angular momentum requires L = 2 and the parity carried by the alpha particle must be positive.

---

99. R. Poppe, Physics <u>50</u>, 48 (1970).
100. N. Tanner, Phys. Rev. <u>107</u>, 1203 (1957).

The decay $2^- \xrightarrow{\alpha} 0^+$ is therefore forbidden by
parity conservation. The search for such a decay
has been performed in ^{16}O. The relevant decay
scheme is shown in Fig. 4.23. The 8.88 MeV level
in ^{16}O has a spin-parity assignment 2^-, the ground
state of ^{12}C one of 0^+. Energetically, the 2^-

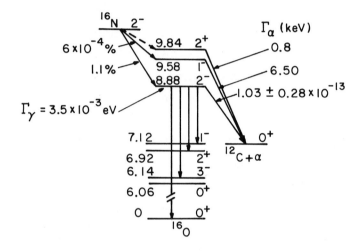

Fig. 4.23. Essential features of the energy level
            diagram of the A = 16 isobars. Only
            relevant states are included. All the
            states shown in ^{16}O have isospin I = 0.
            Energies are given in MeV.

state can decay only by alpha emission to the ground
state of ^{12}C; the first excited state lies too high.
Experimentally, the excited state of ^{16}O is reached
via a beta decay from ^{16}N, produced by the reaction
d $^{15}N \rightarrow p \; ^{16}N$. The spectra of ^{16}N and ^{16}O show
that the 2^- state in ^{16}O decays primarily by gamma
emission, but that a very weak alpha branch does

occur, with a partial width of $\Gamma_\alpha = (1.03 \pm 0.28) \times 10^{-10}$ eV. [101] Comparison of this width with that of the allowed α decay of the nearby 2^+ state yields a value

$$|\mathcal{F}|^2 \approx 10^{-13} \tag{4.171}$$

and thus provides evidence for a very small parity violation in a hadronic nuclear decay.

H. Parity Violation and Weak Nuclear Forces

Although the value of $|\mathcal{F}|$ determined above is very small, it is of the order of magnitude expected from the presence of the weak interaction. It is interesting to note that the role of the weak interaction in nuclear forces was investigated during the early development of nuclear physics. In 1934, when the nature of the force between nucleons was unknown, Tamm and Iwanenko tried to explain the hadronic nuclear force through the virtual exchange of electron-neutrino pairs, as shown in Fig. 4.24 a. [102] The forces calculated from such an exchange turned out to be much too

101. H. Hättig, K. Hünchen, and H. Wäffler, Phys. Rev. Letters 25, 941 (1970); K. Neubeck, H. Schober, and H. Wäffler, Phys. Rev. C 10, 320 (1974).
102. I. Tamm and D. Iwanenko, Nature 133, 981 (1934); H. A. Bethe and R. F. Bacher, Rev. Mod. Phys. 8, 201 (1936).

weak to fit the experimental facts. It is clear
today that the dominant part of nuclear forces
comes from the hadronic exchange of mesons, as
indicated for instance in Fig. 1.3a. Nevertheless,
a weak nucleon-nucleon force should also be
present[103] and the parity violation in the weak
interaction makes its detection possible. Observation of the contribution from the diagram 4.24a
is hopeless because the weak interaction occurs
twice, once at each vertex, and thus leads to
$|\mathcal{F}| \approx 10^{-12}$, or $|\mathcal{F}|^2 \approx 10^{-24}$ (See Table 1.3). If,

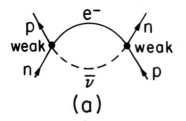
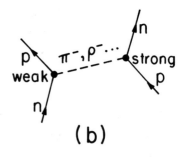

Fig. 4.24.   Weak contribution to the nuclear
             force.

however, the weak interaction acts but once, a

---

103.   R. P. Feynman and M. Gell-Mann, Phys. Rev.
       109, 193 (1958).

value of $|\mathcal{F}|^2 \sim 10^{-12}$ can be expected: the corre-
sponding contribution to the nucleon-nucleon force
is illustrated in Fig. 4.24 b.

Three different classes of effects can provide
information about the parity-violating contribution
to the force between nucleons[104,105]:

I.   Violation of the parity selection rule,

II. Nonvanishing helicity (longitudinal polarization) of a particle or photon emitted by an
unpolarized system in an electromagnetic or
strong transition,

III. Asymmetry in the angular distribution of a
particle or photon emitted by a polarized
system ("cosθ term" relative to the polarization direction).

Class I effects are proportional to $|\mathcal{F}|^2$, class II
and III to $\mathcal{F}$.

In the previous subsection we have discussed
a class-I experiment and the result expressed in
Eq. (4.171) indicates a parity violation of the
expected order of magnitude. Such experiments are,
however, exceedingly difficult because of the
extremely small effect and it indeed took many years
before a positive result was obtained. Even class
II and III experiments are very difficult and

104. D. H. Wilkinson, Phys. Rev. 109, 1603, 1610
(1958).
105. E. M. Henley, Ann. Rev. Nuclear Sci. 19,
367 (1969).

success so far has been achieved mainly because
nuclear enhancement effects increase the magnitude
of the observable helicity.

The first evidence for a parity-violating
component in the nuclear force was found by Abov
and co-workers[106] (class III) and by Lobashov and
co-workers[107] (class II). We sketch Lobashov's
approach here because it is beautifully simple and
ingenious. The goal of the experiment was to
determine the circular polarization of the gamma
rays emitted by ^{181}Ta which is the daughter of
^{181}Hf. The ^{181}Ta gamma rays from a 500 Ci source
of ^{181}Hf were passed through magnetized iron and
then observed in a scintillation counter. In order
to obtain sufficient statistics, the integral
current of the detector, rather than individual
counts, was measured. Now, the cross section of
scattering of polarized gamma rays by polarized
electrons is different for states with parallel
and antiparallel spins of the electron and
photon.[95] Therefore, if the photons are polarized,
the count rate in the detector changes with the
sign of the polarization of the magnet. In Lobashov's experiment, the magnetic field was reversed

106.   Yu. G. Abov, P. A. Krupchitsky, Yu. A. Ora-
       tovskii, Phys. Letters 12, 25 (1964) and
       Yadern. Fizika 1, 479 (1965) [Transl. Sov. J.
       Nucl. Phys. 1, 341 (1965)]; Yu. G. Abov, P.
       A. Krupchitsky, M. I. Bulgakov, O. N. Yerma-
       kov, and I. L. Karpikhin, Phys. Letters 27B,
       16 (1968).
107.   V. M. Lobashov, V. A. Nazarenko, L. F. Saenko,
       L. M. Smotritskii, and G. I. Kharkevich, JETP
       Letters 3, 47, 173 (1966); 5, 59 (1967);
       Phys. Letters 25B, 104 (1967).

every second, causing an alternating component with
a frequency of 1/2 Hz to appear in the detector
current. To detect the a.c. component, an astro-
nomical pendulum was used, which had a proper
frequency of 1/2 Hz, a Q value of 10^5, and a period
constant within 10^{-7}. The signal from the detector
was amplified and transformed with an electromagnet
into mechanical forces which caused the pendulum
to oscillate. The pendulum was tuned to the magnetic field switching frequency with an accuracy
of 10^{-6}. The measured photon helicity,
$P_\gamma = -(6 \pm 1) \times 10^{-6}$, clearly demonstrates the parity-
violating component in the nuclear force. Since
this pioneering effort, many more decays have been
studied, not only by the "pendulum technique," but
also with sophisticated electronic arrangements,[108]
and the existence of the parity-violating component
is well established.

To understand the experiments just discussed,
we consider a hypothetical gamma decay, $2^- \rightarrow 0^+$
between two bound states, as shown in Fig. 4.25.
If parity is not conserved, then the excited 2^-
state can have a small 2^+ component, primarily due
to admixtures of nearby 2^+ states, so that it is
described by

$$|m\rangle = |2^-\rangle + \mathcal{F}|2^+\rangle \quad .$$

---

108. F. Boehm, in Angular Correlations in Nuclear
      Disintegration, H. van Krugten and B. van
      Noojen, eds. (Rotterdam University Press,
      Gröningen, 1971).

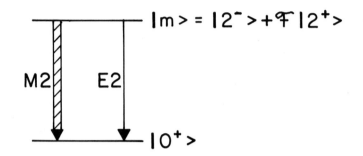

Fig. 4.25. Hypothetical example of a gamma decay
            giving rise to circular polarization
            $(|\mathcal{F}|<<1)$.

Similarly the lower state will generally have a
small 0^- component admixed to the dominant 0^+ part.
For simplicity we assume the lower state to be
pure 0^+ here. With $\mathcal{F} = 0$, only magnetic quadrupole
radiation is emitted; $\mathcal{F} \neq 0$ permits a small admixture of electric quadrupole radiation. The
circular polarization is given by

$$P_\gamma = \frac{2<0|M2|m><0|E2|m>}{<0|M2|m>^2+<0|E2|m>^2} \quad , \qquad (4.172)$$

where all matrix elements have been chosen to be
real. (The justification for this choice will be
given in Section 4.6).

        Since $\mathcal{F}$ is very small, P_γ becomes in a good approximation

$$P_\gamma = 2\mathcal{F}\frac{<0|M2|2^-><0|E2|2^+>}{<0|M2|2^->^2} = 2\mathcal{F}\frac{<0|E2|2^+>}{<0|M2|2^->} \quad ,$$

or

$$P_\gamma = 2\mathcal{F}R \quad , \qquad\qquad\qquad (4.173)$$

where R is the ratio of matrix elements of multipole operators of opposite parity.

Equation (4.173) shows how transitions can be chosen that make the experimentally observed circular polarization reasonably large: If nuclear structure effects hinder the regular transition (M2 in Fig. 4.25) much more than the irregular one (E2 in Fig. 4.25), then R can become rather large and the search for P_γ becomes easier. Experiments thus are done primarily on nuclides where theory predicts a large value for R.

While the circular polarization experiments indicate unambiguously that a parity-violating component exists, detailed comparison between theory and experiment is far from satisfactory. The experimentally determined values are considerably larger than theory can explain.[109,110] It is

---

109. F. C. Michel, Phys. Rev. 133B, 329 (1964);
     S. Wahlborn, Phys. Rev. 138B, 530 (1965).
     B. H. McKellar, Phys. Rev. Letters 20, 1542
     (1968).
110. E. Fischbach and D. Tadic, Physics Reports
     6C, 123 (1973). M. Gari, Physics Reports
     6C, 317 (1973).

possible that part of the problem lies in the calculation of R; nearly all experiments have been performed in heavy nuclei where it is difficult to calculate R accurately. We therefore describe an experiment performed on a light nucleus where theory is more reliable.

The photon helicity measurements discussed above are analogous to the electron helicity experiment described in Subsection F. In that subsection, we have shown that the observation of an asymmetry of electron emission with respect to the direction of a nuclear spin proves parity non-conservation in the weak interaction. Similarly, the observation of the asymmetry of photon emission with respect to an oriented nucleus demonstrates violation of mirror invariance in the nuclear interaction.

Such measurements have been performed with nuclei polarized through the capture of polarized neutrons[106], through reactions induced with polarized protons[111], or by using a dilution refrigerator.[112] Comparison with theory is easiest in ^{19}F, where the parity-violating admixture occurs primarily between the ground and the first excited states.[111] The relevant states are shown

111. E. G. Adelberger, H. E. Swanson, M. D. Cooper, J. W. Tape, and T. A. Trainor, Phys. Rev. Letters 34, 402 (1975).
112. K. S. Krane, C. E. Olsen, J. R. Sites, and W. A. Steyert, Phys. Rev. C 4, 1906, 1942 (1971).

in Fig. 4.26. The first excited state of ^{19}F is produced through the reaction ^{22}Ne$(\vec{p},\alpha)^{19}$F*, using

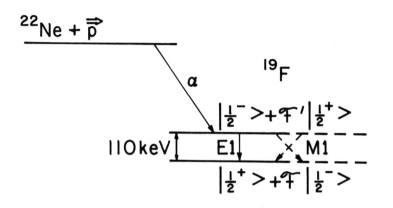

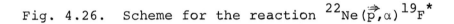

Fig. 4.26. Scheme for the reaction ^{22}Ne$(\vec{p},\alpha)^{19}$F*

polarized protons from the University of Washington Van de Graaff. The arrow $(\vec{p})$ indicates that the incident protons are polarized. A large fraction (~75%) of the polarization is transferred from the incident beam to the excited ^{19}F. [The degree of polarization transfer was measured through the spin correlations of the 110 keV photon and the $1/2^-$ state.] The parity-violating signal is a directional correlation between the momentum of the 110 keV gamma-ray and the polarization vector. If the rate of transition from the unpolarized excited state is w_o, then for a polarization $\vec{P}$, the rate can be written as

$$w = w_o(1 + \delta\langle\vec{P}\cdot\hat{k}\rangle) \quad , \tag{4.174}$$

where $\hat{k}$ is a unit vector in the direction of the photon and δ is a measure of the parity-violation. This coefficient was measured by reversing the direction of polarization at a rate of ~3Hz: it was found to be

$$\delta = -(1.8 \pm 0.9) \times 10^{-4} \quad .$$

Numerous checks were made to ascertain that no instrumental asymmetries were introduced.

How does the parity-violation come about? As we saw, in the presence of a parity-violating interaction, a state function no longer has a definite parity. Since the parity-violating interaction is a scalar under rotations, the $1/2^+$ ground state becomes

$$|1/2^+> + \mathcal{F}|1/2^->$$

and the excited state becomes

$$|1/2^-> + \mathcal{F}'|1/2^+> \quad .$$

For a weak interaction, the coefficients $\mathcal{F}$ and $\mathcal{F}'$ can be computed in perturbation theory. If H_N is the nuclear Hamiltonian with a complete set of eigenstates $|n>$,

$$H_N|n> = E_n |n> \quad ,$$

then in the presence of an additional parity-violating force, $H = H_N + H_{PV}$, the mixing into the ground $(1/2^+)$ state of energy E_o becomes

$$\mathcal{F}\,|1/2^-> \;=\; \Sigma_n |n,1/2^-> \;\frac{<n,1/2^-\,|H_{PV}|1/2^+>}{E_o - E_n} \quad .$$

Because there are no states of angular momentum 1/2 other than the 110 keV state until an excitation of about 5.3 MeV is reached, the sum over n can be approximated by a single term, namely that due to the first excited state n = 1 or $|1,1/2^->$ at 110 keV which we abbreviate as $|1/2^->$,

$$\mathcal{F}\,|1/2^-> \;=\; |1/2^-> \;\frac{<1/2^-\,|H_{PV}|1/2^+>}{-110\ keV}$$

or

$$\mathcal{F} = -\;\frac{<1/2^-\,|H_{PV}|1/2^+>}{110\ keV} \quad .$$

Similarly, for the first excited state, we readily find $\mathcal{F}' = -\mathcal{F}$.

   The normal decay of the 110 eV state occurs through an electric dipole transition. Owing to the admixed states of opposite parity, there occur additional contributions from magnetic dipole transitions (see Fig. 4.26). Since we can neglect the square of $\mathcal{F}$, we find for the rate

$$w = |<1/2^+|E1|1/2^->|^2\{1$$

$$+ 2\;\mathcal{F}\left[\frac{<1/2^-|M1|1/2^-> - <1/2^+|M1|1/2^+>}{<1/2^+|E1|1/2^->}\right]\}$$

(4.176)

The coefficient δ is thus proportional to the ratio of the difference of the two magnetic dipole transition amplitudes to the electric dipole one, multiplied by the coefficient $\mathcal{F}$. It is gratifying that the magnetic dipole matrix elements are given by the static dipole moments of the appropriate states; these have the opposite sign so that the two matrices in the square bracket of Eq. (4.176) add. Since the electric dipole transition rate is measured one obtains $|\mathcal{F}|$ from the measured asymmetry. It is

$$|\mathcal{F}| = (8.2 \pm 4.1) \times 10^{-6}\quad.$$

This measured coefficient is in reasonable agreement with theory[113], which gives

$$|\mathcal{F}| \approx (2\text{-}4) \times 10^{-6}\quad,$$

or

$$|\delta| \approx (5\text{-}8) \times 10^{-5}\quad.$$

---

113.   B. H. J. McKellar and M. Box (to be published); M. Gari, A. Huffman, J. B. McGrory and R. Offerman (to be published).

All experiments discussed so far suffer from two shortcomings: they are performed on complex systems and are limited in the energy range over which studies can be made. The first shortcoming has been avoided by Lobashov and his group who measured the circular polarization of the gamma ray emitted in thermal np capture and found[114]

$$n \; p \rightarrow d \; \gamma \quad , \; P_\gamma = (1.3 \pm 0.45) \times 10^{-6} \quad .$$

This value is roughly two orders of magnitude larger than predicted by theory[115] and thus invites further experiments with the same system. A second approach that eliminates both shortcomings involves the determination of $A = (\sigma_R - \sigma_L)/(\sigma_R + \sigma_L)$ where $\sigma_R (\sigma_L)$ is the total cross section for scattering of nucleons with positive (negative) helicity by unpolarized nucleons. Theory[116] predicts values of A of roughly a few parts in 10^7. Experiment[117]

---

114. V. M. Lobashov, D. M. Kaminker, G. I. Kharkevich, V. A. Kniazkov, N. A. Lozovoy, V. A. Nazarenko, L. F. Sayenko, L. M. Smotritsky, and A. I. Yegorov, Nucl. Phys. A197, 241 (1972).

115. E. Hadjimichael and E. Fischbach, Phys. Rev. D3, 755 (1971); G. S. Danilov, Yad. Fiz. 14, 788 (1971) [Sov. J. Nucl. Phys. 14, 443 (1972)].

116. M. Simonius, Phys. Letters 41B, 415 (1972). V. R. Brown, E. M. Henley, and F. R. Krejs, Phys. Rev. Letters 30, 770 (1973) and Phys. Rev. C9, 935 (1974). E. M. Henley and F. R. Krejs (to be published).

117. J. M. Potter, J. D. Bowman, C. F. Hwang, J. L. McKibben, R. E. Mischke, D. E. Nagle, P. G. Debrunner, H. Frauenfelder, and L. B. Sorensen, Phys. Rev. Letters 33, 1307 (1974).

presently gives a limit of less than 10^{-6} at 15 MeV; improvements will hopefully lead to un- ambiguous comparison between theory and experiment at this and other energies.

4.6   TIME REVERSAL

CLEMENTI

Some musical works, notably by Bach, Beethoven, and Schönberg, contain passages that appear first in normal and later in time-reversed order (crab- wise movements). The entire work, however, can usually not be played in time-reversed order with- out doing violence to the composer and the lis- tener.* Similarly, very few aspects of life are completely reversible, but the idea of a reversal of time or a return in time has always fascinated people.[118]

---

118.   M. Twain, A Connecticut Yankee in King Arthur's Court, 1889.
        *Schönberg, however, has written a time-rever- sal invariant opera, called "Hin und Zurück."

In physics, the issue concerns the fundamental equations: Are these laws invariant under
a reversal of t into -t? Could a system, starting
from a given point in time, develop equally in
both time directions? Newton's and Maxwell's
equations are unchanged under time reversal, and
Boltzmann used time-reversal invariance, or microscopic reversibility, in his derivation of the H
theorem. Wigner introduced the time-reversal
operation into quantum mechanics.[119] Kramers
had earlier proved that the states of an atomic
system with an odd number of electrons must be
doubly degenerate and that the degeneracy cannot
be removed by an external electric field.[120]
Wigner then showed that Kramer's theorem follows
from invariance under time reversal.

In classical physics, Onsager investigated
the question as to when the time could indeed be
"reversed."[121] However, few doubts concerning
the validity of time-reversal invariance in microscopic properties were expressed before 1956.
When parity was found to be violated, time reversal invariance was also examined, but it emerged
unscathed. In 1964, however, a rare decay mode
of neutral kaons was discovered which cast grave

119. E. Wigner, Nachr. Akad. Wiss. Göttingen,
     Math. Phys. Kl. II A, 31, 546 (1932).
120. H. A. Kramers, Proc. Acad. Amsterdam 33,
     959 (1930).
121. L. Onsager, Phys. Rev. 37, 405 (1931).

(but indirect) doubts on the global validity of
time-reversal invariance. We will discuss these
experiments in Chapter 5.

A.   Time Reversal for Spinless Particles

Classically, we define time-reversal or time-reflection by

$$t \rightarrow -t \quad , \quad \vec{x} \rightarrow +\vec{x} \quad .$$

Newton's laws of motion and Maxwell's equations
are second order differential equations in t and
are therefore invariant under time reflection.
The classical momentum $\vec{p} = m \, d\vec{x}/dt$ and angular
momentum change sign under the time-reversal
operation

$$\vec{p} \longrightarrow -\vec{p} \qquad\qquad \vec{J} \longrightarrow -\vec{J} \quad .$$

In quantum mechanics, we postulate the
operator relations

$$T \; \vec{x}_{op} \; T^{-1} = \vec{x}_{op}$$

$$T \; \vec{p}_{op} \; T^{-1} = -\vec{p}_{op} \tag{4.177}$$

$$T \; \vec{J}_{op} \; T^{-1} = -\vec{J}_{op} \quad .$$

Furthermore, T transforms a wave function $\psi(t)$
into the time-reversed wave function,

$$T \, \psi(t) = \psi_T(-t) \quad . \tag{4.178}$$

Note that the argument t changes sign and that the form of ψ can also change; the change in form is represented by the subscript T.

To find the explicit form for the operation of T , we use the Schrödinger equation,

$$i \, \frac{\partial}{\partial t} \psi(t) = H \, \psi(t) \quad .$$

Time reversal invariance holds if T and H commute:

$$[T \,, H] = 0 \quad . \tag{4.179}$$

$T\psi(t)$ must formally satisfy the same Schrödinger equation as $\psi(t)$, namely,

$$i \, \frac{\partial T \, \psi(t)}{\partial t} = H \, T\psi(t) \quad . \tag{4.180}$$

The simplest ansatz

$$T \, \psi(t) = \psi_T(-t) = \psi(-t)$$

introduced into Eq. (4.180), gives

$$i \, \frac{\partial}{\partial t} \psi(-t) = -i \, \frac{\partial \, \psi(-t)}{\partial (-t)} = H \, \psi(-t) \quad .$$

Since t is just a parameter, we can set $t' = -t$ and compare the equation for t' with the original

Schrödinger equation. It is evident that
$\psi(t') = \psi(-t)$ does <u>not</u> satisfy the Schrödinger
equation; in fact, it leads to a negative energy.
The difficulty was overcome by Wigner[119]: Consider the complex conjugate of the Schrödinger
equation,

$$-i\,\frac{\partial\psi^*(t)}{\partial t} = H^*\,\psi^*(t) \quad \text{or} \quad i\,\frac{\partial\psi^*(-t)}{\partial t} = H^*\,\psi^*(-t).$$

We notice that $\psi^*(-t)$ satisfies the same Schrödinger equation as $\psi(t)$ if H is real. We can
therefore set

$$T\,\psi(\vec{x},t) = K\,\psi(\vec{x},-t) = \psi^*(\vec{x},-t) \qquad , \qquad (4.181)$$

where K is the operator of complex conjugation
satisfying

$$K^2 = 1, \qquad K^{-1} = K \quad . \tag{4.182}$$

The complex conjugate of an operator $\mathcal{O}$ is defined
as $\mathcal{O}^* \equiv K\,\mathcal{O}\,K^{-1} = K\,\mathcal{O}\,K$. The operator T is
<u>antilinear</u>,

$$T\{a|\Phi> + b|\psi>\} = a^*\,T|\Phi> + b^*\,T|\psi> \quad ,(4.183)$$

and it preserves the norm.

As an example, consider a free particle with
momentum $\vec{p}$, given by a wave function
$\psi = \exp[i(\vec{p}\cdot\vec{x} - Et)]$. The time reversed wave
function is

$$T\psi(\vec{x},t) = \psi^*(\vec{x},-t) = e^{-i(\vec{p}\cdot\vec{x}+Et)} = e^{i(-\vec{p}\cdot\vec{x}-Et)},$$

and it describes a particle with momentum $-\vec{p}$. For a spinless free particle of momentum $\vec{p}$, we therefore write

$$T|\vec{p}\rangle = |-\vec{p}\rangle \ . \tag{4.184}$$

As a second example, we treat a spinless particle bound to a fixed center and possessing an orbital angular momentum ℓ with respect to that center. Its wave function then is given by $Y_\ell^m(\theta,\phi)$. With Eq. (4.181), we get

$$TY_\ell^m(\theta,\phi) = KY_\ell^m(\theta,\phi)$$

$$= Y_\ell^{m*}(\theta,\phi) = (-1)^{-m} Y_\ell^{-m}(\theta,\phi) \ ,$$

where the last step follows from the explicit expression for Y_ℓ^m and can be verified with the examples given in Table 3.2. For many applications, and for the generalization to particles with spin, it is more convenient to introduce spherical harmonics with a different phase,

$$\mathcal{Y}_\ell^m(\theta,\phi) = (i)^\ell Y_\ell^m(\theta,\phi) \ . \tag{4.185}$$

We then find

$$T\mathcal{Y}_\ell^m(\theta,\phi) = (-1)^{\ell-m} \mathcal{Y}_\ell^{-m}(\theta,\phi) \ . \tag{4.186}$$

B.   General Formulation of Time Reversal

The simple form Eq. (4.181) does not always suffice for the description of the time-reversal operation as we will see below in the case of a spin 1/2 particle.   Indeed, even for a spinless particle, the choice $T = K$ is correct only in some representations; a more general time reversal operator is

$$T \psi(t) = \psi_T(-t) = U_T K \psi(-t) = U_T \psi^*(-t) \quad ,$$

$$(4.187)$$

where U_T is a unitary operator.   T is still anti-linear and we get for the scalar product of two states (with $\langle U\psi| = \langle\psi|U^\dagger$)

$$\langle T \psi(t) | T \Phi(t)\rangle = \langle\psi_T(-t)|\Phi_T(-t)\rangle$$

$$= \langle U_T\psi^*(-t)|U_T\Phi^*(-t)\rangle$$

$$= \langle\psi^*(-t)|\Phi^*(-t)\rangle$$

$$= \langle\psi(-t)|\Phi(-t)\rangle^*$$

$$= \langle\Phi(-t)|\psi(-t)\rangle \quad .$$

This relation connects the scalar products of the two time-reversed states to the scalar product of the two original states:

$$\langle \psi_T | \Phi_T \rangle = \langle \psi | \Phi \rangle^* = \langle \Phi | \psi \rangle \quad . \tag{4.188}$$

T preserves the norm and leaves scalar products invariant to within a phase. Operators transform under T as

$$\sigma_T = T \ \sigma \ T^{-1} = U_T \ \sigma^* \ U_T^\dagger \quad . \tag{4.189}$$

The transformation properties of <u>matrix elements</u> under time reversal follow with Eq. (4.187) as

$$\langle T \ \psi(t) | T \ \sigma \ T^{-1} | T \ \Phi(t) \rangle$$

$$= \langle U_T \psi^*(-t) | U_T \ \sigma^* \ U_T^\dagger | U_T^* \ -t) $$

$$= \langle \psi(-t) | \sigma | \Phi(-t) \rangle^*$$

or

$$\langle \psi_T | \sigma_T | \Phi_T \rangle = \langle \psi | \sigma | \Phi \rangle^* = \langle \Phi | \sigma^\dagger | \psi \rangle \quad . \tag{4.190}$$

If σ is Hermitian, as for any observable, Eq. (4.190) can be written as

$$\langle \psi_T | \sigma_T | \Phi_T \rangle = \langle \Phi | \sigma | \psi \rangle \quad . \tag{4.191}$$

These relations connect the matrix elements of σ between time-reversed states to the matrix elements of σ between the original states; they differ from the corresponding relations for

parity-reversed states by the complex conjugation
on the right-hand side. We emphasize that <u>no</u>
invariance arguments are involved in the derivation of Eqs. (4.190) and (4.191); such arguments
enter only through a relation between σ_T and σ.
If σ_T and σ are related, for instance $\sigma_T = \pm \sigma$,
then a phase relation between matrix elements can
be established. If in addition the states $|\Phi\rangle$
and $|\psi\rangle$ are eigenstates of T, for instance
$T|\Phi\rangle = |\Phi\rangle$, $T|\psi\rangle = |\psi\rangle$, then Eq. (4.191) shows
that $\langle\Phi|\sigma|\psi\rangle$ must be real or pure imaginary.

As an example of how U_T is determined, we
consider non-relativistic particles of spin 1/2.
In Eq. (4.177), we have characterized the behavior
of the operators $\vec{x}$, $\vec{p}$, and $\vec{J}$ under T. In configuration space K satisfies the first two relations
(4.177); for instance we have $T \vec{p}_{op} T^{-1} =$
$K \vec{p}_{op} K^{-1} = \vec{p}_{op}^* = -\vec{p}_{op}$. The matrix U_T thus can
be chosen to operate only in spin space; it is
then a 2 × 2 matrix and with Eq. (4.189) the
transformation for the spin operator $\vec{s} = \vec{\sigma}/2$
becomes

$$\vec{s}_T = U_T \vec{s}^* U_T^\dagger = -\vec{s} \quad . \tag{4.192}$$

In the standard representation, Eq. (3.123), the
operators σ_x and σ_z are real, and σ_y is pure
imaginary. Hence, Eq. (4.192) gives

$$U_T \, \sigma_x^* \, U_T^\dagger = U_T \, \sigma_x \, U_T^\dagger = -\sigma_x \ ,$$

$$U_T \, \sigma_z^* \, U_T^\dagger = U_T \, \sigma_z \, U_T^\dagger = -\sigma_z \ ,$$

but

$$U_T \, \sigma_y^* \, U_T^\dagger = -U_T \, \sigma_y \, U_T^\dagger = -\sigma_y \quad .$$

In order to satisfy Eq. (4.192), U must anti-commute with σ_x and σ_z, but commute with σ_y. The only operator that fulfills these conditions is $\varepsilon\sigma_y$, where ε is an arbitrary phase. Without loss of generality we select ε to be $-i$ and write for a two-component spinor

$$U_T = -i\sigma_y = \begin{pmatrix} 0 & -1 \\ 1 & 0 \end{pmatrix} \ ,$$

$$(4.193)$$

$$T \, \psi(\vec{x},t) = -i\sigma_y K \, \psi(\vec{x},-t) \quad .$$

The choice $\varepsilon = -i$ makes U_T real, and U_T and K hence commute,

$$K(-i \, \sigma_y) = (-i \, \sigma_y) K \quad .$$

As an example, consider a free non-relativistic spin 1/2 particle; its wave function is given by

$$\psi(\vec{x},t) = \begin{pmatrix} a \\ b \end{pmatrix} e^{i(\vec{p}\cdot\vec{x} \, - \, Et)} \quad .$$

The time-reversed wave function,

$$
T\,\psi = \begin{pmatrix} 0 & -1 \\ 1 & 0 \end{pmatrix} K \begin{pmatrix} a \\ b \end{pmatrix} e^{i(\vec{p}\cdot\vec{x}\,+\,Et)}
$$

$$
= \begin{pmatrix} -b^* \\ +a^* \end{pmatrix} e^{i(-\vec{p}\cdot\vec{x}\,-\,Et)}
$$

(4.194)

describes a particle with reversed momentum. If we compute the expectation value for the spin vector, $\vec{\sigma}$, for the original and the time-reversed state, we see that the spin has also been reversed. If the particle is in a definite spin magnetic (m) state, then we can summarize the action of the operator T as

$$
T\,|\vec{p},m> \;=\; (-1)^{1/2-m}\;|-\vec{p},-m> \quad .
$$

(4.195)

The state vector on the right-hand side appears without the complex conjugate sign.

Equations (4.193) and (4.195) can be generalized to states of arbitrary angular momentum j. To do so, we first rewrite the operator $-i\,\sigma_y$ as

$$
-i\,\sigma_y = e^{-i\,\pi\,s_y} \quad .
$$

The operator $\exp(-i\,\pi\,s_y)$ induces a rotation by 180° about the y axis and reverses the spin, in agreement with Eq. (4.195). Formally, the relation is established by using the identity

$$e^{ic\ \hat{n}\cdot\vec{\sigma}} = \cos c + i(\hat{n}\cdot\vec{\sigma})\ \sin c\ ,\qquad (4.196)$$

which follows from Eq. (4.8) after evaluation of the sum. In Eq. (4.196), $\hat{n}$ is a unit vector, c a real number, and $\vec{\sigma} = 2\vec{s}$. The expression (4.193) can now be generalized to read

$$T\ \psi(\vec{x},t) = e^{-i\ \pi\ J_y}\ K\ \psi(\vec{x},-t)\ .\qquad (4.197)$$

The state function of a bound state with angular momentum j then can be chosen so that[*]

$$T\ |a,j,m> = (-1)^{j-m}|\ a_T,j,-m>\ .\qquad (4.198)$$

Here a denotes all quantum numbers other than j and a_T denotes the corresponding time-reversed quantum numbers. If $[T,H] = 0$, then $a_T = a$. For a spinless particle, j is equal to the orbital angular momentum and Eq. (4.198) agrees with the earlier expression (4.186). For a free particle at rest, with spin s, the generalization of Eq. (4.195) is

$$T\ |a,s,m> = (-1)^{s-m}|a_T,s,-m>\ .$$

If this particle has total angular momentum j,

---

[*] The phase convention here and in similar equations is not unique. The factor $(-1)^{j-m}$ is sometimes replaced by $(-1)^{2m}$ or $(-1)^{j+m}$.

the state function can be chosen so that Eq.
(4.198) holds. Alternatively, for a particle of
spin s and linear momentum $\vec{p}$, the state function
can be chosen so that

$$T \, |a,\vec{p},s,m> \; = \; (-1)^{s-m} \; |a_T,-\vec{p},s,-m> \quad . \quad (4.199)$$

This relation holds not only for free particles;
the phases of the state functions can be so chosen
that it is also valid for systems. Consider for
instance a nucleus with quantum numbers (j_1,m_1)
and a particle with angular momentum (j_2,m_2) with
respect to the nucleus. Equation (4.199) then
holds, with $j = j_1+j_2$ and $m = m_1+m_2$.

To study time reversal in Dirac theory we
note that under T the charge density $\rho = j^0$ is
unchanged and the current $\vec{j} = \rho\vec{v}$ changes sign:

$$T \, j^0 \, T^{-1} = + \, j^0$$

$$T \, \vec{j} \, T^{-1} = - \, \vec{j} \quad , \quad\quad\quad (4.200)$$

or

$$T \, j^{\mu}(t,\vec{x}) \, T^{-1} = + \, j_{\mu}(-t,\vec{x}) \quad .$$

The Dirac four-current is given by Eq. (3.169) as
$j^{\mu} = \bar{\psi} \, \gamma^{\mu} \, \psi$. With Eqs. (3.157) and (3.160) it is
straightforward to see that the operator

$$T = i\gamma^1 \, \gamma^3 \, K \quad , \quad T^{-1} = iK\gamma^3\gamma^1 \quad\quad (4.201)$$

applied to the current operator j^{μ}, gives the

required transformation properties, Eq. (4.200),
if the standard representation (3.174) is used.
T commutes with the Dirac Hamiltonian, Eq.
(3.159). If $\psi(t)$ is a solution of the Dirac equation (3.159), so is $T \psi(t) = i\gamma^1\gamma^3\psi*(-t)$.

In the presence of electromagnetic fields,
A^μ, the substitution (4.17) changes the Dirac
equation (3.167) to

$$(i\slashed{\partial} - q\slashed{A} - m)\psi = 0 \quad .$$

$T \psi$ is now a solution if the field A^μ transforms
as

$$T A^0 T^{-1} = A^0 \quad , \quad T \vec{A} T^{-1} = -\vec{A} \quad ,$$

(4.202)

$$T A^\mu(t,\vec{x}) T^{-1} = A_\mu(-t,\vec{x}) .$$

With Eqs. (4.200) and (4.202), the electromagnetic
interaction (4.149) is seen to be invariant under
time reversal, $[T,H_\gamma] = 0$.

The operation of complex conjugation, as
used in the Wigner time reversal, Eq. (4.187), is
not the sole way to define a time reversal operator. A procedure that is useful in field theory
has been suggested by Schwinger.[122] In Schwinger's time reversal operation, the complex conjugation is replaced by the interchange of bras
and kets and by taking the transpose of operators.

---

122. J. Schwinger, Phys. Rev. 82, 914 (1951).

For observables, the Schwinger and Wigner time
reversal operations give identical results.

C.   Some Consequences of Time Reversal Invariance

We first show that double time-reversal invariance induces a super-selection rule. The
physical properties of any system must remain
unchanged under a double time reversal operation;
hence we must have

$$T^2 |\psi> = c|\psi> \quad .$$

(4.203)

Any state is thus an eigenstate of T^2. T^2, in
contrast to T, is Hermitian. To find the eigenvalues c of T^2, we use Eq. (4.199) and get

$$c = (-1)^{2s} \quad .$$

(4.204)

This result does not depend on the choice of representation and it shows that the eigenvalue of T^2
is $+1$ for a system with integer spin and -1 for one
with half-integer spin. All possible physical
states thus fall into two classes, which we denote
with $|+>$ and $|->$. If σ is an arbitrary observable,
we have

$$T^2 \sigma T^{-2} \equiv \sigma_{TT} = \sigma \quad ,$$

and hence

$$<+|\sigma|-> \ = \ <+|T^{-2} \ \sigma_{TT} \ T^2|-> \ = \ - \ <+|\sigma|-> \qquad ,$$

or

$$<+|\sigma|-> \ = \ 0 \qquad . \qquad\qquad (4.205)$$

No matrix elements connect states with integer and half-integer spins and we have thus justified the superselection rule stated in Subsection 4.5.B.

Next we consider a state Φ containing an odd number of fermions. Equation (4.204) states that the eigenvalue of T^2 for such a state is -1. Introducing $\psi = T \ \Phi$ into Eq. (4.188) then yields

$$<\Phi|\psi> \ = \ <\Phi| \ T \ \Phi> \ = \ <T^2 \ \Phi| \ T \ \Phi> \ = \ -<\Phi| \ T \ \Phi>$$

or

$$<\Phi|\psi> \ = \ <\Phi| \ T \ \Phi> \ = \ 0 \qquad . \qquad\qquad (4.206)$$

The states Φ and $\psi = T \ \Phi$ are orthogonal; if the Hamiltonian of the system is time-reversal invariant, $[T,H] = 0$, then Φ and $T \ \Phi$ belong to the same energy and the state is doubly degenerate. The degeneracy cannot be lifted by a field, such such as the electric one, that commutes with T. We have thus derived Kramers' theorem[120] which finds primary application in solid state physics. However, it also restricts nuclear wave functions. For an odd-A nucleus, it predicts that any eigenstate is at least doubly degenerate. The symmetry

between K and -K shown in Eq. (4.77) is an example
of this degeneracy.

The application of time reversal invariance
to transitions between two states $|\alpha\rangle$ and $|\beta\rangle$
yields some interesting results. We assume that
the interaction responsible for the transition
is weak enough so that pertubation theory is
applicable. This assumption is valid for electromagnetic and weak decays, (Fig. 4.27). The
transition rate $w_{\beta\alpha}$ is then given by the <u>golden</u>

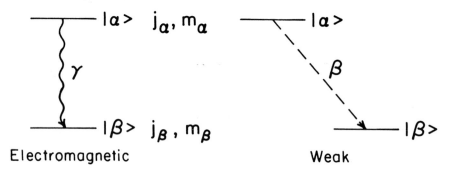

Electromagnetic   Weak

Fig. 4.27.   Decays that can be described by
             perturbation theory.

<u>rule</u>,

$$w_{\beta\alpha} = 2\pi |\langle \beta, \vec{k} | H | \alpha \rangle|^2 \rho_\beta \quad .\tag{4.207}$$

Here, $\vec{k}$ is the momentum of the particle in the

final state, H the Hermitian interaction Hamiltonian, and ρ_β is the density-of-states factor.
For a Hermitian Hamiltonian we get from Eq. (3.31)

$$<\beta,\vec{k}|H|\alpha> = <\alpha|H|\beta,\vec{k}>* \quad ,$$

and with Eq. (4.207), we find the <u>principle of</u>
<u>detailed balance</u>,

$$w_{\beta\alpha} \, \rho_\alpha = w_{\alpha\beta} \, \rho_\beta \quad . \tag{4.208}$$

Hermiticity alone suffices to relate the rates of
a transition $|\alpha> \to |\beta>$ and its inverse $|\beta> \to |\alpha>$.

 <u>Time-reversal invariance</u> provides additional
information about the matrix element $<\beta|H|\alpha>$;
with $[H,T] = 0$ or

$$H_T \equiv T \, H \, T^{-1} = H,$$

Eq. (4.190) gives

$$<\beta_T|H|\alpha_T> = <\beta|H|\alpha>* \tag{4.209}$$

for arbitrary states $|\beta>$ and $|\alpha>$. If these states
are eigenstates of the total angular momentum
operators J^2 and J_z, we can write them as
$|\alpha> = |a,j_a,m_a>$ and $|\beta> = |b,j_b,m_b>$, where a and
b denote all quantum numbers not connected with
the rotational properties of the states. If,
furthermore, these states are eigenstates of a
Hamiltonian that commutes with T, then $a_T = a$,

$b_T = b$, and we obtain with Eq. (4.198)

$$\langle \beta_T | H | \alpha_T \rangle = (-1)^{j_a + j_b - m_a - m_b}$$

$$\times \langle b, j_b, -m_b | H | a, j_a, -m_a \rangle \quad .$$

If in addition H is a <u>scalar operator</u>, Eq. (4.79) shows that it connects only states of the same j and m and is independent of m so that the matrix element becomes

$$\langle \beta_T | H | \alpha_T \rangle = (-1)^{2(j_a - m_a)} \tag{4.210}$$

$$\times \langle b, j_b | H | a, j_a \rangle \, \delta_{j_b j_a} \, \delta_{m_b m_a} \quad .$$

Since $2(j-m)$ is even, we finally get from Eqs. (4.209) and (4.210)

$$\langle \beta | H | \alpha \rangle = \langle \beta | H | \alpha \rangle^* \quad . \tag{4.211}$$

Thus, the matrix elements of a time-reversal even and rotationally invariant Hermitian operator between the states α and β are real.[123]

The above argument can be generalized for arbitrary operators; it is then useful for electromagnetic transitions of arbitrary multipolarity,

123. F. Coester, Phys. Rev. __89__, 619 (1953); L. C. Biedenharn and M. E. Rose, Rev. Mod. Phys. 25, 736 (1953). S. P. Lloyd, Phys. Rev. __81__, 161 (1951).

and we shall need it in Subsection D. Thus, consider an arbitrary operator, which we assume to be a spherical tensor $T_q^{(k)}$, where k denotes the tensor order and q the spherical component (Subsection 4.3 E). Under the time reversal transformation, we assume that $T_{q\ T}^{(k)}$ is related to $T_q^{(k)}$ by $T_{q\ T}^{(k)} = \pm T_{-q}^{(k)}$. In particular, we take

$$T\ T_q^{(k)}\ T^{-1} = \eta_k (-1)^{k-q}\ T_{-q}^{(k)} \ , \tag{4.212}$$

with $\eta_k = \pm 1$, depending on the operator $T^{(k)}$. (The phase choice here is similar to the one given in Eq. (4.198) for a state.)

Repeating the steps from Eq. (4.209) to Eq. (4.210), and using the Wigner-Eckart theorem, Eq. (4.84), we obtain

$$\langle \beta_T |\, T\ T_q^{(k)}\ T^{-1} |\, \alpha_T \rangle$$

$$= \eta_k (-1)^{j_a+j_b+k-m_a-m_b-q} \langle b, j_b, -m_b | T_{-q}^{(k)} | a, j_a, -m_a \rangle$$

$$= \eta_k (-1)^{j_a+j_b+k-m_a-m_b-q} \frac{\langle j_a -m_a\ k-q | j_b -m_b \rangle}{\sqrt{2j_b + 1}}$$

$$\times \langle b, j_b || \ T^{(k)}\ || a, j_a \rangle \ . \tag{4.213}$$

The Wigner-Eckart theorem can also be applied to $\langle \beta | T_q^{(k)} | \alpha \rangle$; the resulting expression can be compared to Eq. (4.213). With the symmetry relation,

$$\langle j_a m_a kq | j_b m_b \rangle = (-1)^{j_a+k-j_b} \langle j_a -m_a \, k-q | j_b -m_b \rangle \, ,$$

$$(4.214)$$

and using $m_a + q = m_b$, $2(j_b - m_b)$ even, we obtain with Eq. (4.190)

$$\langle \beta_T | T \, T_q^{(k)} \, T^{-1} | \alpha_T \rangle$$

$$= \eta_k \langle \beta | T_q^{(k)} | \alpha \rangle = \langle \beta | T_q^{(k)} | \alpha \rangle^* \quad . \qquad (4.215)$$

Thus, depending on the phase $\eta_k = \pm 1$, the matrix elements of the tensor operators are either purely real or purely imaginary. Indeed, it is always possible to select the phases so that all matrix elements are real. (See subsection D).

If perturbation theory is no longer applicable, the approach must be generalized; time reversal invariance then leads to the principle of reciprocity.[124] To discuss the principle, we introduce the S matrix here in a superficial way. We define S as the operator that transforms the initial state into the final state. Initial and final states can be the results of a decay as in the previous paragraph, or they can be the state of the incoming and the outgoing particles in a

124.   F. Coester, Phys. Rev. 84, 1259 (1951).

scattering experiment (Fig. 4.28).   In either case, we write

$$|out> = S|in> \quad . \tag{4.216}$$

S is a linear and rotationally invariant operator which preserves the normalization; it is hence unitary,

$$S^\dagger S = S S^\dagger = 1 \quad . \tag{4.217}$$

In order to apply the S operator to a scattering problem we note that the incident and the outgoing particles far away from the target do not interact and thus can be described by using free-particle states.   Scattering then leads from a given initial state $|\alpha>$ to <u>all</u> possible final states $|out>$:

$$|out> = S|\alpha> \quad .$$

The probability of finding the particles in a <u>particular</u> final state $|\beta>$ is proportional to $|<\beta|out>|^2$; thus

$$w_{\beta\alpha} = const|<\beta|S|\alpha>|^2 \rho_\beta \quad . \tag{4.218}$$

The element $S_{\beta\alpha}$ of the <u>S matrix</u> gives the transition rate from a given initial state $|\alpha>$ to a particular final state $|\beta>$.

We now turn to <u>time-reversal invariance</u>. Without proof we state that S satisfies the oper-

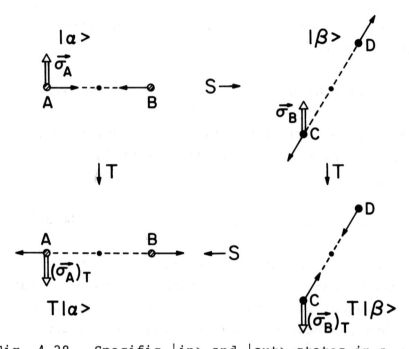

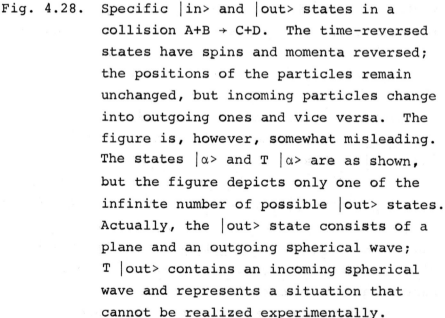

Fig. 4.28.   Specific |in> and |out> states in a
            collision A+B → C+D. The time-reversed
            states have spins and momenta reversed;
            the positions of the particles remain
            unchanged, but incoming particles change
            into outgoing ones and vice versa. The
            figure is, however, somewhat misleading.
            The states |α> and T |α> are as shown,
            but the figure depicts only one of the
            infinite number of possible |out> states.
            Actually, the |out> state consists of a
            plane and an outgoing spherical wave;
            T |out> contains an incoming spherical
            wave and represents a situation that
            cannot be realized experimentally.

ator relation

$$S_T \equiv T S T^{-1} = S^{\dagger} \tag{4.219}$$

if T and H commute, i.e., if time-reversal invariance holds.[124] Equations (4.219) and (4.190) together show that

$$<\beta|S|\alpha> = <\alpha_T|S|\beta_T> \; ; \tag{4.220}$$

the matrix element for the transition $|\alpha> \rightarrow |\beta>$ is equal to that for the inverse transition <u>between time-reversed states</u>, $|\beta_T> \rightarrow |\alpha_T>$. The rates for the reactions $|\alpha> \rightarrow |\beta>$ and $|\beta_T> \rightarrow |\alpha_T>$ are therefore connected by the <u>reciprocity</u> relation

$$\frac{w_{\beta\alpha}}{\rho_\beta} = \frac{w_{\alpha_T \beta_T}}{\rho_{\alpha_T}} \tag{4.221}$$

if time-reversal invariance holds. At first sight it could be expected that time-reversal invariance demands equality of rates between transitions $|\alpha> \rightarrow |\beta>$ and $|\beta> \rightarrow |\alpha>$. However, time-reversal invariance yields no a-priori connection between a transition and its <u>inverse</u>.

A question is raised by the results in Eqs. (4.220) and (4.221): Are the measurable consequences of time-reversal invariance and of reciprocity identical? Time-reversal invariance demands equality of the relevant matrix elements whereas reciprocity only implies equality of the absolute

values of these quantities. The problem has been
studied by T. D. Lee and applied to the lambda
decay, $\Lambda^0 \rightarrow N\pi$, and its reversed process $\pi N \rightarrow \Lambda^0$.[125)]
It turns out that the measurable consequences of the
two principles are indeed the same.

We apply Eq. (4.220) first to <u>decay processes</u>.
Since S is a scalar under rotations, we can take
over Eq. (4.210) if the states $|\alpha\rangle$ and $|\beta\rangle$ are
eigenstates of angular momentum, and we get
$\langle\alpha_T|S|\beta_T\rangle = \langle\alpha|S|\beta\rangle$. From Eq. (4.220) it then
follows that S matrix elements between states
$|\alpha\rangle \equiv |a,j,m\rangle$ and $|\beta\rangle \equiv |b,j,m\rangle$ are symmetric[123)]

$$\langle b,j,m|S|a,j,m\rangle = \langle a,j,m|S|b,j,m\rangle \quad . \qquad (4.222)$$

For the discussion of <u>scattering processes</u>, we
use plane wave states, e.g. $|\alpha\rangle = |a,\vec{p}_a,s_a,m_a\rangle$,
where m_a is the z component of the spin s_a in the
state $|\alpha\rangle$. With Eq. (4.199) and with $a_T = a$,
$b_T = b$, the matrix element becomes

$$\langle\alpha_T|S|\beta_T\rangle$$
$$= (-1)^{s_a+s_b-m_a-m_b}\langle a,-\vec{p}_a,s_a,-m_a|S|b,-\vec{p}_b,s_b,-m_b\rangle .$$

The factor before the matrix element can only be

125. T. D. Lee, in <u>Preludes in Theoretical Physics</u>,
      A. DeShalit, H. Feshbach, and L. Van Hove,
      eds. (North-Holland Publishing Co., Amsterdam,
      1966).

$\pm$ 1, and Eq. (4.220) leads to

$$<b,\vec{p}_b,s_b,m_b|S|a,\vec{p}_a,s_a,m_a>$$

$$= \pm <a,-\vec{p}_a,s_a,m_a|S|b,-\vec{p}_b,s_b,-m_b> \quad . \tag{4.223}$$

We have shown earlier (Eq. (4.208)) that the principle of detailed balance, which connects the transitions $|\alpha> \rightarrow |\beta>$ and $|\beta> \rightarrow |\alpha>$, holds if perturbation theory is valid and if the transition operator is Hermitian. We have also noted that time-reversal invariance leads only to reciprocity, Eq. (4.221), but not to detailed balance. Equation (4.223) now provides a tool to bridge the gap partially by establishing a relation that is often called principle of <u>semi-detailed balance</u>. To derive this principle, the right-hand side of Eq. (4.223) is changed by noting that the relation

$$<a,-\vec{p}_a,s_a,-m_a|S|b,-\vec{p}_b,s_b,-m_b>$$

$$= <a,\vec{p}_a,s_a,-m_a|S|b,\vec{p}_b,s_b-m_b>$$

holds if parity is conserved and a and b are invariant under the parity transformation. The transition rate between states m_b and m_a is proportional to the square of the relevant matrix element. The various transitions $m_b \rightarrow m_a$ do not interfere and the total rate $|\beta> \rightarrow |\alpha>$ is proportional to the sum over all component transitions. If all m states are equally populated, i.e., if the incident

particles and the target are not polarized and if
the polarization in the final state is not measured,
then it does not matter whether the sum is performed
over m or -m so that

$$\sum_{m_a m_b} |<a,\vec{P}_a,s_a,-m_a|S|b,\vec{P}_b,s_b,-m_b>|^2$$

$$=\sum_{m_a m_b} |<a,\vec{P}_a,s_a,m_a|S|b,\vec{P}_b,s_b,m_b>|^2 \quad.$$

With Eq. (4.223) and the last two relations, the
principle of semi-detailed balance is obtained:

$$\sum_{m_a m_b} |<b,\vec{P}_b,s_b,m_b|S|a,\vec{P}_a,s_a,m_a>|^2$$

$$\tag{4.224}$$

$$=\sum_{m_a m_b} |<a,\vec{P}_a,s_a,m_a|S|b,\vec{P}_b,s_b,m_b>|^2 \quad.$$

In contrast to Eq. (4.208), this principle holds
even if perturbation theory is not applicable and
S is not Hermitian.

Time reversal invariance also leads to conditions on <u>moments and form factors</u> of particles and
nuclei. As a simple example, we consider here the
dipole moments of particles. The relevant part
of the electromagnetic interaction Hamiltonian
is given by

$$H'_\gamma = -\vec{\mu}_m \cdot \vec{B} - \vec{\mu}_e \cdot \vec{E} \quad, \tag{4.225}$$

where $\vec{B}$ and $\vec{E}$ are magnetic and electric field
strengths and $\vec{\mu}_m$ and $\vec{\mu}_e$ are the magnetic and
electric dipole moment operators. The Hermicity
of H'_γ requires $\vec{\mu}_m$ and $\vec{\mu}_e$ to be Hermitian. To find
their transformation properties under P and T,
we first get the transformation properties of $\vec{B}$
and $\vec{E}$ from Eqs. (3.6), (4.148), and (4.202):

$$P \vec{B} P^{-1} = + \vec{B} \quad , \quad T \vec{B} T^{-1} = -\vec{B}$$

$$\quad (4.226)$$

$$P \vec{E} P^{-1} = -\vec{E} \quad , \quad T \vec{E} T^{-1} = + \vec{E} \quad .$$

Assuming that H_γ commutes with P and T, we get
from Eq. (4.225)

$$P\vec{\mu}_m P^{-1} = \vec{\mu}_m \quad , \quad T \vec{\mu}_m T^{-1} = -\vec{\mu}_m \quad \quad (4.227)$$

$$P\vec{\mu}_e P^{-1} = -\vec{\mu}_e \quad , \quad T \vec{\mu}_e T^{-1} = \vec{\mu}_e \quad . \quad \quad (4.228)$$

Equations (4.132) and (4.133) show that for a
particle described by a state $|\alpha\rangle$ of <u>well-defined
parity</u>, no condition on $\langle\alpha|\vec{\mu}_m|\alpha\rangle$ results, but that
the expectation value of $\vec{\mu}_e$ vanishes,

$$\langle\alpha|\vec{\mu}_e|\alpha\rangle = 0 \quad . \quad \quad (4.229)$$

To show that time-reversal invariance also leads
to a vanishing electric dipole moment, we consider
the particle of spin s (s $\geq$ 1/2) to be at rest
and in a state $|\alpha\rangle = |a,s,m\rangle$. The spherical

components of the Hermitian vector operator $\vec{\mu}_e$
follow from Eq. (4.83) as

$$T_0^{(1)} = \mu_z \quad , \quad T_{\pm 1}^{(1)} = \mp (\mu_x \pm i\mu_y)/\sqrt{2}$$

and Eq. (4.228) becomes

$$T\,T_0^{(1)}\,T^{-1} = +\,T_0^{(1)}$$

$$\tag{4.230}$$

$$T\,T_{\pm 1}^{(1)}\,T^{-1} = -\,T_{\mp}^{(1)} \quad .$$

The transformation properties of a general tensor
operator under T are given by Eq. (4.212); comparing
it with Eq. (4.230) gives a value $\eta_k = -1$ for the
phase in Eq. (4.212). With $\eta_k = -1$ and Eq. (4.215),
the expectation value of $T^{(1)}$ for a non-degenerate
state turns out to be pure imaginary,

$$\langle \alpha |\, T_q^{(1)}\, | \alpha \rangle = -\langle \alpha |\, T_q^{(1)}\, | \alpha \rangle^* \quad .$$

However, since $\vec{\mu}_e$ and therefore also $T^{(1)}$ are
measurable and thus Hermitian, the expectation value
of $\vec{\mu}_e$ should be real and it must therefore be zero:
Time reversal invariance alone rules out static
electric dipole moments. This fact can be understood by the following physical argument: The only
direction along which an electric dipole moment can
point is that of the spin. The spin changes sign
under time reversal but the operator $\vec{\mu}_e$ does not
do so and its expectation value must therefore
vanish. This argument can be generalized to all

odd electric and even magnetic moments.

The proofs for the vanishing of the electric dipole moment assume definite spins for the particles. For a system with degenerate eigenstates differing in parities or spins, a permanent electric dipole moment is not ruled out; indeed polar molecules with electric dipole moments exist.

D. Time Reversal Invariance Tests

Does time reversal invariance hold? We will see in Chapter 5 that there is evidence for a small T violation. Clearly, direct and sensitive tests are important. The results of the previous subsection suggest some experiments and we discuss a few examples here.

We first consider semi-detailed balance. Equation (4.224) states that the squares of the transition matrix elements, summed over spins, are the same for a nuclear reaction and its inverse if time-reversal invariance holds. However, some pitfalls must be avoided; Eq. (4.224) can hold for reasons other than time-reversal invariance, and some nuclear reactions are more sensitive to time-reversal violations than others.[126,127]

126. E. M. Henley and B. A. Jacobsohn, Phys. Rev. 113, 225 (1959).
127. P. A. Moldauer, Phys. Rev. 165, 1136 (1968); Phys. Letters 26B, 713 (1968); D. Robson, Phys. Letters 26B, 117 (1968); T. E. O. Ericson, Phys. Letters 23, 97 (1966); C. Malhaux and H. A. Weidenmüller, Phys. Letters 23, 100 (1966).

The reactions

$$d + {}^{24}Mg \rightleftharpoons p + {}^{25}Mg$$

are primarily direct ones; they have been studied
very carefully with deuterons and protons from a
tandem Van De Graaff accelerator. It is extremely
difficult to determine absolute cross sections to
accuracies of better than 1%; Weitkamp et al.[128]
therefore measured, as a function of energy, the
ratio of differential cross sections simultaneously
at two angles. Reciprocity requires these ratios
to be equal for the two reactions when compared
at the same pair of c.m. angles and the same ${}^{26}Al$
excitation energy. The result is shown in Fig.
4.29. The ratios of the two reactions agree to
within an over-all uncertainty of 0.3%. Lack of
a good model for the reaction makes a quantitative
interpretation difficult, but simple estimates based
on a distorted wave approximation show that the
ratio of T-noninvariant to T-invariant forces,
implied by the experiment, is also of the order
of 0.3%.[129] A similar limit has been obtained
in an investigation of the reactions $\alpha + {}^{25}Mg \rightleftharpoons p + {}^{27}Al$,
which proceed primarily via the compound nucleus
${}^{28}Si$.[130]

---

128. W. G. Weitkamp, D. W. Storm, D. C. Shreve,
     W. J. Braithwaite, and D. Bodansky, Phys. Rev.
     165, 1233 (1968).
129. E. M. Henley and A. H. Huffman, Phys. Rev.
     Letters 20, 1191 (1968); A. H. Huffman,
     Phys. Rev. D1, 882, 890 (1970).
130. W. von Witsch, A. Richter, and P. von Bren-
     tano, Phys. Letters 22, 631 (1966); Phys. Rev.
     Letters 19, 524 (1967); Phys. Rev. 169, 923
     (1968).

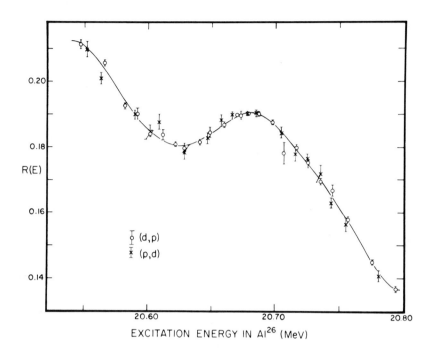

Fig. 4.29. Excitation function for the ratio $\sigma(\theta_{1m}, E)/\sigma(\theta_{2m}, E)$ for the $^{24}Mg(d,p)^{25}Mg$ and $^{25}Mg(p,d)^{24}Mg$ reactions at $\theta_{1m} = 29.7°$ and $\theta_{2m} = 119.2°$ (c.m.). The solid curve is a least-squares fit of an eight-term Fourier cosine series to the two sets of data. [From reference 128].

The experiments described in references 128 and 130 were performed at particle energies between 10 and 20 MeV. Tests at higher energies have been suggested[131], and the reactions $\gamma d \rightleftarrows np$ and $\gamma n \rightleftarrows \pi^- p$ have been investigated[132] in the region of the first nucleon resonance (200-300 MeV photon energies). No evidence for a time-reversal violation is found.

The tests discussed so far check the validity of semi-detailed balance, Eq. (4.224). This relation holds if the operator S is rotationally invariant and satisfies the relation $T S T^{-1} = S^\dagger$. If S is also Hermitian, other checks become possible. With $S^\dagger = S$, Eqs. (4.222) and (3.31) show that all matrix elements of S between angular momentum eigenstates can be chosen to be real.[123] The same result follows from Eq. (4.211). (Our phase choice that leads to real matrix elements is given by Eq. (4.198).) Although the reality of one matrix element is not measurable, the ratio of the matrix elements for any two competing modes between the same states must be real, regardless of phase convention. The phase between two competing matrix elements must therefore be 0 or 180°, and this phase can be determined experimentally.

131. N. Christ and T. D. Lee, Phys. Rev. 148, 1520 (1966); S. Barshay, Phys. Rev. Letters 17, 49 (1966).

132. D. F. Bartlett, C. E. Friedberg, P. E. Goldhagen, and K. Goulianos, Phys. Rev. Letters 27, 881 (1971); C. Joseph in High Energy Physics and Nuclear Structure, G. Tibell, ed. (North-Holland Publishing Co., Amsterdam, and Almqvist and Wiksell, Stockholm, 1974), pp. 33-40.

In order to explain the idea underlying measurements of the relative phase of two matrix elements, we consider the neutron decay, $n \rightarrow p \, e^- \, \bar{\nu}$. In this decay, electron and antineutrino can be emitted with antiparallel spins (Fermi decay), or with parallel spins (Gamow-Teller decay); we denote the corresponding transition operators with H_F and H_{GT}, the matrix elements with M_F and M_{GT}. For a time-reversal experiment, polarized neutrons are produced. The number of neutrons with spin $\vec{s}$, giving rise to electrons with momentum $\vec{p}_e$ and anti-neutrinos with momentum $\vec{p}_{\bar{\nu}}$ is then determined and compared to the corresponding number for the "time-reversed" configuration $(-\vec{s}, -\vec{p}_e, -\vec{p}_{\bar{\nu}})$. (In an actual experiment, the antineutrino momentum is inferred from a measurement of the momentum of the recoiling proton.) The directions of $\vec{s}$, $\vec{p}_e$ and $\vec{p}_{\bar{\nu}}$ are so chosen that the scalar observable

$$\mathcal{O} = \vec{s} \cdot (\vec{p}_e \times \vec{p}_{\bar{\nu}}) \tag{4.231}$$

is a maximum. $\mathcal{O}$ is even under P, but odd under T. A nonvanishing expectation value of $\mathcal{O}$ will therefore be evidence for time-reversal violation. To connect this fact with the relative phase of matrix elements, we write for the ratio of decays discussed above

$$R = \frac{|<\vec{p}_e,\vec{p}_{\bar{\nu}}|H_F + H_{GT}|\vec{s}>|^2}{|<-\vec{p}_e,-\vec{p}_{\bar{\nu}}|H_F + H_{GT}|-\vec{s}>|^2}$$

(4.232)

$$= \frac{|<-\vec{p}_e,-\vec{p}_{\bar{\nu}}|T(H_F + H_{GT})T^{-1}|-\vec{s}>|^2}{|<-\vec{p}_e,-\vec{p}_{\bar{\nu}}|H_F + H_{GT}|-\vec{s}>|^2}.$$

In the second step, Eq. (4.190) has been used. If $H_w = H_F + H_{GT}$ commutes with T, R is unity. If, however,

$$H_w = H_F + e^{i\phi} H_{GT}, \quad H_F, \ H_{GT} \text{ Hermitian,} \quad (4.233)$$

the ratio of matrix elements of H_F and H_{GT} is no longer real, H_w and T no longer commute, and R can be different from one. Computation of the matrix elements gives an expression for the ratio R of the form

$$R = 1 + \text{const} <\mathcal{O}> \sin \phi \quad , \quad (4.234)$$

where $\mathcal{O}$ is given by Eq. (4.231); the constant can be evaluated explicitly.[133] The ratio R for the neutron decay has been measured[134] and the best

133. J. D. Jackson, S. B. Treiman, and H. W. Wyld, Jr., Phys. Rev. 106, 517 (1957); Nuclear Phys. 4, 206 (1957). K. Alder, B. Stech, and A. Winther, Phys. Rev. 107, 728 (1957). C. G. Callan, Jr. and S. B. Treiman, Phys. Rev. 162, 1494 (1967).

134. M. T. Burgy, V. E. Krohn, T. B. Novey, G. Ringo and V. L. Telegdi, Phys. Rev. 120, 1829 (1960); R. I. Steinberg, P. Liaud, B. Vignon and V. W. Hughes, Phys. Rev. Letters 33, 41 (1974).

result for the phase is

$$\phi = (180.14 \pm 0.22)° \quad .$$

Within the limits of error the ratio of the two
matrix elements is real and no evidence for a
violation of time-reversal invariance is obtained.

Another class of tests involves electromagnetic transitions between bound states.[135]
The idea is similar to the one discussed above:
Two competing matrix elements, usually E2 and M1,
are involved and their relative phase is determined. The most accurate measurements have been
performed by using the Mössbauer effect.[136]
Experiments with ^{99}Ru and ^{193}Ir indicate that the
phase ϕ is less than about 2×10^{-3} radians.

One word of caution must be added to the
experiments that test the reality of matrix
elements. The arguments are based on first-order
perturbation theory and are only valid if final
state interactions between the particles can be
neglected. In the neutron decay, for instance,
the Coulomb interaction between the proton and

135. B. A. Jacobsohn and E. M. Henley, Phys. Rev.
     113, 234 (1959); E. M. Henley and B. A.
     Jacobsohn, Phys. Rev. Letters 16, 706 (1966);
     W. A. Steyert and K. S. Krane, Phys. Letters
     47B, 294 (1973).
136. O. C. Kistner, Phys. Rev. Letters 19, 872
     (1967). (^{99}Ru). M. Atac, B. Chrisman, P.
     Debrunner and H. Frauenfelder, Phys. Rev.
     Letters 20, 691 (1968). (^{193}Ir). See also
     M. I. Bulgakov et. al., Yad. Fiz. 18, 12
     (1973) [Sov. J. Nucl. Phys. 18, 6 (1974)].

the electron in the final state must be shown to
be negligible or it must be included in the computation.[133]   In the Mössbauer experiments, nuclear
radiative corrections[135] and the interaction of the
emitted photon with atomic electrons have to be
taken into account.[137]

As a final example, we discuss the search for
the electric dipole moment of the neutron. As
indicated by Eq. (4.229), the electric dipole moment
of nuclei and particles must vanish if parity is
conserved.[87]   In 1952, Smith, Purcell, and Ramsey
tried to test parity conservation by setting an
upper limit to the electric dipole moment of the
neutron.[138]   With the advent of parity nonconservation, the arguments were reanalyzed and Landau
pointed out that a permanent electric dipole moment
can only exist if parity and time reversal invariance are violated.[139]   Now parity is violated
and particles could have small electric dipole
moments if time-reversal invariance is also violated. The parity breaking admixture $\mathcal{F}$ to the
nuclear interaction is of the order of 10^{-7}. An
electric dipole moment of the order of $e\mathcal{F}r$,
where r is a measure of the particle radius, could
therefore be expected. With $\mathcal{F} = 10^{-7}$, $r = 10^{-13}$ cm,

137.   J. P. Hannon and G. T. Trammell, Phys. Rev.
       Letters 21, 726 (1968).
138.   J. H. Smith, E. M. Purcell and N. F. Ramsey,
       Phys. Rev. 108, 120 (1957).
139.   L. Landau, Nuclear Phys. 3, 127 (1957).

a crude estimate is

$$\langle \mu_e \rangle \stackrel{\sim}{\scriptstyle\sim} 10^{-20} \text{ e cm.} \tag{4.235}$$

To test for a possible T violation, $\langle \mu_e \rangle$ of some
particle must be measured to a higher accuracy
than the estimate (4.235). The best candidate for
such an experiment is the neutron; for charged
particles, the interaction of electromagnetic
fields with the charge makes very accurate measurements of $\langle \mu_e \rangle$ extremely difficult.

Experiments to detect a possible electric
dipole moment of the neutron have been performed[138,140] and we sketch here the one by
Ramsey and coworkers. A simplified diagram of
their arrangement is shown in Fig. 4.30. Very slow
neutrons from a reactor are polarized by total
reflection from a magnetized mirror,[141] as indicated in Fig. 4.31. The polarization results from
the fact that neutrons can be totally reflected
from mirrors[142] and that the angle of total
reflection depends on the sign of $\vec{\mu}_m \cdot \vec{B}$. The

---

140. W. B. Dress, J. K. Baird, P. D. Miller, and
     N. F. Ramsey, Phys. Rev. 170, 1200 (1968);
     179, 1285 (1969). C. G. Shull and R. Nathans,
     Phys. Rev. Letters 19, 384 (1967). V. W.
     Cohen, R. Nathans, H. B. Silsbee, E. Lipworth,
     and N. F. Ramsey, Phys. Rev. 177, 1942 (1969).
     W. P. Dress, P. D. Miller and N. F. Ramsey,
     Phys. Rev. D7, 3147 (1973).
141. D. J. Hughes and M. T. Burgy, Phys. Rev. 81,
     498 (1951).
142. E. Fermi and L. Marshall, Phys. Rev. 71, 666
     (1947).

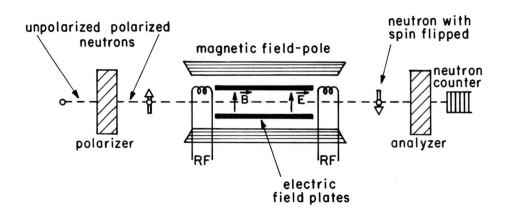

Fig. 4.30.   Experimental arrangement used by Ramsey
             and coworkers[140) in the search for an
             electric dipole moment.

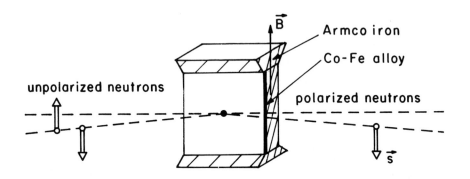

Fig. 4.31.   Polarization of neutrons by total
             reflection from a magnetized mirror
             (Co-Fe alloy).

polarized neutrons then pass through a resonance
spectrometer, an analyzer and finally are recorded
in a counter. The analyzer, also a magnetized
Co-Fe mirror, is adjusted in such a way that neutrons pass through it easily if they have undergone
a spin-flip in the spectrometer. The spectrometer
consists of a magnetic field $\vec{B}$(9 G), radio-frequency
coils, and an electric field $\vec{E}$, parallel or antiparallel to $\vec{B}$. The energy and hence the precession
frequency of the neutron in the spectrometer follows
from Eq. (4.225) as

$$\omega = <\mu_m> \; |\vec{B}| \; \pm <\mu_e> \; |\vec{E}| \quad .$$

If the electric field is zero, a maximum number of
neutrons will flip their spin at the resonance
frequency $\omega_R = <\mu_m> \; |\vec{B}|$; the counting rate will
then be a maximum. With electric field present,
the resonance frequency is shifted by the amount

$$\Delta\omega = \pm <\mu_e> \; |\vec{E}| \quad ,$$

where the + and - sign refer to $\vec{E}$ parallel or antiparallel to $\vec{B}$. A shift in resonance frequency
results in a change in the neutron counting rate.
In the experiment, a field of 140 kV/cm was applied
and its direction was reversed every five minutes.
No significant counting rate change was observed
and a limit on the electric dipole moment of

$$|<\mu_e>/e| \lesssim 1.0 \times 10^{-23} \text{ cm} \qquad (4.236)$$

was set. This value results in a limit on time-
reversal violation roughly one order of magnitude
smaller than the limit imposed by the nuclear
measurements discussed earlier.[143]

4.7   CHARGE CONJUGATION

In addition to the space-time symmetries that
we discussed in the last sections, there exist
internal symmetries that are independent of the
space-time continuum. An example is charge conjugation which was suggested by the particle-hole
symmetry in the Dirac theory, and the particle-
antiparticle symmetry in the Stueckelberg-Feynman
description. Invariance under charge conjugation
requires that the laws of physics remain invariant
under replacement of all particles by their antiparticles. Formally, charge conjugation was first
applied to electrons and positrons[11,144], but
it was later generalized to all particles.

A.   Particle-Antiparticle Conjugation

We characterize a particle by the values of
its additive quantum numbers: baryon number A,

143. D. T. Broadhurst, Phys. Rev. D5, 1228 (1972).
144. W. Pauli, Inst. H. Poincaré Ann. 6, 130
      (1936). H. A. Kramers, Proc. Amst. Akad.
      Sci. 40, 814 (1937).

hypercharge Y, electric charge Q, lepton number L, and muon number L_μ. These quantum numbers do not have equal standing: A and Y are determined by the hadronic interaction, Q is given through inter- actions with electromagnetic fields, and L and L_μ are only defined through weak processes. We then define a charge conjugation or particle-antiparticle operator C by

$$C|A,Y,q,L,L_\mu; \vec{p},\vec{s}\rangle$$

$$= \eta_c^*|-A,-Y,-q,-L,-L_\mu; \vec{p},\vec{s}\rangle \quad . \tag{4.237}$$

The operator C reverses the sign of all additive internal quantum numbers of a particle; η_c is a phase factor. Equations (4.46) and (4.108) show that I_3 and the strangeness S also change sign under C. Space-time properties, such as spin and momentum, remain unchanged.

Some properties of C are important for later applications. In order to insure invariance of the norm, C, just as P, is taken to be unitary,

$$C^\dagger C = C C^\dagger = 1 \quad . \tag{4.238}$$

C does not commute with the electric charge oper- ator Q; to find the value of the commutator, we write

$$Q|q\rangle = q|q\rangle \tag{4.239}$$

and then get

$$CQ|q\rangle = Cq|q\rangle = qC|q\rangle = q|-q\rangle \quad ,$$

and

$$QC|q\rangle = Q|-q\rangle = -q|-q\rangle \quad .$$

Hence

$$\{C,Q\} = 0 \quad , \quad CQC^{-1} = -Q \quad , \tag{4.240}$$

and

$$\left[C,Q\right]|q\rangle = 2q|-q\rangle = 2CQ|q\rangle \quad ,$$

or

$$\left[C,Q\right] = 2CQ \quad . \tag{4.241}$$

C and Q do not commute. Since the electric current is the time derivative of the electric charge, we generalize Eq. (4.240) to

$$Cj^{\mu}C^{-1} = -j^{\mu} \quad . \tag{4.242}$$

This transformation is intuitively the correct one since C leaves the momentum vector unchanged; the reversal in charge then also reverses the direction of the electric current $q\vec{v}$. (The electric current actually is given by ej^{μ}, but e is a constant and

is not affected by C.)   Similarly, because the
electromagnetic potential is caused by electric
charges, we postulate

$$C A^\mu C^{-1} = -A^\mu \quad . \tag{4.243}$$

With Eqs. (4.242) and (4.243), the electromagnetic
interaction $j^\mu A_\mu$ is invariant under C.

The arguments that have led from Eq. (4.239)
to Eq. (4.241) apply also to other additive quantum
numbers and to I; they show that

$$\left[C, A\right] \neq 0, \quad \left[C, Y\right] \neq 0, \quad \left[C, I\right] \neq 0,$$

$$\tag{4.244}$$

$$\left[C, L\right] \neq 0, \quad \left[C, L_\mu\right] \neq 0 \quad .$$

On the other hand, C commutes with momentum and
angular momentum operators, and hence also with
the helicity operator,

$$\left[C, \vec{p}\right] = \left[C, \vec{s}\right] = \left[C, \hat{p} \cdot \vec{s}\right] = 0 \quad . \tag{4.245}$$

We have introduced the operator C without much
justification, but we now take a second look at
Eq. (4.237).   It really makes a strong statement:
The hadronic, the electromagnetic, and the weak
interactions appear intimately linked.   If we go
from a particle with electric charge q to its antiparticle with charge - q (charge conjugation), we

also change the sign of the hadronic and the leptonic additive quantum numbers (hadronic and leptonic particle-antiparticle conjugation). Are charge conjugation and particle-antiparticle conjugation, as defined in the previous sentence, always identical? For all known particles, these operations indeed coincide and Eq. (4.237) is valid. However, it is possible that particles exist that do not obey Eq. (4.237).[145] A related question is the validity of Eq. (4.242). Bernstein, Feinberg, and Lee[146] have suggested that the electromagnetic current of <u>hadrons</u> could be of the form

$$\mathcal{Y}^\mu = j^\mu + K^\mu ,$$

with j^μ satisfying Eq. (4.242) and K^μ obeying

$$C K^\mu C^{-1} = + K^\mu . \tag{4.246}$$

An interaction, $\mathcal{Y}^\mu A_\mu$, then would contain terms like $K^\mu A_\mu$ that do not commute with C. Such an interaction could explain the experimentally observed T violation (Chapter 5). However, no evidence for the existence of currents K^μ has been found.

---

145. T. D. Lee, Phys. Rev. <u>140</u>, B 959, B 967 (1965).
146. J. Bernstein, G. Feinberg, and T. D. Lee, Phys. Rev. <u>139</u>, B 1650 (1965).

B. Charge Conjugation Formalism

Time reversal and parity operations can be described in a satisfactory way in ordinary quantum mechanics because they relate for instance spin and momentum of a particle with spin and momentum of the same particle in the time-reversed or the space-reflected state. Charge conjugation, however, relates a particle to its antiparticle and a one-particle description is inadequate. The language of field theory is better suited for such a situation than the language of ordinary quantum mechanics.

We first discuss a non-Hermitian Klein-Gordon field, as given by Eq. (3.115), and define the operator C by

$$C \Phi(x) C^{-1} = \eta_C \Phi^\dagger(x) \quad . \tag{4.247}$$

Here, η_C is an unobservable phase factor with

$$\eta_C^* \eta_C = 1 \quad . \tag{4.248}$$

Applying C to the field (3.115) and comparing terms gives

$$C a_p C^{-1} = \eta_C a_{\bar{p}} \quad , \quad C a_p^\dagger C^{-1} = \eta_C^* a_{\bar{p}}^\dagger \quad , \tag{4.249}$$

$$C a_{\bar{p}} C^{-1} = \eta_C^* a_p \quad , \quad C a_{\bar{p}}^\dagger C^{-1} = \eta_C a_p^\dagger \quad .$$

To see that the definition (4.247) is in agreement with the earlier one, Eq. (4.237), we use the charge operator Q as given by Eq. (3.118). We easily find

$$CQC^{-1} = -Q \quad ,$$

in agreement with Eq. (4.240). Furthermore, we get for the Klein-Gordon current Eq. (3.108)

$$Cj^{\mu} C^{-1} = - j^{\mu} \quad ,$$

in agreement with Eq. (4.242). Finally we define for the vacuum state

$$C|0> = |0> \quad , \tag{4.250}$$

and then get for a one-particle state $|p>$

$$C|p> = Ca_p^{\dagger} |0> = n_C^* a_{\underline{p}}^{\dagger}|0> = n_C^*|\bar{p}> . \tag{4.251}$$

C changes a particle into its antiparticle, in agreement with Eq. (4.237).

Next, we turn to a <u>Hermitian Klein-Gordon field</u>, as given by Eq. (3.105). With $\Phi = \Phi^{\dagger}$, we find from Eq. (4.247) $n_C^* = n_C$; the phase n_C must be real and by Eq. (4.248)

$$n_C = \pm 1 \quad . \tag{4.252}$$

The neutral one-particle state $|p^o> = a_{p^o}^{\dagger}|0>$ then is an eigenstate of C with eigenvalue η_C:

$$C|p^o> = \eta_C |p^o> \qquad \eta_C = \pm 1 \quad . \tag{4.253}$$

The eigenvalue η_C is now an observable and it is called charge parity. We will discuss it in more detail in Subsection C.

The properties of the electromagnetic field under C are given by Eq. (4.243). With the expansion (3.54) we get for the destruction and creation operators

$$C\ a_{k\lambda}\ C^{-1} = -\ a_{k\lambda}\ ,\ C\ a_{k\lambda}^\dagger\ C^{-1} = -\ a_{k\lambda}^\dagger , \tag{4.254}$$

and then find for a one-photon state $|k\lambda> = a_{k\lambda}^\dagger |0>$ with Eq. (4.250)

$$C|k\lambda> = -\ |k\lambda> \quad . \tag{4.255}$$

The charge parity η_C of a photon is negative. For an n-photon state then follows

$$C|\gamma_1 \cdots \cdots \gamma_n> = (-1)^n |\gamma_1 \cdots \cdots \gamma_n> \quad . \tag{4.256}$$

The charge parity of an n-photon state is $(-1)^n$. A consequence of this result is Furry's theorem[147]:

$$\left.\begin{array}{c} <n\ \text{photons}\ |\ H_\gamma\ |\ m\ \text{photons}> = 0 \\[1em] \text{if n+m odd, and } [H_\gamma, C] = 0 \quad . \end{array}\right\} \tag{4.257}$$

147. W. H. Furry, Phys. Rev. 51, 125 (1937).

A state with an even number of photons cannot connect to a state with an odd number of photons if
no external influences are present.

Charge conjugation was forced on physicists
by the <u>Dirac equation</u>. We have outlined the
physical aspects in Section 3.4. To close the
circle, we sketch here some formal aspects. The
Dirac equation for an electron with charge -e
follows from Eqs. (3.167) and (3.87) as

$$(i\not\partial + e\not A - m)\psi = 0 \quad .$$

Had we started with positrons, the equation would
read

$$(i\not\partial - e\not A - m)\psi_C = 0 \quad .$$

Note that the negative-energy solutions of the
first equation are positrons, that of the second
one electrons. In the standard representation,
Eq. (3.174), the wavefunctions ψ_C and ψ are connected by

$$\psi_C = i \gamma^2 \psi^* \quad , \tag{4.258}$$

where the phase factor i has been chosen to make
the matrix $i \gamma^2$ real.

Equation (4.258) describes charge conjugation
in the one-particle Dirac theory; its interpretation involves hole states and is therefore somewhat
artificial. In field theory, charge conjugation

can be handled in a more natural way through the operator equation

$$C \psi C^{-1} = i \gamma^2 \gamma^0 \bar{\psi} \quad , \tag{4.259}$$

where the field operators ψ and $\bar{\psi}$ are given by Eq. (3.190). For calculations, it is sometimes best to display matrix indices explicitly:

$$C \psi_\alpha C^{-1} = (i \gamma^2 \gamma^0)_{\alpha\beta} \bar{\psi}_\beta \quad .$$

Relations similar to Eqs. (4.240) can then be established and used.

C. Charge Parity

The operator C is unitary; if we apply it twice, we return to the original state. C is therefore of the type of Eq. (4.12) and may have observable eigenvalues. However, states found in nature are characterized by the eigenvalues of the operators Q, A, and Y. Equations (4.241) and (4.244) show that C does not commute with these operators; particles will in general not be eigenstates of C. However, fully neutral particles or systems, i.e., states with eigenvalues zero of Q, A, and Y, form an exception; they can be eigenstates of C. The photon, neutral nonstrange mesons such as the π^0 and the η^0, and particle-antiparticle systems such as e^+e^-, $p\bar{p}$, $n\bar{n}$, $K\bar{K}$ qualify.

For a <u>fully neutral</u> system, we can write

$$C|\psi> = \eta_C|\psi> \quad , \qquad \eta_C = \pm 1 \quad . \tag{4.260}$$

This equation agrees with our earlier Eq. (4.253) for the Hermitian Klein-Gordon field: as stated there η_C is called the <u>charge parity</u>. The phase is fixed by Eq. (4.250), $C|0> = |0>$.

By Eq. (4.255) we have already found the charge parity of the photon - it is negative. To find the charge parities of the neutral pion and the eta, we note that both decay electromagnetically into two gamma rays,

$$\pi^{\circ} \rightarrow 2\,\gamma \quad , \quad \eta^{\circ} \rightarrow 2\,\gamma \quad .$$

Assuming that C is conserved in electromagnetic decays, we find with Eq. (4.256)

$$\eta_C(\pi^{\circ}) = +1 \quad , \quad \eta_C(\eta^{\circ}) = +1 \quad . \tag{4.261}$$

C parity is also used to classify states of particle-antiparticle systems. As first example, we consider the two-pion state $|\pi^{+}(\vec{x}_1), \pi^{-}(\vec{x}_2)> \equiv |\pi^{+}, \pi^{-}>$. Charge conjugation interchanges the two pions,

$$C|\pi^{+}\pi^{-}> = |\pi^{-}\pi^{+}> \quad . \tag{4.262}$$

For a system of two spinless bosons, the interchange of the two particles in the c.m. system is identical to the parity operation

$$P|\pi^+\pi^-> = C|\pi^+\pi^-> \quad .$$

We know that

$$P|\pi^+\pi^-> = (-1)^\ell |\pi^+\pi^->$$

and hence

$$C|\pi^+\pi^-> = (-1)^\ell |\pi^+\pi^-> \quad , \quad \eta_C(\pi^+\pi^-) = (-1)^\ell \quad ,$$

$$(4.263)$$

where ℓ is the c.m. orbital angular momentum. For two neutral pions, the same argument applies, but since the two particles are identical, only even values of ℓ can appear and η_C is always $+1$:

$$\eta_C(\pi^0\pi^0) = +1 \quad .$$

This result is in agreement with the fact that a single neutral pion already has $\eta_C = +1$.

The states of a fermion-antifermion system $|f\bar{f}>$ can be classified similarly. The result is[148]

$$C|f\bar{f}> = (-1)^{\ell+s}|f\bar{f}> \quad , \quad \eta_C(f\bar{f}) = (-1)^{\ell+s} \quad ,$$

$$(4.264)$$

148. R. Laverriere, J. Proviol, and C. Thevenet, Am. J. Phys. 36, 258 (1968).

where s is the spin of the system.

D.   Charge Conjugation Invariance Tests

The fact that antiparticles to each known particle have been observed is already an indication that invariance under C is at least a good approximation. We discuss here a few experiments that give more quantitative limits on the validity of C invariance.

Selection rules based on C invariance give information on the electromagnetic interaction. Consider first positronium. In the $^1S_0$ state ($\ell = 0$, $s = 0$), Eq. (4.264) shows that it has $\eta_C = +1$. It can therefore decay into two, but not into three gamma rays. If C invariance holds, then the selection rule is absolute,[149] and the decay is not only slower by a factor $e^2/4\pi$. Conservation of C parity also forbids the decays

$$\pi^0 \to 3\,\gamma\,, \quad \eta^0 \to 3\,\gamma\quad.$$

Experiments have been performed to search for the forbidden decays of the neutral pion and the singlet positronium and the following limits have been set[150]

150.   J. Duclos, D. Freytag, K. Schlüpmann, V. Soergel, J. Heintze, and H. Rieseberg, Phys. Letters 19, 253 (1965). D. Cline and R. M. Dowd, Phys. Rev. Letters 14, 530 (1965). ($\pi^0 \to 3\gamma$). A. P. Mills and S. Berko, Phys. Rev. Letters 18, 420 (1967). ($^1S_0 \to 3\gamma$).

See also K. Marko and A. Rich, Phys. Rev. Letters 33, 980 (1974).

$$R = \frac{w(\pi^0 \to 3\gamma)}{w(\pi^0 \to 2\gamma)} \leq 5 \times 10^{-6} \quad ,$$

$$R' = \frac{w(^1S_0 \to 3\gamma)}{w(^1S_0 \to 2\gamma)} \leq 3 \times 10^{-6} \quad ,$$

where w is the appropriate decay rate. What do these ratios mean in terms of a possible C violation? Bernstein, Feinberg, and Lee[146] have estimated the rate R under the assumption that the C-even current K^μ exists; they find $R \sim 3 \times 10^{-6}$, less than the present experimental limit. The experimental evidence for the validity of invariance under conjugation in the electromagnetic decay of the neutral pion and the singlet positronium is therefore not very strong.

However, other tests of C can be and have been performed. An example is the symmetry required by Eq. (4.262) in the electromagnetic decays (see Subsection E) of the η: $\eta \to \pi^+\pi^-\pi^0$ and $\eta \to \pi^+\pi^-\gamma$.[151]   The asymmetry, A,

$$A = \frac{N_+ - N_-}{N_+ + N_-} \quad ,$$

---

151. J. J. Thaler, J. A. Appel, A. Kotlewski, J. G. Layter, W. Lee, and S. Stein, Phys. Rev. Letters 29, 313 (1972); J. G. Layter, J. A. Appel, A. Kotlewski, W. Lee, S. Stein, and J. J. Thaler, Phys. Rev. Letters 29, 316 (1972). See also M. R. Jane et. al., Phys. Letters 48B, 260, 265 (1974).

where $N_+(N_-)$ is the number of decay events for which the $\pi^+(\pi^-)$ has a greater energy than the $\pi^-(\pi^+)$, is found to be

$$A = -0.005 \pm 0.0022$$

in the decay mode $\eta \to \pi^+\pi^-\pi^0$, and

$$A = 0.005 \pm 0.006$$

in the decay $\eta \to \pi^+\pi^-\gamma$. There is thus no indication of a violation of C in the electromagnetic decays of this hadron. The invariance

$$[H_\gamma, C] = 0 \tag{4.265}$$

is thus established.

The most direct test for charge conjugation invariance in hadronic interactions would be the comparison of a reaction with the corresponding antiparticle reaction. Such an experiment is difficult to perform and it is easier to follow a suggestion by Pais[152] and compare reactions

$$p\bar{p} \to 1+2+ \ldots$$

$$p\bar{p} \to \bar{1}+\bar{2}+ \ldots \ .$$

If charge conjugation invariance holds, the energy and momentum distributions of the charge-conjugate channels should be the same. Such experiments have

152. A. Pais, Phys. Rev. Letters 3, 242 (1959).

been performed with $\bar{p}p$ annihilation in flight and at rest.[153]   The result is

$$\frac{\text{C-nonconserving amplitude}}{\text{C-conserving amplitude}} \leq 0.01 \quad .$$

These experiments are not sensitive enough to test the electromagnetic or the weak interaction; the limit applies to the hadronic interaction and we express this fact as

$$\left[H_h , C \right] = 0 \quad . \tag{4.266}$$

The weak interaction is easy to discuss. Already the first experiments that showed the breakdown of parity conservation also indicated violation of charge conjugation.[89,90]   The violation is manifest in the helicity of electrons and muons emitted in weak decays. Eq. (4.245) shows that C commutes with the helicity operator. We then expect the helicity of electrons and positrons to be the same. If electrons and positrons are produced in states related by $C|e^+> = |e^->$, then

---

153.   N. H. Xuong, G. R. Lynch, and C. K. Hinrichs, Phys. Rev. 124, 575 (1961). C. Baltay, N. Barash, P. Franzini, N. Gelfand, L. Kirsch, G. Lütjens, J. C. Severiens, J. Steinberger, D. Tycko, and D. Zanello, Phys. Rev. Letters 15, 591 (1965).

$$<e^-|\vec{\sigma}\cdot\hat{p}|e^-> = <e^-|C^{-1}C\,\vec{\sigma}\cdot\hat{p}\,C^{-1}C|e^->$$

$$= <e^+|\vec{\sigma}\cdot\hat{p}|e^+>\quad.$$

The experiments, however, show just the opposite! Eq. (4.168) states that electrons have negative, positrons positive helicity. The same holds for negative and positive muons. Charge conjugation invariance does not hold in the weak interactions,

$$\left[H_w, C\right] \neq 0\quad.\tag{4.267}$$

We summarize this subsection by stating that charge conjugation invariance is violated in the weak interactions, but appears to hold in the electromagnetic and the hadronic interactions.

E. G Conjugation and G Parity

In Subsection C we noted that charge parity can only be assigned to fully neutral states. However, it is possible to generalize the concept of charge parity to states with nonzero electric charge; the new quantum number is called G parity.[154]

---

154. K. Nishijima, Progr. Theor. Phys. 6, 614, 1027 (1951); L. Michel, Nuovo Cimento 10, 319 (1953); D. Amati and B. Vitale, Nuovo Cimento 2, 719 (1955); C. Goebel, Phys. Rev. 103, 258 (1956); T. D. Lee and C. N. Yang, Nuovo Cimento 3, 749 (1956).

To find an operator that can be used to classify states with arbitrary electric charge we consider systems with $A = Y = 0$. Equation (4.108) then gives $Q = eI_3$ and we hence demand that the new operator G commute with I_3,

$$[G, I_3] = 0 \quad . \tag{4.268}$$

Since C changes the sign of Q and hence also of I_3, we must add an operation that returns $-I_3$ to I_3. A rotation in isospin space by 180° about the I_2 axis will do it. Equation (4.93) shows that such a rotation is given by $\exp(-i \pi I_2)$. We thus set

$$G = C e^{-i \pi I_2} \quad . \tag{4.269}$$

G conjugation is a 180° rotation in isospin space about I_2, followed by charge conjugation. The phase factors associated with C can be so chosen that G commutes not only with I_3, but with all components of $\vec{I}$,

$$[G, \vec{I}] = 0 \quad . \tag{4.270}$$

G can thus be diagonalized simultaneously with I^2 and I_3 if $A = Y = 0$; the eigenvalues of G are independent of I_3 or Q, and eigenvalues of G can be found even for electrically charged particles. However, $\exp(-i \pi I_2)$ acts only in isospin space and does not affect the baryon number or the hypercharge; as with C conjugation we have

$$[G,A] \neq 0 \quad , \quad [G,Y] \neq 0 \quad . \tag{4.271}$$

Only states with $A = Y = 0$ can be eigenstates of
G. For such states, the eigenvalue equation is

$$G \ |\psi> = \eta_g \ |\psi> \quad , \tag{4.272}$$

and η_g is called G parity.
When is G conserved?   C commutes with H_h and
H_γ, but Eq. (4.90) shows that I_2 is not a good
quantum number in the presence of electromagnetic
interactions.   Hence

$$\left[G,H_h \right] = 0 \quad , \tag{4.273}$$

$$\left[G,H_\gamma \right] \neq 0 \quad . \tag{4.274}$$

G can only be used for hadronic processes, whereas
C applies to hadronic and electromagnetic ones.

Before giving some examples of the use of G
parity, we derive an expression for the action of
the operator $\exp(-i \ \pi \ I_2)$. This operator is related to charge symmetry and it can be used to
derive selection rules in nuclear reactions.[155]
For a state with integral isospin, Eqs. (4.65),

155.   N. M. Kroll and L. L. Foldy, Phys. Rev. 88,
       1177 (1952).

(4.70), and (3.15) give

$$e^{-i\pi I_2} |I,0> = \sum_{I_3} d^I_{I_3 0}(\pi) |I,I_3>$$

$$\text{(4.275)}$$

$$= (-1)^I |I,0> \quad .$$

Since G commutes with $\vec{I}$, it is independent of I_3 and we have for integral isospin

$$G = (-1)^I C \quad . \tag{4.276}$$

The G parities of the pion and the eta follow from Eqs. (4.276) and (4.261)

$$\eta_g(\pi) = -1 \quad , \qquad \eta_g(\eta^0) = +1 \quad . \tag{4.277}$$

The G parity is the same for all three charge states of the pion. Next we consider a system of n pions. Unlike C, G commutes with the charge operator Q; since each pion is in a definite charge state, the G parity of the system is the product

$$\eta_g(n\pi) = (-1)^n \quad . \tag{4.278}$$

The G parity for the $p\bar{p}$ system, and hence for any nucleon-antinucleon system, follows from Eqs. (4.264) and (4.275) as

$$\eta_g(N\bar{N}) = (-1)^{\ell+s+I} \quad . \tag{4.279}$$

Equations (4.273), (4.278), and (4.279) lead to many selection rules. We give three examples here.

(a) Under the strong interaction, an even number of pions cannot transform into an odd number. This selection rule is similar to Furry's theorem for photons.

(b) The η^o has G-parity +1, but decays into three pions. This decay is forbidden by conservation of G parity; why does it occur? It turns out that the transition $\eta^o \to 2\pi$ is forbidden by other selection rules; since G parity forbids the decay $\eta^o \to 3\pi$, this observed mode must proceed via the electromagnetic interaction. Indeed, the observed width of the η^o, (2.63 ± 0.58) keV, is much smaller than the widths of mesons that decay hadronically. Moreover, the mode $\eta^o \to 2\gamma$ competes effectively with that into three pions, giving another indication that the electromagnetic interaction is responsible for the eta decay.

(c) For decays $N\bar{N} \to n\pi$, Eqs. (4.278) and (4.279) show that states with even $\ell+s+I$ can decay hadronically only into an even number of pions.

4.8 TCP

A. TCP Conjugation and TCP Invariance

So far, we have treated the operations T, C, and P individually, but combinations, such as CP or CT, can be useful. The most important combined operation is TCP conjugation*, usually denoted by Θ, and given by

$$\Theta = TCP \quad . \tag{4.280}$$

The operator Θ changes a particle with momentum $\vec{p}$ and spin $\vec{s}$ into an antiparticle with the same momentum, but opposite spin:

$$\Theta|a,\vec{p},s,m\rangle = \eta_\Theta|-a,\vec{p},s,-m\rangle \quad . \tag{4.281}$$

Here a stands for all additive quantum numbers, A, Y, q, L, and L_μ, and η_Θ is a phase factor. Since C and P are unitary, but T is antiunitary, Θ is also antiunitary and we get with Eq. (4.188)

$$\langle\Theta\psi|\Theta\Phi\rangle = \langle\Phi|\psi\rangle \quad ; \tag{4.282}$$

---

*Yes, Virginia, there exists an "official" ordering of T, C, and P. There are 3! possibilities and nearly all appear in the literature, but Lüders and Zumino checked that their choice agreed with the name of a well-advertised gasoline additive. (G. Lüders, Physikalische Blätter 22, 421 (1966).

the properties of the Θ-reversed states are given
by Eq. (4.281). Θ - conjugation conserves the norm
of a state. Matrix elements of an operator σ
transform in analogy to Eq. (4.190);

$$\langle \Theta\psi | \Theta\sigma\Theta^{-1} | \Theta\Phi \rangle = \langle \psi | \sigma | \Phi \rangle *$$

$$= \langle \Phi | \sigma^\dagger | \psi \rangle \quad .$$

(4.283)

A system is <u>invariant under Θ conjugation</u>
if its Hamiltonian commutes with Θ,

$$[\Theta, H] = 0 \quad .$$

(4.284)

In terms of the Hamiltonian density, Eq. (3.199),
we write the invariance condition as

$$\Theta \, \mathcal{H}(t, \vec{x}) \Theta^{-1} = \mathcal{H}(-t, -\vec{x}) \quad .$$

(4.285)

With Eqs. (4.283) and (4.284), we obtain

$$\langle \Theta\Psi | H | \Theta\Phi \rangle = \langle \Psi | H | \Phi \rangle * \quad .$$

(4.286)

The behavior of the S matrix under Θ,

$$\Theta \, S \, \Theta^{-1} = S^\dagger \quad ,$$

(4.287)

parallels that of S under T, as given in Eq.
(4.219). Eq. (4.283) then yields

$$\langle \Theta\Phi | S | \Theta\psi \rangle = \langle \psi | S | \Phi \rangle \quad ,$$

(4.288)

Equations (4.284) to (4.288) assume Θ - invariance. Of course, only experiments can decide whether nature is Θ invariant and in this respect Θ is on an equal footing with T, C, and P and all other symmetries. There exists, however, a profound difference with regard to the theoretical situation. At present it is unknown why, for instance, the hadronic and electromagnetic, but not the weak, interactions conserve parity. It is as easy to write down interactions that violate parity as it is to find some that do not. The situation for Θ is different; as we will discuss in the next subsection, Eq. (4.284) must be valid for a very wide class of theories. In fact, it is difficult to write down interactions or field equations that do not conserve Θ.

The importance of studying combined operations was first observed by Schwinger and by Lüders;[156] the correct form of a TCP theorem was then suggested by Zumino to Lüders.[157] Pauli[158] and Bell[159] generalized the earlier work to show that a Lorentz invariant theory with the normal spin-statistics connection is invariant under Θ. Pauli's proof was simplified by Lüders,[160] and Jost finally demon-

156. J. Schwinger, Phys. Rev. <u>82</u>, 914 (1951); <u>91</u>, 713 (1951). G. Lüders, Z. Physik <u>133</u>, 325 (1952).
157. G. Lüders, Kgl. Danske Vidensk. Selsk. Mat.-Fys. Medd. <u>28</u>, No. 5 (1954).
158. W. Pauli, in <u>Niels Bohr and the Development of Physics</u> (Pergamon Press, London, 1955).
159. J. S. Bell, Proc. Roy. Soc. London A <u>231</u>, 479 (1955).
160. G. Lüders, Ann. Phys. (New York) <u>2</u>, 1 (1957).

strated the validity of the TCP or Schwinger-
Lüders-Pauli theorem under very weak assumptions.[161]

Before the discovery of parity non-conservation,
the TCP theorem appeared to be an interesting but
useless luxury. However, with the realization that
the individual operations T,C, and P can be violated
in some interactions, it became a cornerstone of
theoretical arguments.[162] We will discuss the
TCP theorem in Subsection B and experimental tests
in Subsection C.

B.   The TCP or Schwinger-Lüders-Pauli Theorem

The TCP theorem states that $\Theta \equiv TCP$ is a
conserved symmetry operation in any field theory
that satisfies the following four conditions:
(i)   The field equations are invariant under
        continuous Lorentz transformations.
(ii)   The field equations are local.
(iii)   The usual connection between spin and
        statistics holds.
(iv)   The Hamiltonian is Hermitian.
The first assumption shows that invariance under
inversions (T or P) is not required. The second
assumption means that all field quantities are
spinors or tensors of finite rank and that the
interaction is local and contains derivatives only

---

161.   R. Jost, Helv. Phys. Acta 30, 409 (1957).
162.   T. D. Lee, R. Oehme, and C. N. Yang, Phys.
       Rev. 106, 340 (1957).

of finite order. The third assumption states that field operators of a field with integer spin commute, those of a field with half-integer spin anticommute. Finally we note that the order of T, C, and P in Θ is irrelevant.

General proofs of the TCP theorem, both in the Hamiltonian formalism and in axiomatic field theory can be found in the references 160 and 161, and in books and reviews listed at the end of the chapter. Here we demonstrate the TCP theorem with one example.

If the Hamiltonian density $\mathcal{H}(t,\vec{x})$ is a Hermitian Lorentz invariant linear combination of products of fields and of finite-order derivatives of fields, it satisfies the conditions (i), (ii), and (iv). In order to satisfy also conditions (iii), the products must be symmetrized with respect to Bose fields and antisymmetrized with respect to Fermion fields.[*] (An example of antisymmetrization is given in Eq. (3.191).) As an example we take

$$\mathcal{H}(t,\vec{x}) \equiv \mathcal{H}(x) = g\Phi(x)\,\psi^{\dagger}(x) + \text{H.c.} \quad , (4.289)$$

where Φ and ψ are field operators for spinless bosons, g is a coupling constant, and H.c. denotes the Hermitian conjugate added to insure Hermiticity. Symmetrization gives

---

[*]The same result can be achieved by using Wick's time-ordered product (: :).

$$\mathcal{H}(x) = \tfrac{1}{2}g\Big(\Phi(x)\;\psi^{\dagger}(x) + \psi^{\dagger}(x)\;\Phi(x)\Big) + \text{H.c.}\qquad.$$

The transformation properties of Φ are given by Eqs. (4.139), (4.247), and Problem 4.61 as

$$P\Phi(t,\vec{x})P^{-1} = \eta_P\;\Phi(t,-\vec{x})\qquad,$$

$$P\Phi^{\dagger}(t,\vec{x})P^{-1} = \eta_P^*\Phi^{\dagger}(t,-\vec{x})\qquad,$$

$$C\Phi(t,\vec{x})C^{-1} = \eta_C\Phi^{\dagger}(t,\vec{x})\qquad,$$

$$C\Phi^{\dagger}(t,\vec{x})C^{-1} = \eta_C^*\Phi(t,\vec{x})\qquad,\qquad\qquad(4.290)$$

$$T\Phi(t,\vec{x})T^{-1} = \eta_T\;\Phi(-t,\vec{x}),$$

$$T\Phi^{\dagger}(t,\vec{x})T^{-1} = \eta_T^*\Phi^{\dagger}(-t,\vec{x})\qquad.$$

Analogous relations, with phases `η', hold for ψ. These transformations make no assumptions about invariances. With Eq. (4.290) we get (with $TgT^{-1} = g*$)

$$\Theta\;\mathcal{H}(x)\Theta^{-1} = \tfrac{1}{2}\eta_T^*\eta_C^*\eta_P^*\;\eta_T'\eta_C'\eta_P'\;g*\Big(\Phi^{\dagger}(-x)\;\psi(-x)$$

$$+\;\psi(-x)\;\Phi^{\dagger}(-x)\Big) + \text{H.c.}\qquad.$$

If we select the arbitrary phases such that for Φ and ψ

$$\eta_T\eta_C\eta_P = 1\qquad\qquad\qquad\qquad(4.291)$$

we find that $\mathcal{H}(x)$ satisfies Eq. (4.285). If we

use T, C, and P in a different permutation, the
condition on the arbitrary phases changes, but Θ
invariance still holds. The proof is unchanged if
derivatives appear in the expression for $\mathscr{H}$,
because differentiation with respect to x^μ commutes
with T, C, and P. For spinor fields, and for
boson fields with spin, the relations corresponding
to Eq. (4.290) must be established first and the
behavior of products under Θ can then be studied
similarly.

C. TCP Tests; Consequences[163,164)]

Because of the few and weak assumptions underlying the TCP theorem, experimental tests are
particularly important. If a breakdown were found,
it would lead to far-reaching revisions of present
ideas. We discuss some tests in the present subsection.

We consider first the mass m of a free stable
particle and the mass $\bar{m}$ of the corresponding antiparticle. If

$$H = H_h + H_\gamma + H_w$$

is the total Hamiltonian and if $|\Phi\rangle$ is the eigen-
state of the particle at rest, m and $\bar{m}$ are given by

---

163.   L. B. Okun, Comm. Nucl. Part. Phys. **2**, 116
       (1968).
164.   L. Wolfenstein, Nuovo Cimento **63A**, 269 (1969).

$$m = <\Phi|H|\Phi> \quad , \quad \bar{m} = <\Theta\Phi|H|\Theta\Phi> \quad . \qquad (4.292)$$

The fact that the antiparticle state $|\Theta\Phi>$ has spin $-\vec{J}$ is irrelevant because the mass of a free particle does not depend on its orientation. Eq. (4.286) shows that[162,165)]

$$m = \bar{m} \quad . \qquad (4.293)$$

TCP invariance guarantees that the masses of particle and antiparticle are the same to all orders in H. For particles such as the electron, the muon, and the pion, the equality (4.293) has been checked to an accuracy of between 10^{-3} and 10^{-4}. The best test comes from comparing the masses of the K^O and $\bar{K}^O$. We will see in Chapter 5 that

$$(m_{K^O} - m_{\bar{K}^O})/m_{K^O} \lesssim 10^{-14} \quad . \qquad (4.294)$$

This limit indicates that the ratio of the Θ-noninvariant to that of the Θ-invariant amplitude is less than $\sim 10^{-14}$ for H_h, less than $\sim 10^{-12}$ for H_γ, and less than about 10^{-8} for the hypercharge conserving part of H_w.

TCP invariance also requires the electromagnetic moments of particles and antiparticles to be equal in magnitude and opposite in sign. Experimentally, the magnetic dipole moments of negative

---

165.   G. Lüders and B. Zumino, Phys. Rev. 106, 385
       (1957).

and positive electrons, and of negative and positive
muons have been measured accurately. The magnitudes
of the dipole moments for leptons and antileptons
agree to better than 1 part in 10^5.[166]

Next we consider the total <u>lifetimes</u> of parti-
cles decaying from a state $|\alpha>$ to a state $|\beta>$
through the weak interaction H_w. With Eqs. (4.286)
and (4.207) we find for $\tau = 1/w_{\beta\alpha}$

$$\tau = \bar{\tau} \quad .$$ (4.295)

Particles and antiparticles that are stable under
H_h and H_γ have the same total lifetime to first
order in H_w.[162]   (Actually, the result holds under
less restrictive assumptions.[165]) Experimentally,
the lifetimes of a few particle-antiparticle pairs
have been determined accurately. Of particular
interest is the ratio for the charged kaons, since
the decay involves a change in hypercharge. The
result[167]

$$\tau(K^+)/\tau(K^-) = 1.0005 \pm 0.001 \quad ,$$

shows that the ratio of the Θ noninvariant to the
Θ invariant amplitude is less than 0.03.

---

166.   Review of Particle Properties, Particle Data
       Group, Physics Letters Vol. <u>50B</u>, No. 1,
       April 1974.
167.   F. Lobkowicz, A. C. Melissinos, Y. Nagashima,
       S. Tewksbury, H. von Briesen, Jr., and J. D.
       Fox, Phys. Rev. Letters <u>17</u>, 548 (1966).

The situation for <u>partial lifetimes</u> is more
complicated.[*] TCP invariance insures the equality
of partial decay rates only if there are no hadronic
or electromagnetic interactions in the final state
that connect different decay channels.[165,168]

Once TCP invariance is assumed, a number of
<u>consequences</u> follows. We discuss here two examples.
The first consequence is obvious: If in a certain
interaction one of the operators T, C, or P is not
conserved, at least one other must also be violated.
We have already seen an example in the weak interaction, where both P and C fail. In Chapter 5, we
will encounter another example. The decays of the

---

*The notion of a <u>partial lifetime</u> is sometimes
misinterpreted. A state (particle or nucleus)
has only <u>one</u> lifetime, the total lifetime τ.
If $N(0)$ particles are created at time $t = 0$,
the number that survive at time t is given
by $N(t) = N(0) \exp(-t/\tau) = N(0) \exp(-\Gamma t)$,
regardless of how many decay modes exist.
$\Gamma = 1/\tau$ is the total decay constant. If more
than one decay mode exists, partial decay rates
Γ_i are introduced, with $\sum_i \Gamma_i = \Gamma$. A fraction
(Γ_i/Γ) of all particles decays into mode i;
this fraction is independent of time and all
modes decay with the same lifetime τ. Partial
lifetimes τ_i are now <u>defined</u> as $\tau_i = 1/\Gamma_i$.
A fraction (τ/τ_i) of all particles decays into
mode i, but τ_i does not represent an observed
lifetime.

168.   S. Okubo, Phys. Rev. <u>109</u>, 984 (1958). Y. Ueda
       and S. Okubo, Phys. Rev. <u>139</u>, B1591 (1965).
       F. Strocchi, Phys. Rev. Letters <u>19</u>, 1456
       (1967).

neutral kaons show that CP is violated; the TCP
theorem then predicts that T must be violated
also, but so far it is unclear how the violation
occurs.

TCP invariance also leads to a classification
of particle multiplets.[169]   As one example, we
consider self-conjugate isospin multiplets, where
particles and anti-particles belong to the same
multiplet. The $I = 1$ multiplet (π^+, π^o, π^-), for
instance, is identical to its antimultiplet
(π^-, π^o, π^+), whereas the $I = \frac{1}{2}$ multiplets (K^+, K^o)
and $(\overline{K}^o, K^-)$ are distinct. Carruthers has shown
that self-conjugate bosons of half-integer isospin
lead to a violation of causality and locality.[170]
Zumino and Zwanziger,[169] and Steinman[171] later
demonstrated that Carruthers' theorem follows
from TCP invariance for bosons and fermions. We
will derive the theorem here for bosons. To do so,
we first note that the phase choice $\eta_T\, \eta_C\, \eta_P = 1$,
given in Eq. (4.291), leads to Θ-invariance. With
Eq. (4.290) then follows that for bosons

$$\Theta^2 \Phi(t,\vec{x}) \Theta^{-2} = +\Phi(t,\vec{x}) \quad . \qquad (4.296)$$

Similarly we can obtain for fermions

---

169. B. Zumino and D. Zwanziger, Phys. Rev. 164,
      1959 (1967).
170. P. Carruthers, Phys. Rev. Letters 18, 353
      (1967).
171. O. Steinmann, Phys. Letters 25B, 234 (1967).

$$\Theta^2 \psi(t,\vec{x}) \Theta^{-2} = -\psi(t,\vec{x}) \quad . \tag{4.297}$$

The eigenvalues of Θ^2 thus are (c.f. Eq. (4.204))

$$\Theta^2 = (-1)^{2s} \quad , \tag{4.298}$$

where s is the spin of the particle.[172]

A self-conjugate isospin multiplet can exist only for particles with A = Y = 0. For a member of such a multiplet, Eq. (4.281) can be written as

$$\Theta|I,I_3,s,m\rangle = \eta_\Theta|I,-I_3,s,-m\rangle \quad . \tag{4.299}$$

For bosons, Eq. (4.298) gives $\Theta^2 = 1$; applying Θ a second time to Eq. (4.299) yields with $\Theta\eta\Theta^{-1} = \eta^*$

$$|I,I_3,s,m\rangle = \eta_\Theta^* \Theta|I,-I_3,s,-m\rangle \quad . \tag{4.300}$$

The state $-I_3$ can also be obtained from the state I_3 by a rotation in isospin space about the 2-axis by the angle π; Equations (4.65), (4.70), and (B.15) together give for the action of the isospin rotation operator $U(\pi) = e^{-i\pi I_2}$:

---

172.   T. D. Lee and G. C. Wick, Phys. Rev. 148, 1385 (1966).

$$U(\pi)|I,I_3,s,m\rangle$$

$$= \sum_{I_3'} D^I_{I_3'I_3}(0,\pi,0)|I,I_3',s,m\rangle$$

(4.301)

$$= (-1)^{I-I_3}|I,-I_3,s,m\rangle \quad .$$

With Eqs. (4.299) - (4.301), $n_\Theta^* n_\Theta = 1$, and $[U,\Theta] = 0$, it follows that

$$|I,I_3,s,m\rangle = n_\Theta^* \Theta |I,-I_3,s,-m\rangle$$

$$= n_\Theta^* (-1)^{I_3-I} \Theta U|I,I_3,s,-m\rangle$$

$$= n_\Theta^* (-1)^{I_3-I} U\Theta |I,I_3,s,-m\rangle$$

$$= n_\Theta^* n_\Theta (-1)^{I_3-I} U|I,-I_3,s,m\rangle$$

$$= (-1)^{I_3-I} (-1)^{I+I_3} |I,I_3,s,m\rangle \quad .$$

Thus $(-1)^{2I_3} = 1$; I_3 and hence I must be integer and Carruthers' theorem has been proved for bosons. The theorem agrees with the empirical result expressed by the Gell-Mann-Nishijima relation, Eq. (4.108): For $Y = A = 0$, the relation gives $I_3 = Q/e$; the observation that only particles with integer values of Q/e exist then limits I to integers also.

PROBLEMS

4.1   Derive Eq. (4.1).

4.2   Devise experiments that distinguish between
       additive and multiplicative quantum numbers.
       What experiments indicate that the electric
       charge obeys an additive rather than a multi-
       plicative conservation law?

4.3   Derive the expressions for the Schrödinger,
       the Klein-Gordon, and the Dirac currents by
       using Eq. (4.24) and the proper Lagrangians.

4.4   Assume that a Lagrangian density $\mathcal{L}$ is not
       invariant under a transformation of the type
       (4.21). Relate the change in $\mathcal{L}$ to the diver-
       gence of the associated current.

4.5   It has been postulated that the magnetic
       fields of the earth and the sun are caused
       by a small difference δe between the magni-
       tudes of the charges of the electron and the
       proton. (A. Einstein, reported in A. Piccard
       and E. Kessler, Arch. Sci. Phys. et Nat.
       (Genève) $\underline{7}$, 340 (1925).) R.Lyttleton and H. Bondi
       (Proc. Roy. Soc. $\underline{A252}$, 313 (1959)) have
       conjectured that the expansion of the uni-
       verse could be due to such a difference.
       Estimate the fractional difference $\delta e/e$ that
       would lead to the observed magnetic fields
       and the observed rate of expansion. Do
       present experimental data rule out these
       hypotheses?

4.6   Assume that the entire heat flow at the
       surface of the earth is caused by the decay
       of nucleons. Estimate the limit on the
       nucleon lifetime.

4.7   Devise an experiment that would allow an
       improvement of the limits quoted in Table
       4.1 by two orders of magnitude.

4.8   a)   Formulate the gauge invariance that leads
            to the conservation law of lepton number.
      b)   If the gauge invariance were of the type
            Eq. (4.27), what measurable consequences
            would ensue?

4.9   Estimate the magnitude of the cross section
      for antineutrino capture

$$\bar{\nu}\, p \rightarrow e^+ n$$

      for neutrinos of energy 3 MeV and 1000 MeV.
      (Use the Hamiltonian given in Eq. (1.4)).

4.10 Double beta decay is capable of giving infor-
      mation on lepton conservation.   Discuss
      experimental arrangements that can be used
      to search for such a decay.   What are the
      main problems?

4.11 How can electrons, muons, and pions of the
      same momentum be distinguished in a bubble
      chamber?   in a spark chamber?

4.12 Calculate the rate of decay for $\mu \rightarrow e\gamma$ if the
      matrix element of the interaction $j_\nu A^\nu$ is
      given by

$$\langle e| j_\nu A^\nu |\mu \rangle = \frac{1}{\sqrt{2q2E_{p'}E_p}}$$

$$\times \; \bar{u}_e(p') \; \sigma_{\nu\omega} \frac{q^\nu}{2m_\mu} F(q^2) \; u_\mu(p) \; \varepsilon^\omega$$

      where $F(q^2)$ is a form factor, $q = p'-p$, and
      ε^ω is the photon polarization vector.   Assume
      $F(q^2) = F(0) = 1$, and a muon at rest; for the
      radiation gauge $\varepsilon^\omega = (0, \vec{\varepsilon})$.   Compare the
      calculated rate to the measured muon lifetime.
      (The matrix element as given assumes a quan-
      tization volume V = 1; see Eq. (3.190).)

4.13 Discuss the evidence that indicates that the negative muon is a particle (i.e. has L = +1) and not an antiparticle. (The negative electron is defined to be a particle.)

4.14 What experiments could decide whether the lepton number is an additive or a multiplicative quantum number? [J. J. Amato et al., Phys. Rev. Letters 21, 1709 (1968).]

4.15 Kaons can be produced weakly. What is the threshold energy for kaon production

a) by p-p collisions with free protons,
b) from heavy nuclei (assume a Fermi momentum of 25 MeV/c),
c) in pion-nucleon collisions.
d) Assume a proton beam of 1 mA and a proton kinetic energy of about 1 GeV. How many kaons per sec can be expected?

4.16 Discuss reactions that determine the strangeness of the neutral cascade particle and of the antisigmas.

4.17 Discuss the interaction of K^+ and K^- with matter:

a) List reactions that have been observed.
b) Discuss the differences between K^+ and K^- interactions.
c) Kaons (+ and -) of momentum 5 GeV/c pass through a 30 cm liquid hydrogen target. Calculate the decrease in intensity for both types due to decay, scattering, and absorption.

4.18   Which of the following reactions are possible?
       If forbidden, state by what selection rule(s).
       If allowed, indicate through which interaction
       the reaction will proceed.

a) $\pi^+ n \rightarrow K^0 p$,

b) $\pi^- p \rightarrow \Sigma^- K^+$,

c) $\pi^- p \rightarrow p \; \pi^+ \pi^- K^- K^0$,

d) $e^- \mu^+ \rightarrow \bar{\nu}_e \nu_\mu$

e) $K^- p \rightarrow K^+ K^- \pi^+ \pi^- \pi^0$,

f) $\gamma \; n \rightarrow p \; \pi^- \pi^0$,

g) $\bar{\nu}_\mu p \rightarrow \mu^+ n$,

h) $K^+ p \rightarrow K^0 \pi^+ \pi^- p$,

i) $p \; \bar{p} \rightarrow \Sigma^- p \; \pi^+ \pi^-$,

j) $\nu_\mu n \rightarrow \overline{\Sigma^+} \Sigma^+ \pi^+ \pi^- \pi^0 p \; \mu^-$

k) $\bar{p} \; n \rightarrow \pi^+ \pi^+ \pi^- \pi^- \pi^- K^+ K^-$,

l) $\Xi^- p \rightarrow \Sigma^0 \Lambda^0$,

m) $\alpha \; {}^{12}C \rightarrow d \; {}^{14}N \, (J^P = 1^+)$,

n) $\alpha \; {}^{12}C \rightarrow d \; {}^{14}N \, (J^P = 0^+)$,

o) $d \; d \rightarrow {}^4He \; \gamma$.

4.19   Consider rotations about the x,y, and z axes.
       By means of a Taylor expansion show that
       J_x, J_y, and J_z correspond to the classical
       operators for angular momenta, i.e.,
       $J_z = xp_y - yp_x$.

4.20 Prove Eq. (4.61).

4.21 Prove Eq. (4.62).

4.22 Use the commutation relations for angular momentum and the knowledge that the magnetic quantum number m is bounded from below by m_{min} and from above by m_{max} to show that j can only take integral and half-integral values. Do not use Eq. (4.58); determine the constant $j(j+1)$.

4.23 Prove Eq. (4.80).

4.24 Derive Eqs. (4.82), (4.83).

4.25 Discuss the experimental evidence for charge independence of hadronic forces in nuclear and particle physics.

4.26 Calculate the ratio of cross section,

$$\sigma(p \ d \rightarrow \pi^+ \ {}^3H)/\sigma(p \ d \rightarrow \pi^0 \ {}^3He),$$

assuming charge independence and neglecting electromagnetic corrections.

4.27 Give an example each for isospin singlet, doublet, triplet, quadruplet, and quintuplet. Justify in each case why you think that all members of the multiplet have been found. Compare the computed Coulomb energy differences with the experimentally observed values. How much do the energy differences calculated on the basis of the simple classical model deviate from the observed and from the correctly calculated values?

4.28 Calculate the following branching ratios:

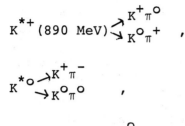

$$K^{*+} (890 \text{ MeV}) \begin{cases} \to K^+ \pi^0 \\ \to K^0 \pi^+ \end{cases} ,$$

$$K^{*0} \begin{cases} \to K^+ \pi^- \\ \to K^0 \pi^0 \end{cases} ,$$

$$N^{*+} (1236) \begin{cases} \to p \ \pi^0 \\ \to n \ \pi^+ \end{cases} .$$

4.29 Discuss selection rules and reactions that test the following isospin assignments:

$$\Lambda(I = 0), \ \Sigma(I = 1), \ \Xi(I = \tfrac{1}{2}).$$

4.30 Assume validity of the Gell-Mann-Nishijima relation, Eq. (4.108). List and identify all baryons predicted by the relation if you do not introduce multiply charged particles.

4.31 Justify the entries in Table 4.4.

4.32 Construct a table showing the allowed angular momentum states for a system of two pions, with isospin I = 0,1,2.

4.33 Repeat problem 4.32 for three pions.

4.34 Discuss the description of the nucleon field operators including isospin.

4.35 Show that $\vec{I}$, Eq. (4.113), satisfies the commutation relations characteristic for an angular momentum.

4.36 Verify Eq. (4.114).

4.37 If two neutrons in a $^1S_0$ state were bound with the same energy as the deuteron (2.2 MeV), estimate the binding energy of the np (I = 1) and pp $^1S_0$ states. Take into account a) Coulomb effects, b) $m_n - m_p$. What other effects might be important?

4.38   Compute $[I_3, \Phi_1 \pm i\Phi_2]$; use the result to show
       that $|\pi^+> = \Phi_+|0>$ is a state with charge +1.

4.39   The non-leptonic weak interaction satisfies
       an approximate isospin selection rule
       $\Delta I = \pm\frac{1}{2}$. $(I_f = I_i \pm \frac{1}{2})$. Which of the
       following decay modes are allowed:
       $K^+ \rightarrow \pi^+\pi^0$, $K^0 \rightarrow \pi^0\pi^0$, $\overline{K^0} \rightarrow \pi^+\pi^-$, $K^- \rightarrow \pi^-\pi^0$?

4.40   Show that the intrinsic parity of a fermion-
       antifermion pair depends on their relative
       orbital angular momentum ℓ and is given by
       $(-1)^{\ell+1}$.

4.41   Describe the experimental determination of
       the spin of the charged pions.

4.42   Would the assignment $\eta_p = -\eta_n = \eta_\Lambda = 1$ lead
       to observable effects that are different from
       the ones predicted by the assignment Eq.
       (4.158)?

4.43   a) Discuss the experimental determination
           of the relative parities of K and Σ.
       b) How can the parity of the Ξ be found?
           Relative to what particles(s) is the
           parity determined?

4.44   Find an explicit expression in field theory
       for the operator P for a complex Klein-Gordon
       field and show that it has the properties
       appropriate for a parity operator.

4.45   a) Find the expansion for a Hermitian free
           Klein-Gordon field in terms of creation
           and destruction operators of a given
           angular momentum, $a^\dagger_{k\ell m}$ and $a_{k\ell m}$. Use the
           definition

$$a_{k\ell m} = \text{const } i^\ell \int d\Omega_k \, k \, Y^m_\ell \, a_k \quad .$$

       b) Determine the commutation relations
           between the various combinations of the

$a_{k\ell m}$ and $a_{k\ell m}^{\dagger}$.

c) Find the Hamiltonian in terms of the $a_{k\ell m}$.

d) Show that in an angular momentum representation

$$P = \exp\left\{-i\pi \sum_{k\ell m} \ell a_{k\ell m}^{\dagger} a_{k\ell m}\right\}$$

$$\text{or} \quad \exp\left\{-i\pi \sum_{k\ell m} (\ell+1) a_{k\ell m}^{\dagger} a_{k\ell m}\right\}.$$

e) Show that $H' = \int d^3x\, \rho(\vec{x})\Phi$ is even

under P (nonrelativistic scalar coupling),

and $H'' = \int d^3x\, \rho(\vec{x})\vec{\sigma}\cdot\vec{\nabla}\Phi$ is odd under P

(nonrelativistic pseudoscalar coupling).
Assume $P\,\Phi\,P^{-1} = +\,\Phi$; $\rho(\vec{x})$ is a scalar
source function, invariant under space
reflection.

4.46 a) Show that for a stationary system of
well-defined parity, the electric dipole
moment must be zero.

b) Show that for a stationary system, which
is an eigenfunction of a time-reversal
even Hamiltonian, H, the electric dipole
moment must be zero.

c) If $PHP^{-1} = H$, does it necessarily follow
that every stationary state must have a
definite parity?

d) Is the polarization vector of a particle
a vector or a pseudovector? Explain.

4.47 Møller and Bhabha scattering can be used to
determine the helicity of electrons and

positrons, respectively.

a) Sketch the theory underlying the method.
b) Sketch a simple set-up.
c) The helicity of positrons of kinetic
    energy 2 MeV is measured.
      What is the optimum lab angle between
    the scattered positron and the knocked-
    out electron?
d) How has the helicity of muons, resulting
    from the decay of pions, been determined?
    What are the values of the helicity for
    the positive and for the negative muon?

4.48 The neutral pion field Φ could interact with
      the electromagnetic field through terms of
      the form

$$g \int \Phi(\vec{x}) \left\{ \vec{E}^2(\vec{x}) - \vec{H}^2(\vec{x}) \right\} d^3x$$

or

$$g \int \Phi(\vec{x}) \vec{E}(\vec{x}) \cdot \vec{H}(\vec{x}) \, d^3x \quad ,$$

where g is a coupling constant. Assume invariance of the pion-electromagnetic field
interaction under space inversion. Discuss
the transformation properties of the two
terms under the parity operation. Which
term would you expect to find realized in
nature?

4.49 Verify Eq. (4.190).

4.50 Compute $\langle\vec{\sigma}\rangle$ for the original and the time-
      reversed spinor in Eq. (4.194) and show that
      time-reversal reverses the spin.

4.51 Show that

$$\langle T\phi | \psi \rangle = \pm \langle \phi | T\psi \rangle^* = \pm \langle T\psi | \phi \rangle$$

where the plus sign applies to systems with
integer, the minus sign to systems with
half-integer spin.

4.52   Verify Eq. (4.197).

4.53   Derive Eq. (4.199).

4.54   a)   Show that if $\psi(t)$ is a solution of the
              Dirac equation, so is
              $$T \psi(t) = \gamma^1 \gamma^3 K \psi(-t).$$
        b)   Show that the Dirac Hamiltonian commutes
              with T.
        c)   Discuss the effect of T on a Dirac plane
              wave solution.

4.55   a)   Show the equivalence of the Schwinger
              and Wigner time reversal operations for
              Hermitian operators.
        b)   Show that the commutator $[x, p_x] = i$ is
              invariant under Wigner and Schwinger
              time reversal.

4.56   Show that the helicity is invariant under
        time reversal.

4.57   Show that the eigenvalues of the operator
        T^2 (Eq. 4.203) do not depend on the choice
        of representation.

4.58   Discuss Kramers' degeneracy for a para-
        magnetic substance in an external electric
        field.

4.59   Show that the matrix elements of an arbitrary
        vector operator (between states of well-
        defined angular momentum and z-component of
        the angular momentum) are proportional to the
        corresponding matrix elements of the angular
        momentum operator $\vec{J}$.

4.60   Discuss the production of polarized neutrons
        by reflection from magnetized mirrors. Is
        the spin direction indicated in Fig. 4.31
        correct?

4.61   Discuss time reversal in field theory, using
        as an example a non-Hermitian Klein-Gordon
        field, (Eq. (3.115)). Use as definition of
        time reversal

$$T \, \Phi(t,\vec{x}) \, T^{-1} = \eta_T \, \Phi(-t,\vec{x}) \quad .$$

Find the result of T on the expansion coefficients and on a one-particle and a one-antiparticle state.

4.62 Time reversal tests with electromagnetic transitions can be "spoiled" by effects that simulate terms of the type of Eq. (4.233). (E. M. Henley and B. A. Jacobsohn, Phys. Rev. Letters 16, 706 (1966). J. P. Hannon and G. T. Trammell, Phys. Rev. Letters 21, 726 (1968).)

    a) Draw some Feynman diagrams for such higher-order effects.
    b) Estimate the order-of-magnitude of the spoiling.

4.63 a) In principle, the decay $\eta \to \mu^+\mu^-$ can be used to search for a neutral weak current through a parity-violating effect. How would you search for this effect experimentally?
    b) The effect sought occurs through an interference of the weak and electromagnetic interaction. What would you expect for its order of magnitude?
    c) Show that if time reversal symmetry is valid, the effect sought must vanish.

4.64 Show that the transformation Eq. (4.258) represents charge conjugation in Dirac theory in the standard representation.

4.65 Show that the Lagrangians given in Table 3.1 are invariant under C.

4.66 Construct an explicit expression for the operator C for a non-Hermitian Klein-Gordon field in terms of the creation and annihilation operators.

4.67 Verify Eq. (4.264).

4.68 Why would detection of a charge symmetry (i.e. a different momentum spectrum for π^+ and π^-) in the decay $\eta^0 \to \pi^+\pi^-\pi^0$ indicate violation of C invariance? Discuss the experiments performed to search for such an asymmetry and the limits obtained.

4.69 Prove Eq. (4.270).

4.70 Find the relation of the operator $\exp(-i \pi I_2)$ and charge symmetry.

4.71 What selection rule forbids the decay $\eta^0 \to 2\pi$?

4.72 There is a resonance in the 3-pion system with mass $\sim$ 780 MeV called "ω," having quantum numbers A = 0, Y = 0, J = 1, I = 0, G = -1, P = -1. The width of this resonance is $\sim$ 10 MeV.

   a) Give an estimate for the minimum life-time of this state.
   b) The dominant decay mode of the ω is $\omega \to \pi^+\pi^-\pi^0$. Is this a weak decay?
   c) Would you expect the following reactions to occur, assuming that the above quantum numbers are "good?" If so, give a rough estimate of the branching ratio. If not, why not?

$$\omega \to \gamma \, \pi^0 \qquad \omega \to \pi^+\pi^-$$

   d) Calculate the threshold kinetic energy for the production of the ω in the reaction $K^-p \to \omega^0\Lambda^0$.

4.73 In the capture of antiprotons by protons, the annihilation takes place with 99% probability from the states $^1S_0$ and $^3S_1$.

   Construct a table that shows which of the annihilation modes $p\bar{p} \to n\pi$, $n \leq 4$ are allowed for the four possible initial states.

4.74 Repeat problem 4.73 for decays $p\bar{p} \to \pi \, \omega$, $p\bar{p} \to \pi \, \eta$.

4.75 Find the phase factor conditions corresponding to Eq. (4.291) if $\Theta' = PCT$ and if $\Theta'' = CTP$.

4.76 Assume that ψ describes a spin $\frac{1}{2}$ field. Find the transformation properties under T, C, P and $\Theta = TCP$ for

a) $\bar{\psi} \psi$

b) $\bar{\psi}_1 \gamma^\mu (1+\gamma^5) \psi_2 \ \bar{\psi}_3 \gamma_\mu (1+\gamma^5) \psi_4$.

4.77 Show that the equal magnitude and opposite sign of the charge and the magnetic moment of an antiparticle compared to its particle requires only invariance under TCP, but not necessarily under C.

4.78 Show that TCP invariance does not guarantee equality of the partial decay rates for

$$\Sigma^+ \to p \ \pi^0$$

$$\overline{\Sigma^+} \to \bar{p} \ \pi^0$$

4.79 Show that TCP invariance does not guarantee equality of the decay rates for

$$K^+ \to \pi^+ \pi^0 \gamma \ ,$$

$$K^- \to \pi^- \pi^0 \gamma \ .$$

Establish a relation among the various decay modes of the positive and negative kaon that does follow from TCP invariance.

4.80 Show that TCP invariance can hold, but C invariance must be violated if the muons from the decay

$$\pi^+ \to \mu^+ \nu \ , \quad \pi^- \to \mu^- \bar{\nu}$$

have nonvanishing helicity and if the helicity of the μ^+ is opposite to that of the μ^-.

4.81 Derive Eq. (4.297).

4.82 Prove Carruthers' theorem for fermions.

ADDITIONAL REFERENCES

A beautiful introduction into the subject of
symmetries is given by Hermann Weyl in his book
Symmetry (Princeton University Press, Princeton,
New Jersey, 1952). A short history of some discoveries, particularly of parity nonconservation,
is presented by C. N. Yang, Elementary Particles
(Princeton University Press, Princeton, New Jersey,
1962). E. P. Wigner provides an exciting account
of the present situation in "Violations of Symmetry
in Physics," Scientific American 213, No. 6, p. 38
(1965).

The level of field theory that we have pre-
sented in Chapter 3 and used in Chapter 4 can be
characterized as "stone age." A more sophisti-
cated and complete approach exists; it leads for
instance to a very satisfactory proof of the TCP
theorem. This approach is described in two books:
R. F. Streater and A. S. Wightman, PCT, Spin and
Statistics, and All That (W. A. Benjamin, Inc.,
Reading, Mass., 1964); Res. Jost, The General
Theory of Quantized Fields, Boulder Lectures in
Applied Mathematics (American Mathematical Society,
Providence, Rhode Island, 1966), Vol. 4.

More details on the symmetries treated in the
present chapter can be found in many reviews; we

list a few here: G. C. Wick, "Invariance Principles of Nuclear Physics," Ann. Rev. Nucl. Sci. <u>8</u>,
1 (1958).

N. Kemmer, J. C. Polkinghorne, and D. L.
Pursey, "Invariance in Elementary Particle Physics,"
Reports Progr. Physics <u>22</u>, 368 (1959).

G. Grawert, G. Lüders, and H. Rollnik, "The
TCP Theorem and its Applications," Fortschritte
Physik <u>7</u>, 291 (1959).

M. I. Shirokov, "Tests of PC and PCT Invariance in Decay Processes," Soviet Physics Uspekhi
<u>5</u>, 913 (1963).

G. C. Wick, "Group Theory, Invariance Princi-
ples, Symmetries," in <u>High Energy Physics</u>, C. de-
Witt and M. Jacob, eds. (Gordon and Breach, New
York, 1965).

T. D. Lee and C. S. Wu, "Weak Interactions,"
Ann. Rev. Nuclear Sci. <u>15</u>, 381 (1965); <u>16</u>, 471
(1966).

E. M. Henley, "Parity and Time-Reversal
Invariance in Nuclear Physics," Ann. Rev. Nucl.
Sci. <u>19</u>, 367 (1969).

W. D. Hamilton, "Parity Violation in Electromagnetic and Strong Interaction Processes," Progr.
Nucl. Phys. <u>10</u>, 1 (1969).

Useful for many applications are <u>rotations in
ordinary space</u> and the <u>quantum theory of angular
momentum</u>, together with the corresponding gymnastics.
For a first introduction, we recommend the texts
by K. Gottfried, A. Messiah, and G. Baym. The
subject is treated in more detail in the following

monographs: M. E. Rose, Elementary Theory of
Angular Momentum (John Wiley, New York, 1957).
A. R. Edmonds, Angular Momentum in Quantum Mech-
anics (Princeton University Press, Princeton,
New Jersey, 1957). U. Fano and G. Racah, Irredu-
cible Tensorial Sets (Academic Press, New York,
1959). A. P. Yutsis, J. B. Levinson, and V. V.
Vanagas, Mathematical Apparatus of the Theory of
Angular Momentum (Israel Program for Scientific
Translations, Jerusalem, 1962). A. De-Shalit
and I. Talmi, Nuclear Shell Theory (Academic
Press, New York, 1963). D. M. Brink and G. R.
Satchler, Angular Momentum (Oxford University Press,
Oxford, 1968). J. D. Louck, Am. J. Phys. $\underline{38}$, 3
(1970). The most important papers are reprinted
in L. C. Biedenharn and H. Van Dam, Quantum Theory
of Angular Momentum (Academic Press, New York,
1965).

Unfortunately many of these books contain
misprints; before relying on any equation it is
best to check the equation and the basic definitions.

Isospin has become of great interest in
nuclear physics in the past few years because of
the discovery of the isobaric analog states. A
number of excellent reviews exist. In addition to
the ones quoted in the text, we mention first the
classic paper describing the early ideas: E. Feenberg and E. P. Wigner, Repts. Progr. in Physics $\underline{8}$,
274 (1941). The state of the subject before the
discovery of analog states is summarized in W. E.
Burcham, Progr. Nucl. Phys. $\underline{4}$, 171 (1955). The
theory of analog states is described by B. F.

Bayman, in <u>Nuclear Structure and Nuclear Reactions</u>,
M. Jean and R. A. Ricci, eds. (Academic Press, New
York, 1969), Proceedings of the International
School of Physics "Enrico Fermi." More details
concerning theory and experiments are given in
the Proceedings of the various Isospin-Meetings:
<u>Isobaric Spin in Nuclear Physics</u>, J. D. Fox and
D. Robson, eds. (Academic Press, New York, 1966).
<u>Isospin in Nuclear Physics</u>, D. H. Wilkinson, ed.
(North-Holland Publishing Co., Amsterdam, 1969).
<u>Nuclear Isospin</u>, J. D. Anderson, S. D. Bloom,
J. Cerny, and W. W. True, eds. (Academic Press,
New York, 1969).

Finally we note that a brief description of
many important books and papers on symmetries in
physics are collected in D. Park, "Resource
Letter SP-1 on Symmetry in Physics," Am. J.
Phys. <u>36</u>, 577 (1968).

Chapter FIVE   SUPERPOSITION AND INTERFERENCE

Interference effects occur in a variety of
nuclear and particle physics phenomena. An example
is the coherent diffraction scattering from a
nucleus which takes place if the wavelength of
the incident particle is comparable to, or smaller
than, the interparticle spacing. A particularly
dramatic example is the interference of the K^O
and $\overline{K^O}$. Since investigations of the neutral kaon
system have led to far-reaching conclusions, and
because the interference effects in this system
occur between only two states, we treat the
neutral kaons in some detail.

5.1   THE NEUTRAL KAONS

The only quantum number that distinguishes
the neutral kaon, K^O, from its antiparticle, $\overline{K^O}$,
is the hypercharge Y. The hypercharge of the K^O
is Y = 1, and that of the $\overline{K^O}$ is opposite, $Y(\overline{K^O}) =$
-1. The K^O and $\overline{K^O}$ appear as two distinct and

different particles in all experiments that involve only the strong and electromagnetic interactions $H_h + H_\gamma$, since they conserve hypercharge.
However, the weak interaction does not respect
hypercharge and can therefore induce transitions
between the two particles. Although the masses
of the K^O and $\overline{K^O}$ are identical by the TCP theorem,
the weak interactions of these particles will
give rise to an energy (mass) splitting and to
time-dependent transition rates between them.[1,2]
These effects have been observed in many experiments and they provide spectacular evidence for
interference between particle states. In the
following subsections we expound the most important
effects.

For a first introduction, see:

1. H. Frauenfelder and E. M. Henley, Subatomic
    Physics (Prentice-Hall, Inc., Englewood Cliffs,
    New Jersey, 1974), pp. 214-217.
2. R. P. Feynman, R. B. Leighton, and M. Sands,
    The Feynman Lectures on Physics (Addison-Wesley,
    Reading, Massachusetts, 1965), Vol. III,
    Chapters 8-11.

A. K_1^O and K_2^O

In 1954, Gell-Mann and Pais[3] pointed out that neutral kaons should have remarkable properties. Although the K^O and $\overline{K^O}$ are quite different with respect to hadronic and electromagnetic interactions, which conserve hypercharge, they are connected by the weak interactions. The K^O and $\overline{K^O}$ have the same weak decays, for instance $K^O \to 2\pi$, $\overline{K^O} \to 2\pi$, and are therefore linked by virtual weak interactions such as

$$K^O \rightleftarrows 2\pi \rightleftarrows \overline{K^O} \quad . \tag{5.1}$$

Since the connection is a weak one, perturbation theory can be used to describe the resulting effects; however, because the K^O and $\overline{K^O}$ are degenerate eigenstates of

$$H_s = H_h + H_\gamma \quad , \tag{5.2}$$

it is necessary to use degenerate perturbation theory. The transition described by Eq. (5.1) and shown in Fig. 5.1a is then a second order one in the perturbation, H_w.

The stationary states of the total Hamiltonian

$$H = H_s + H_w \tag{5.3}$$

---

3. M. Gell-Mann and A. Pais, Phys. Rev. <u>97</u>, 1387 (1955).

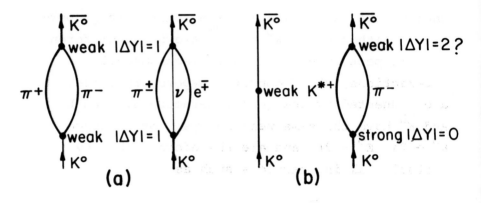

Fig. 5.1. Virtual transitions $K^O \rightarrow \overline{K^O}$.

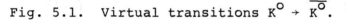

are determined by the solution of

$$H \ |\psi> = E \ |\psi> \quad .$$
(5.4)

As usual with degenerate perturbation theory, the lowest order solution of $|\psi>$ is a linear super-position of the states K^O and $\overline{K^O}$. The correct combination can be obtained by considering symmetries that are maintained in the presence of H_w. Although both P and C are violated in the weak interactions, the combined CP transformation is conserved to roughly 10^{-3}. If we neglect this small effect, then the total Hamiltonian commutes with the combined parity operator CP

$$[H,CP] = [H_s + H_w, CP] = 0 \quad . \qquad (5.5)$$

The eigenstates of H should thus be chosen to be simultaneous eigenstates of CP. With the conventional phase choice, we have

$$C \; |K^o\rangle = |\overline{K^o}\rangle \quad , \quad C \; |\overline{K^o}\rangle = |K^o\rangle \quad . \qquad (5.6)$$

The intrinsic parity of the kaons is negative (Subsection 4.5E),

$$P \; |K^o\rangle = -|K^o\rangle \quad , \quad P \; |\overline{K^o}\rangle = -|\overline{K^o}\rangle \quad , \qquad (5.7)$$

so that the effect of the combined parity operator is

$$CP \; |K^o\rangle = -|\overline{K^o}\rangle \quad , \quad CP \; |\overline{K^o}\rangle = -|K^o\rangle \quad . (5.8)$$

The first-order eigenstates $|\psi\rangle$ of the total Hamiltonian H are thus the symmetric and antisymmetric superpositions of the K^o and $\overline{K^o}$,

$$|K_1^o\rangle = \sqrt{\tfrac{1}{2}} \; \left\{|K^o\rangle - |\overline{K^o}\rangle\right\}$$

$$\qquad (5.9)$$

$$|K_2^o\rangle = \sqrt{\tfrac{1}{2}} \; \left\{|K^o\rangle + |\overline{K^o}\rangle\right\} \quad .$$

with

$$CP \; |K_1^o> \; = \; +|K_1^o> \quad , \quad CP \; |K_2^o> \; = \; -|K_2^o> \quad . \quad (5.10)$$

The K_1^o has a combined parity η_{CP} of +1, the K_2^o one of -1.* If the states $|K^o>$ and $|\overline{K^o}>$ are orthonormal, then so are the states $|K_1^o>$ and $|K_2^o>$. The masses of the K_1^o and K_2^o, which correspond to the eigenstates E of Eq. (5.4), will not be identical, unlike those of the K^o and $\overline{K^o}$. We will discuss this mass splitting in Subsection C. In addition, if a K^o is produced by a hadronic inter-action, for instance $\pi^- p \rightarrow K^o \Lambda^o$, then the weak interactions will induce oscillatory transitions between the K^o and $\overline{K^o}$ through reactions such as those given by (5.1). At the time of production, t = 0, the K particle is in a well-defined state of hypercharge, Y = 1, but not in an eigenstate of CP. Since the eigenstates of the total Hamil-tonian are linear combinations of the K^o and $\overline{K^o}$, as given by Eqs. (5.9), time-dependent transitions occur between the states of Y = 1 and Y = -1. These oscillations can be and have been observed, and we will discuss them in Subsections D and E. Here we treat yet another manifestation of the existence of the states $|K_1^o>$ and $|K_2^o>$, namely the occurrence of a long-lived neutral kaon.

---

*The freedom allowed by the arbitrary phases in the definitions of C and P has led to many dif-ferent ways of writing the linear combinations (5.9). The observable consequences are un-changed by the phase choice.

Kaons have numerous weak decay modes. For the neutral kaons one of the principal decay modes is into two or three pions. The kaon spin is zero, so that the total angular momentum of the pions in the final state must also be zero. The relevant properties of the globally neutral two- and three-pion systems with spin 0 are given in Table 5.1.

Table 5.1 Parities of the neutral two- and three-pion states with zero total angular momentum. The angular momentum ℓ is defined in Fig. 4.20.

State	$\pi^+\pi^-, \pi^0\pi^0$	$\pi^+\pi^-\pi^0$	$\pi^0\pi^0\pi^0$
Parity η_P	+1	-1	-1
Charge parity η_C	+1	$(-1)^\ell$	+1
Combined parity $\eta_{CP} = \eta_P\eta_C$	+1	$(-1)^{\ell+1}$	-1

The angular momentum ℓ that appears in the table is defined in Fig. 4.20. As mentioned above, we here assume that the total Hamiltonian conserves CP; the decay of the K_1^0 and of the K_2^0 hence must also conserve CP. K_1^0, with eigenvalue +1 of CP, can decay into both two and three pions, but the K_2^0 with $\eta_{CP} = -1$ can decay only into three pions.

The kinetic energy available in the two-pion decay
mode is roughly 220 MeV, but it is only 90 MeV in
the three-pion mode. The mean life of the K_1^o,
which decays almost solely into two pions, is thus
expected to be much smaller than that of the K_2^o.
This follows from the larger density of states
[See Eq. (4.207)] available in the two-pion decay
mode.

It is interesting to follow the lifeline of
a neutral kaon. Consider for instance a K^o pro-
duced at t = 0 by the hadronic reaction $\pi^- p \rightarrow K^o \Lambda^o$.
At birth, the neutral kaon has a definite hyper-
charge, Y = 1; with Eq. (5.9) we write this
initial state as

$$|\psi(t = 0)> = |K^o>$$

(5.11)

$$= \sqrt{\frac{1}{2}} \left\{ |K_1^o> + |K_2^o> \right\} \quad .$$

If the particle suffers no encounters with matter,
then it will decay freely through a weak interaction. However, it is K_1^o and K_2^o that are eigenstates of H and have exponential decays with unique
lifetimes. The K^o will not decay with a distinct
lifetime, but will behave like a particle mixture
with two different lifetimes. According to Eq.
(5.11) only 1/2 of the K^o's produced by the
hadronic reaction will decay into two pions. As
pointed out by Gell-Mann and Pais, if the word
"particle" is to apply to an object with a unique

lifetime, then it does not apply to the K^O or $\overline{K^O}$ but only to the K_1^O or K_2^O.

B. Experimental Verification of the K_2^O

The prediction of Gell-Mann and Pais was soon followed by an experimental verification. Although some preliminary evidence for a kaon with a longer lifetime than the then normal one of 10^{-10} sec came from cosmic ray investigations,[4] these "anomalous" decays were not understood. A clear verification came from a Columbia-Brookhaven collaboration using a cloud chamber.[5] This cloud chamber was positioned at a distance of 6 m from the target, corresponding to approximately 100 mean lifetimes for the "normal" decay component. The investigation revealed many "V events" in the chamber which could only be understood by postulating a long-lived three-pion decay mode of the neutral kaon. Other experiments verified this conclusion. The mean life of the K_2^O is $T_2 = 51.8$ nsec, that of the K_1^O is only $T_1 = 88.6$ psec.

The observation of the K_1^O-K_2^O particles has another consequence: It shows that the masses

4. C. Franzinetti and G. Morpurgo, Nuovo Cimento, Supplemento 6, No. 2, Chapter 3 (1957).
5. K. Lande, E. T. Booth, J. Impeduglia, L. M. Lederman, and W. Chinowsky, Phys. Rev. 103, 1901 (1956); 105, 1925 (1957).

of the K^O and the $\overline{K^O}$ must be the same to within
the line width of the K_1^O; otherwise the interference effects would be smaller or absent. The
line width of the K_1^O is given by $\Gamma(K_1^O) = 1/\tau_1$ or
in conventional units

$$\Gamma(K_1^O) = \tau_1^{-1} \simeq 10^{-5} \text{ eV} \quad . \tag{5.12}$$

This limit for the mass difference K^O-$\overline{K^O}$ is the
strongest evidence for TCP conservation, as we
have stressed in Section 4.8.

C. The Mass Splitting $K_1^O - K_2^O$

Gell-Mann and Pais remarked in passing that
the K_1^O and the K_2^O should have slightly different
masses.[3] We discuss here some aspects of this
mass splitting and estimate its magnitude. If
TCP invariance holds, Eq. (4.293) shows that the
K^O and the $\overline{K^O}$ must have the same mass. This
argument does not apply to the K_1^O and K_2^O, because
the K_2^O is not the antiparticle of the K_1^O; indeed,
the energies of these states are split by the
perturbing interaction H_w of the total Hamiltonian,
Eq. (5.3).

To discuss the $K_1^O - K_2^O$ mass splitting, we
assume that we can first switch off H_w, and we
write for the eigenstates of the strong part of
the Hamiltonian, $H_s \equiv H_h + H_\gamma$,

$$H_s \; |K_o> \; = \; m_o \; |K_o> \quad ,$$

$$(5.13)$$

$$H_s \; |\overline{K}_o> \; = \; m_o \; |\overline{K}_o> \quad .$$

The particle K_o (with <u>subscript</u> o!) exists only on paper and in the minds of theoretical physicists; it is not identical to the real K^o. The linear combinations $|K_o^1> \; = \; \{|K_o> \; - \; |\overline{K}_o>\}/\sqrt{2}$ and $|K_o^2> \; = \; \{|K_o> \; + \; |\overline{K}_o>\}/\sqrt{2}$ (note that o is subscript) are also eigenstates of H_s, with the same mass m_o. The difference between $(K_o, \overline{K}_o)$ and (K_o^1, K_o^2) lies in the quantum numbers Y and CP: The first two are eigenstates of Y whereas the last two are eigenstates of CP. Now we switch on H_w. If the total Hamiltonian H conserves CP, we can take for its eigenstates those given by Eq. (5.9),

$$H \; |K_1^o> \; = \; m_1 |K_1^o> \quad , \quad H \; |K_2^o> \; = \; m_2 |K_2^o> \; . \quad (5.14)$$

To find the mass shift $\Delta m_1 \; = \; m_1 \; - \; m_o$ that results from switching on H_w, we use degenerate perturbation theory and write

$$\Delta m_1 \; = \; m_1 \; - \; m_o \; = \; <K_o^1|H_w|K_o^1>$$

$$(5.15)$$

$$+ \; {\sum_{\alpha}}' \; \frac{<K_o^1|H_w|\alpha><\alpha|H_w|K_o^1>}{m_o \; - \; E_\alpha} \quad ,$$

where K_o^1 is the <u>unperturbed</u> state. The sum

$\sum_{\alpha}'$ in the second-order term goes over all acces-

sible intermediate states of the system, except $E_\alpha = m_o$.

With $|K_o^1> = \sqrt{\frac{1}{2}}\left\{|K_o> - |\overline{K}_o>\right\}$, and with TCP

symmetry, the first-order term can be written as

$$(\Delta m_1)^1 = <K_o^1|H_w|K_o^1>$$

$$= \frac{1}{2}[<K_o|H_w|K_o> + <\overline{K}_o|H_w|\overline{K}_o>$$

$$(5.16)$$

$$- <K_o|H_w|\overline{K}_o> - <\overline{K_o}|H_w|K_o>]$$

$$= <K_o|H_w|K_o> - <K_o|H_w|\overline{K_o}> .$$

A similar term appears in the expression for $\Delta m_2 = m_2 - m_o$:

$$(\Delta m_2)^1 = <K_o|H_w|K_o> + <K_o|H_w|\overline{K}_o> . \qquad (5.17)$$

The first term gives rise to a shift of the center of mass of the two neutral kaons from m_o to $m(K^o) = m(\overline{K^o}) = m_o + <K_o|H_w|K_o>$. The shift is unobservable since it is the same for K_1^o and K_2^o. The second term in Eqs. (5.16) and (5.17) produces a mass splitting, $\pm <K_o|H_w|\overline{K}_o>$; this matrix element involves a $|\Delta Y| = 2$ transition, as shown in Fig. 5.1b. Equation (4.48) expresses the observation that such transitions are either absent or strongly

suppressed. The first-order splitting in Eq.
(5.15) is therefore either absent or strongly
reduced and the second-order term must be considered. Justification for the complete neglect of
the first-order term then comes from a comparison
between the measured and calculated mass splitting.

The intermediate states $|\alpha>$ in the second
term consist of virtual particles that satisfy
the relevant selection rules. Since the dominant
decay mode of the K_1^O is that into two pions, we
assume for simplicity that the dominant state $|\alpha>$
is that of two pions, and neglect other states to
obtain an order-of-magnitude estimate for Δm.
Because the state is virtual, the energy of the
pions is not restricted by energy conservation;
the uncertainty principle permits any energy from
$2m_\pi$ to infinity and only the energy m_O is forbidden. To evaluate the term, we change the sum
to an integral and write

$$\Delta m_1 \cong \mathcal{P} \int_{2m_\pi}^{\infty} dn \ \frac{|<E_\alpha|H_w|K_O^1>|^2}{m_O - E_\alpha} \qquad (5.18)$$

Here, $\mathcal{P}$ denotes the principal value, expressing
the fact that the point $E_\alpha = m_O$ has to be omitted
from the integration; n is the density of states.
To compute the integral, we must either know the
matrix element, or compare Δm_1 to some property
of K_1^O that contains the same matrix element. The
decay probability of K_1^O is given by the golden rule,

$$\frac{1}{\tau_1} = 2\pi |<2\pi|H_w|K_1^o>|^2 \frac{dn}{dE} \quad , \tag{5.19}$$

where dn/dE is the number of states per unit
energy (denoted by ρ_β in Eq. (4.207)). Despite
the fact that similar matrix elements appear in
the expressions for Δm_1 and $1/\tau_1$, they cannot be
compared without further assumptions. The two
pions in the final state of Eq. (5.19) are
real and energy conservation must hold
within the uncertainty $1/\tau_1$; we express this
constraint by saying that the matrix element is
evaluated "on the energy shell." In Eq. (5.18),
$|E_\alpha>$ is a virtual state and the corresponding
matrix element is "off the energy shell." The
situation is sketched in Fig. 5.2.

If we now make the crude assumption that the
matrix element $<E_\alpha|H_w|K_1^o>$ is independent of the
energy E_α over a sizable region about m_o and equal
to $<E_\alpha|H_w|K_o^1>$, we can replace the factor
$dn|<E_\alpha|H_w|K_o^1>|^2$ by $dE_\alpha/2\pi\tau_1$ and get

$$m_1 = \frac{1}{2\pi\tau_1} \mathcal{P}\int_{2m_\pi}^{\infty} \frac{dE_\alpha}{m_o - E_\alpha} \quad . \tag{5.20}$$

Since $1/(m_o - E_\alpha)$ is an odd function of $(m_o - E_\alpha)$, the
contribution from $2m_\pi$ to m_o cancels that from m_o
to $2(m_o - m_\pi)$ and we are left with an integral from
$2(m_o - m_\pi)$ to ∞. Since in this energy region E is
always greater than m_o, we get a negative energy

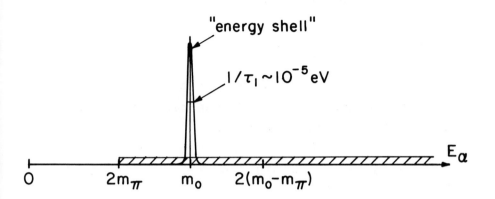

Fig. 5.2. Virtual pion pairs with energies in
           the hatched region contribute to the
           mass splitting Δm_1. The contribution
           to the decay $K_1^o \to 2\pi$ comes only from
           the energy shell, a region of width
           $1/\tau_1$, centered at m_o.

shift. However, because of the assumption of a
constant matrix element the integral diverges
logarithmically, and a <u>cut-off</u> <u>energy</u> E_{max}, must
be introduced. We then get from Eq. (5.20)

$$\Delta m_1 = \frac{-1}{2\pi\tau_1} \ \ell n \left(\frac{E_{max}}{m_o - 2m_\pi} \right) \ . \tag{5.21}$$

Since Δm_1 depends only logarithmically on E_{max},
the actual numerical choice is not important for
an order-of-magnitude estimate. We arbitrarily

take E_{max} = 10 GeV and get[6)]

$$\Delta m_1 \cong - \frac{1}{\tau_1} \quad .$$

An analogous expression holds for Δm_2; but since τ_2 is much larger than τ_1, we can neglect Δm_2 and thus find for the mass splitting

$$\Delta m \cong \Delta m_1 \cong - \frac{1}{\tau_1} \cong -10^{-5} \text{ eV} \quad . \tag{5.22}$$

Calculations using dispersion relations lead to similar, but more reliable, numbers.[7)] Mass shifts and mass splitting, as expected from our crude approach, are shown in Fig. 5.3. The long-lived kaon should be heavier than the short-lived one.

The estimate Eq. (5.22) is based on the assumption that no $|\Delta Y| = 2$ transitions occur. If such transitions exist, then the mass splitting would be first order in H_w, as given by the second terms of Eqs. (5.16) and (5.17) and shown in Fig. 5.1b. The mass splitting should then be about

---

6. Note that the negative sign comes from the assumed constancy of the matrix element. For a matrix element that falls rapidly with energy, the sign would be positive.
7. V. Barger and E. Kazes, Phys. Rev. 124, 279 (1961); J. Nilsson, Nuovo Cimento 22, 414 (1961); R. N. Mohapatra, J. Subba Rao, and R. E. Marshak, Phys. Rev. Letters 20, 1081 (1968). E. V. Gedalin, O. V. Kancheli, and S. G. Matinyan, JETP Letters 7, 136 (1968). R. E. Marshak, Riazuddin and C. P. Ryan, Weak Interactions of Elementary Particles (J. Wiley and Sons, New York, 1969), pp. 618-624.

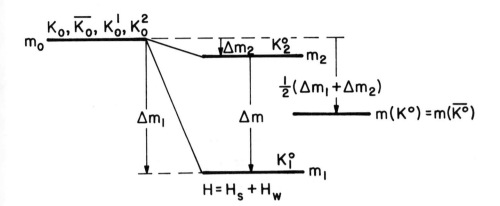

Fig. 5.3. K_1-K_2 mass splitting. The "particles"
K_0, $\overline{K}_0$, K_0^1 and K_0^2 are hypothetical
states that exist if H_w is "switched
off." The mass shifts by H_w are
denoted by Δm_1 and Δm_2. The levels
drawn in the figure indicate only the
centers of mass. The particle K_1^0, for
instance, decays with a lifetime
$\tau_1 \stackrel{\sim}{\sim} 1/\Delta m_1$ and it has therefore a width
comparable to the splitting Δm.

10^5 times bigger than the estimate (5.22) or

$$|\Delta m| \stackrel{\sim}{\sim} 1 \text{ eV} \text{ if } |\Delta Y| = 2 . \tag{5.23}$$

Measurement of Δm hence also gives some information
about the existence of $|\Delta Y| = 2$ transitions.[8]

---

8.   L. Okun and B. Pontecorvo, Soviet Phys. JETP
     5, 1297 (1957). See also S. L. Glashow, Phys.
     Rev. Letters 6, 196 (1961).

Finally we return to the states K^O and $\overline{K^O}$. From Eqs. (5.9) and (5.14) we get

$$H \, |K^O> \, = \sqrt{\tfrac{1}{2}} \left\{ m_1 |K_1^O> \, + \, m_2 |K_2^O> \right\} \quad . \qquad (5.24)$$

Here we see explicitly what we have pointed out earlier: K^O is not an eigenstate of the total Hamiltonian; it does not have a well-defined mass. It is therefore proper to speak of K^O and $\overline{K^O}$ as particle mixtures. However, we can still define a mass as the expectation value of H for the state $|K^O>$ at rest

$$m(K^O) \, = \, <K^O|H|K^O> \, = \tfrac{1}{2}(m_1+m_2) \quad . \qquad (5.25)$$

This mass value is indicated in Fig. 5.3. The TCP test discussed in Section 4.8 has to be interpreted with Eq. (5.25) in mind.

Figure 5.3 is actually somewhat misleading because the energy levels are indicated as sharp. The particles are decaying, and the levels have finite width. In particular, K_1 has a width $1/\tau_1$ that is comparable to the mass splitting Δm_1 if the assumptions leading to Eq. (5.22) are correct.

D. Hypercharge Oscillations

The mass splitting predicted by Eq. (5.33) is very small; in terms of the kaon rest mass it is of the order of 1 part in 10^{14}. Can such a tiny mass splitting be observed and measured?

Indeed it can; moreover, the oscillations between the K^O and the $\overline{K^O}$ can be followed and the sign of the mass difference can be determined! The first suggestion of how to attack this delicate problem came from Pais and Piccioni;[9] their treatment was extended by Case and by Treiman and Sachs.[10]

The idea underlying the determination of the mass splitting can be understood in terms of the time-dependent Schrödinger equation

$$i \frac{d}{dt} |\psi(t)> = H |\psi(t)>$$

$$\qquad (5.26)$$

$$= (H_s + H_w) |\psi(t)> \quad .$$

Owing to the action of the perturbation H_w, a particle produced at time $t = 0$ as a K^O will make transitions to the state $\overline{K^O}$. The time-dependent probability for such transitions depends on the mass splitting, Δm, of the K_1^O and K_2^O; furthermore, by measuring the hypercharge Y we can determine whether a detected particle is a K^O or a $\overline{K^O}$.

Because the rate of hypercharge oscillations between the K^O and $\overline{K^O}$ may be comparable to the lifetimes of the kaons, their decay rates must be taken into account. For the eigenstates of H, K_1^O and K_2^O, of energy E and a proper mean life τ

9. A. Pais and O. Piccioni, Phys. Rev. 100, 1487 (1955).

10. K. M. Case, Phys. Rev. 103, 1449 (1956); S. Treiman and R. G. Sachs, Phys. Rev. 103, 1545 (1956).

(defined in the rest frame), the time dependence
of $\psi(t)$ is given by

$$\psi \propto e^{-iEt-t/2\gamma\tau} \quad , \tag{5.27}$$

where γ is given by Eq. (2.48) or Eq. (2.59). The
exponent is chosen so that the probability $\psi^*\psi$ for
a particle at rest ($\gamma = 1$) decays exponentially
with a mean life τ:

$$\psi^*\psi \propto e^{-t/\tau} \quad .$$

Now we consider a state describing a pure K^o at
rest at time $t = 0$ and write it in terms of
eigenstates of definite energy and exponential
time decay,

$$|\psi(t = 0)> \equiv |K^o> = \sqrt{\tfrac{1}{2}}\left\{|K_1^o> + |K_2^o>\right\} \quad .$$

With the time dependence Eq. (5.27), the state
at time t and at rest becomes

$$|\psi(t)> = \sqrt{\tfrac{1}{2}}\left\{e^{-im_1 t-t/2\tau_1} |K_1^o>\right.$$

$$\left. + e^{-im_2 t-t/2\tau_2} |K_2^o>\right\} \quad . \tag{5.28}$$

The probability of finding a K_2^o at time t is given

by

$$|<K_2^0|\psi(t)>|^2 = e^{-t/\tau_2} \quad ; \qquad (5.29)$$

this equation expresses the appearance of a long-lived kaon component. In order to describe hyper-charge oscillations, we rewrite $|\psi(t)>$ in terms of K^0 and $\overline{K^0}$; with Eq. (5.29), we get

$$|\psi(t)> = \frac{1}{2}\left\{\left(e^{im_1 t - t/2\tau_1} + e^{-im_2 t - t/2\tau_2}\right)|K^0>\right.$$

$$\qquad\qquad\qquad (5.30)$$

$$\left. +\left(e^{-im_2 t - t/2\tau_2} - e^{im_1 t - t/2\tau_1}\right)|\overline{K^0}>\right\} \quad .$$

At time t=0, this expression represents a state with hypercharge Y=1. At later times, a component with Y=-1 appears. The probabilities of observing these components are given by

$$P(K^0,t) = |<K^0|t>|^2 \quad , \quad P(\overline{K^0},t) = |<\overline{K^0}|t>|^2 \quad .$$

These probabilities follow easily from Eq. (5.30); with $\Delta m = m_1 - m_2$ they are

$$P(K^O, t) = \frac{1}{4} \left\{ e^{-t/\tau_1} + e^{-t/\tau_2} \right.$$

$$\left. + 2e^{-(t/2\tau_1 - t/2\tau_2)} \cos(\Delta m\, t) \right\},$$

$$(5.31)$$

$$P(\overline{K^O}, t) = \frac{1}{4} \left\{ e^{-t/\tau_1} + e^{-t/\tau_2} \right.$$

$$\left. - 2e^{-(t/2\tau_1 - t/2\tau_2)} \cos(\Delta m\, t) \right\}.$$

Eqs. (5.31) predict some manifestations of hyper-charge oscillations that can be observed experi-mentally. We discuss two particularly interesting ones.

For times t much greater than τ_1, Eq. (5.31) shows that the kaon beam will consist of equal parts K^O and $\overline{K^O}$. (This fact is already evident from Eq. (5.29).) A kaon beam that has been pro-duced in a pure Y=1 state has changed to one con-taining equal parts Y=1 and Y=-1. Experimentally, the appearance of the $\overline{K^O}$ component has been veri-fied through the observation of hadronic inter-actions such as $\overline{K^O}p \rightarrow \pi^+\Lambda^O$. Since nucleons have Y=1, and the Λ^O has Y=0, it can be produced by $\overline{K^O}$, but not by K^O. Processes such as $\overline{K^O}p \rightarrow \pi^+\Lambda^O$ and $\overline{K^O}p \rightarrow \pi^O\Sigma^+$ have indeed been observed in

emulsions[11] and bubble chambers[12] and have given
additional evidence for the particle mixture
hypothesis. However, these early experiments did
not produce a value for the mass splitting Δm.

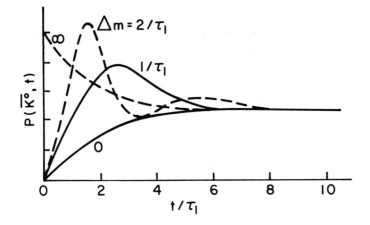

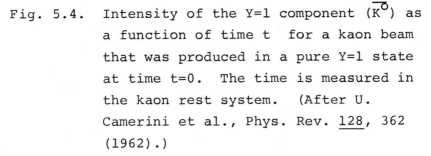

Fig. 5.4. Intensity of the Y=1 component $(\overline{K^0})$ as
          a function of time t for a kaon beam
          that was produced in a pure Y=1 state
          at time t=0. The time is measured in
          the kaon rest system. (After U.
          Camerini et al., Phys. Rev. 128, 362
          (1962).)

11. R. Ammar, J. I. Friedman, R. Levi-Setti, and
     V. L. Telegdi, Nuovo Cimento 5, 1801 (1957).
12. W. B. Fowler, R. L. Lander, and W. M. Powell,
     Phys. Rev. 113, 928 (1959); F. S. Crawford,
     M. Cresti, M. L. Good, K. Gottstein, E. M.
     Lyman, F. T. Solmitz, M. L. Stevenson, and H.
     Ticho, Phys. Rev. 113, 1601 (1959).

Information about the absolute value of the
mass splitting can be obtained if the interference
term in Eq. (5.31) is observed. This term is
appreciable only at times comparable to τ_1.
Figure 5.4 shows $P(\overline{K}^o,t)$ for various values of
$|\Delta m|$. It is clear from this figure that the most
valuable information is contained in the first few
mean lifes. In an actual experiment, the decaying
kaons are not at rest, but are moving and Eq. (5.41)
has to be transformed accordingly. One method for
observing the interference term consists in following the intensity of the Y=-1 component as a
function of the time elapsed after production of
the pure K^o state.[13,14] An example is shown in
Fig. 5.5. A neutral kaon is produced as a K^o in
a charge exchange reaction, $K^+p \rightarrow K^o n$. The state
$|\psi(t)>$ of the neutral kaon then develops according
to Eq. (5.30). The interaction indicated in
Fig. 5.5 results in a Λ^o and a π^+; it is ascribed
to the reaction $\overline{K}^o p \rightarrow \pi^+\Lambda^o$. Energy and momentum
balance, together with the distance between the
two events, permit computation of the time t.
Observation of the intensity of lambda production
as a function of t then yields the desired function
$P(\overline{K}^o,t)$ and hence also $|\Delta m|$.

---

13. W. F. Fry and R. G. Sachs, Phys. Rev. 109,
     2212 (1958).
14. E. Boldt, D. O. Caldwell, and Y. Pal, Phys.
     Rev. Letters 1, 150 (1958).

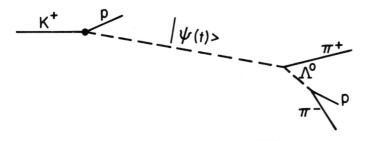

Fig. 5.5.   Observation of the $\overline{K}{}^O$ component of an
            initially pure K^O beam. The state
            denoted by $|\psi(t)\rangle$ is produced as K^O
            and reacts as $\overline{K}{}^O$.

An early experiment[14], performed with a
cloud chamber, already indicated that the mass
splitting is of the order of $1/\tau_1$ and hence agrees
with the estimate Eq. (5.22). Later experiments[15]
with bubble chambers verified this conclusion and
gave a better value for the absolute value $|\Delta m|$.
However, observing the strong interaction of the
$\overline{K}{}^O$ component is not the only method to study Δm;
we discuss in the next subsection an approach
that can be used to measure both sign and magnitude of Δm.

15. V. L. Fitch, P. A. Piroue, and R. B. Perkins,
     Nuovo Cimento 22, 1160 (1961).   U. Camerini,
     W. F. Fry, J. A. Gaidos, H. Huzita, S. V.
     Natali, R. B. Willmann, R. W. Birge, R. P. Ely,
     W. M. Powell, and H. S. White, Phys. Rev. 128,
     362 (1962).   G. W. Meisner, B. B. Crawford,
     and F. S. Crawford, Jr., Phys. Rev. Letters
     16, 278 (1966).

E.   Regeneration[9,10,16)]

> "That the Gell-Mann-Pais theory,
> published in 1955, actually pre-
> ceeded most of the experimental
> evidence known at present is one
> of the most astonishing and grati-
> fying successes in the history of
> the elementary particles."
>
>   R. H. Good et al.[17)]

The games played by neutral kaons go beyond those discussed in the previous subsections. To introduce the next one, we consider the situation shown in Fig. 5.6. Neutral kaons in the state $|K^o\rangle$ are produced, for instance by charge exchange, $K^+n \to K^op$, or by a reaction such as $\pi^-p \to K^o\Lambda^o$. Upon production, the K^o beam will consist of an equal amount of K^o_1 and K^o_2, with a definite relative phase relationship, as given by Eq. (5.11). After a flight path of a few meters, the fast component will have decayed and the beam is in a pure K^o_2 state. If this beam passes through a piece of

---

16.   M. L. Good, Phys. Rev. 106, 591 (1957); Phys.
      Rev. 110, 550 (1958).
17.   R. H. Good, R. P. Matsen, F. Muller, O.
      Piccioni, W. M. Powell, H. S. White, W. B.
      Fowler, and R. W. Birge, Phys. Rev. 124, 1223
      (1961).

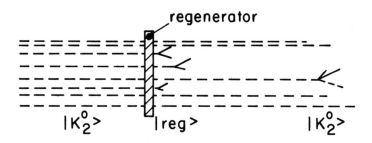

Fig. 5.6.   Regeneration of K_1^0. A pure K_2^0 beam
            passes through a piece of matter
            ("regenerator") and transforms into a
            beam that contains a K_1^0 component
            again. The neutral beam is indicated
            by the dashed line; the decay of the
            short-lived K_1^0 component is charac-
            terized by the two-track events. The
            long lived K_2^0 give rise to three parti-
            cles.

matter, the short lived component, K_1^0 will
reappear; the material used is therefore called a
"regenerator."

     To discuss this effect, we return to the de-
scription in terms of K^0 and $\overline{K^0}$ because the interaction with matter will be predominantly hadronic;
we thus use $|K_2^0\rangle = \sqrt{\frac{1}{2}}\left\{|K^0\rangle + |\overline{K^0}\rangle\right\}$. K^0 and $\overline{K^0}$ in-
teract differently with matter. They both undergo
charge exchange reactions, $K^0 p \rightarrow K^+ n$ and $\overline{K^0} n \rightarrow K^- p$.
The $\overline{K^0}$, however, can produce other reactions

such as

$$\overline{K^o}p \rightarrow \pi^+\Lambda^o \qquad\qquad \overline{K^o}n \rightarrow \pi^o\Lambda^o$$

$$\pi^+\Sigma^o \qquad\qquad\qquad \pi^o\Sigma^o$$

$$\pi^o\Sigma^+ \qquad\qquad\qquad \pi^-\Sigma^+$$

which are forbidden to the K^o because of hypercharge conservation. These inelastic reactions
have larger cross sections than the charge
exchange reactions. We describe the effects of
transmission through the regenerator by two complex numbers, f and $\bar{f}$. Neglecting decay effects,
the amplitude of the beam after the regenerator
becomes

$$|\text{reg}> = \sqrt{\tfrac{1}{2}}\left\{ f\,|K^o> + \bar{f}\,|\overline{K^o}> \right\} \,, \qquad (5.32)$$

or

$$|\text{reg}> = \tfrac{1}{2}\,(f-\bar{f})\,|K_1^o> + \tfrac{1}{2}\,(f+\bar{f})\,|K_2^o> . \qquad (5.33)$$

Because of the difference in the interactions of
the K^o and the $\overline{K^o}$, f and $\bar{f}$ will in general not be
identical and the kaon beam after the regenerator
will again contain a K_1^o component. In Fig. 5.6,
we have indicated the reemergence of this component by two-prong events after the regenerator.
    We treat here the case of coherent forward
regeneration in a plate, as indicated in Fig. 5.7.
If we neglect the decay of the K_2^o, we can write

for the amplitude at the position x: $A_2(x) =$ $\exp(ip_2x)$. (Here, x and p denote components and not four-vectors.) The amplitude for conversion $K_2^O \rightarrow K_1^O$ between x and $x+dx$ is proportional to $A_2(x)$ and $(f-\bar{f})$, as is clear from Eq. (5.33), and we set

$$dA_1(x) \propto (f-\bar{f}) A_2(x) dx \quad . \tag{5.34}$$

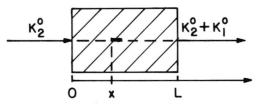

Fig. 5.7.   Regeneration through coherent forward scattering in a plate.

The short mean life of the K_1^O requires that its decay be taken into account; with Eq. (5.27) and with a flight time $t = (L-x)/v_1$ we get for the contribution from $dA_1(x)$ to the K_1^O amplitude at the end of the plate

$$dA_1(L) = e^{ip_1(L-x) - (L-x)/2\Lambda} dA_1(x) \quad , \tag{5.35}$$

Here $\Lambda = \gamma v_1 \tau_1$ is the decay mean free path of the K_1^O and we have neglected secondary regeneration, $K_1^O \rightarrow K_2^O$. The K_1^O amplitude at the end of the plate now becomes

$$A_1(L) = const \ (f-\bar{f}) e^{ip_1 L} e^{-L/2\Lambda}$$

$$\times \int dx \ e^{i(p_2-p_1)x} e^{x/2\Lambda} . \qquad (5.36)$$

To get p_2-p_1 we use the fact that there is no
energy loss to the regeneration plate in forward
scattering so that $p_1^2 + m_1^2 = p_2^2 + m_2^2$. Hence
$p_2^2-p_1^2 = m_1^2-m_2^2$ or $(p_2-p_1)(p_1+p_2)=(m_1-m_2)(m_1+m_2)$, or
in a very good approximation

$$p_2-p_1 = \frac{m}{p} \Delta m, \quad \Delta m = m_1-m_2 , \qquad (5.37)$$

where m and p are mass and momentum of either
meson. After performing the integration in
Eq. (5.36), using Eq. (5.37), and taking the
square of the absolute value, we arrive at the
probability $P(K_1^O;L)$ of finding a K_1^O meson at the
end of a regenerator of length L:

$$P(K_1^O;L) = P_O \left\{ 1 + e^{-L/2\Lambda} \right.$$

$$\left. - 2e^{-L/2\Lambda} \cos\left(\frac{m}{p} \Delta mL\right)\right\} \qquad (5.38)$$

The constant P_O gives the probability of observing
a K_1^O after a thick absorber $(L \gg \Lambda)$. We have
derived Eq. (5.38) under very restrictive
assumptions; more general discussions are given in
references 16 and 17. However, the essential

features of coherent regeneration are contained in
Eq. (5.38): In particular, for absorbers of thickness L comparable to Λ, the decay mean free path of
K_1^o, an oscillatory term appears. This term permits
measurement of the magnitude of the mass splitting.
Values of $|\Delta m|$ have been obtained using this
method.[17,18]

A glance at Eqs. (5.31) and (5.38) shows that
hypercharge oscillations and coherent regeneration
are even functions of the mass difference and
hence cannot be used to determine the sign of the
mass difference. Methods to find the sign have
been suggested by a number of authors. Common to
all is the observation of interference phenomena
in addition to the interference between K^o and $\overline{K^o}$.
This interference can be between two different
leptonic decay modes[19] (e.g. $e^+\pi^-\nu$ and $e^-\pi^+\overline{\nu}$),
between K_1^o coherently regenerated from two
regenerators of dissimilar material[20] (e.g.
carbon and uranium), or between the originally
produced and the regenerated K_1's.[21] Experiments
have been performed and they indicate that the
long-lived meson is heavier than the short-lived

18. T. Fujii, J. V. Jovanovich, F. Turkot, and
     G. T. Zorn, Phys. Rev. Letters 13, 253,324
     (1964). J. H. Christenson, J. W. Cronin,
     V. L. Fitch, and R. Turlay, Phys. Rev. 140B,
     74 (1965).
19. N. Biswas, Phys. Rev. 118, 866 (1960).
20. L. Yu. Kobzarev and L. B. Okun, Soviet Phys.
     JETP 12, 426 (1961).
21. U. Camerini, W. F. Fry, and J. Gaidos, Nuovo
     Cimento 28, 1096 (1963).

one.[22]

Hypercharge oscillations and coherent regeneration are not the only methods that can be used to determine Δm; a systematic survey over all approaches is contained in a review by Lee and Wu.[23] At present, the value for the mass difference is

$$\Delta m = m_1 - m_2 = -(0.540 \pm 0.004) \times 10^{10} \text{sec}$$

$$= -(0.48 \pm 0.01)/\tau_1 \quad . \qquad (5.39)$$

5.2   THE BREAKDOWN OF CP INVARIANCE

Kaons have never ceased to puzzle and surprise us. As we discussed in Subsection 4.5.F., the study of the pionic decay modes of the charged kaons led to the discovery of parity violation. For the neutral kaons, coherence properties were first predicted theoretically. They lead to a mass splitting of two particles, K_1^o and K_2^o, of different lifetimes, to hypercharge oscillations and to regeneration phenomena. The experimental substantiation of these

---

22.   G. W. Meisner, B. B. Crawford, and F. S. Crawford, Jr., Phys. Rev. Letters 17, 492 (1966).   J. Canter, Y. Cho, A. Engler, H. E. Fisk, R. W. Kraemer, C. W. Meltzer, D. G. Hill, D. K. Robinson, and M. Sakitt, Phys. Rev. Letters 17, 942 (1966).   J. V. Jovanovich, T. Fujii, F. Turkot, G. T. Zorn, and M. Deutsch, Phys. Rev. Letters 17, 1075 (1966).
23.   T. D. Lee and C. S. Wu, Ann. Rev. Nucl. Sci. 16, 552 (1966).

phenomena was exciting, indeed. No less exciting, but much more unexpected was the discovery of CP violation, for which the basic cause may not yet have been found.

A. <u>CP Violation in the K_2^0 Decay</u>

We have shown in Subsection 5.1.A that, if CP is conserved, the K_2^0 <u>cannot</u> decay into <u>two</u> pions. In 1964 Christenson, Cronin, Fitch and Turlay carried out an experiment to set a lower limit on the two-pion decay of the K_2^0.[24] The experimental arrangement is shown in Fig. 5.8. Neutral kaons were produced in the Brookhaven AGS; the beam was defined by collimators and charged particles were swept out by magnets. A decay path of about 17m allowed all K_1^0's to decay. Charged pions resulting from the decay of the K_2^0 in a helium bag were

24. J. H. Christenson, J. W. Cronin, V. L. Fitch, and R. Turlay, Phys. Rev. Letters <u>13</u>, 138 (1964). Another experiment, with similar results but smaller statistics, was simultaneously done by an Illinois group, A. Abashian, R. J. Abrams, D. W. Carpenter, G. P. Fisher, B.M.K. Nefkens, and J. H. Smith, Phys. Rev. Letters <u>13</u>, 243 (1964).

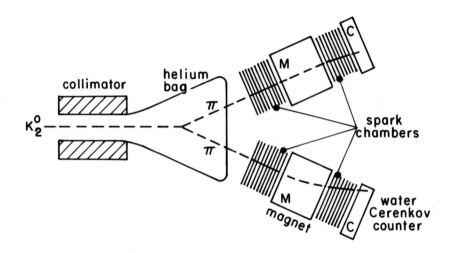

Fig. 5.8.   Observation of the two-pion decay mode
            of the K_2^o.   (Reference 24).

observed by two spectrometers. Each spectrometer
consisted of scintillation counters, a Cerenkov
counter, and two spark chambers separated by a
magnet. The spark chambers were triggered on coincidences between the Cerenkov counter and the
scintillation counters. The vector momentum of
each particle could then be determined from the
curvature in the magnetic field. Assuming both
particles to be pions, the mass of the decaying
particle was calculated. For a two-body decay,
this mass value will be the kaon mass. In a
three-body decay, the neutral particle will also
carry away momentum and the calculated mass
spectrum will not peak at the kaon mass. The

experiment gave the totally unexpected result that
some decays into two pions do occur with an
approximate branching ratio (intensity ratio) of

$$\frac{\text{Int}(K_L \to \pi^+\pi^-)}{\text{Int}(K_L \to \text{all charged modes})} \approx 2 \times 10^{-3}. \quad (5.40)$$

From here on we denote the long-lived
neutral kaon by K_L and the short-lived one by K_S,
to differentiate them from the K_1^o and K_2^o which we
defined to be eigenstates of CP. The decay of K_L
into two pions demonstrates, however, that the
long-lived kaon is not an eigenstate of CP and the
notation K_2^o is therefore inappropriate. We will
discuss notation in more detail in Subsection B.
    The news of the violation of CP was not
readily accepted and other causes for the observations were sought. Skepticism was based on two
prejudices. Firstly, CP violation together with
the TCP theorem predicts a breakdown of time-
reversal invariance and physicists were very
reluctant to give up this attractive symmetry.
Secondly, earlier breakdowns had shown maximal
violations. Parity, for instance, is completely
violated in the weak interactions. CP violation,
on the other hand, appears to be small. Despite
these arguments, the experiments turned out to
be correct and they were verified by other

groups.[25)]

The strongest support for the existence of a CP-violating decay mode of the K_L came from an interference experiment, performed again by a Princeton group.[26)] If an initially pure K_L beam passes through a regenerator, two-pion decays can occur owing to the regeneration $K_L \to K_S$ and the CP-nonconserving K_L decay. If a two-pion decay occurs, it is impossible to say by which process it was produced. The amplitudes for regeneration (A_1) and for CP-nonconserving decay (ε) therefore add coherently and the branching ratio for the two-pion decay in matter is proportional to $|A_1 + \varepsilon|^2$. The regenerator can be so chosen that the two amplitudes are about equal in magnitude and the interference term $A_1\varepsilon^* + \varepsilon A_1^* = 2\,\mathrm{Re}(A_1\varepsilon^*)$ is then at its maximum. The Princeton group found clear evidence for constructive interference between A_1 and ε and thereby proved that the long-lived two-pion component cannot be caused, for instance, by an unknown particle, as had been suggested.[27)]

---

25. X. de Bouard, D. Dekkers, B. Jordan, R. Mermod, T. R. Willitts, K. Winter, P. Scharff, L. Valentin, M. Vivargent, and M. Bott-Bodenhausen, Phys. Letters 15, 58 (1965). W. Galbraith, G. Manning, A. E. Taylor, B. D. Jones, J. Malos, A. Astbury, N. T. Lipman, and T. G. Walker, Phys. Rev. Letters 14, 383 (1965).
26. V. L. Fitch, R. F. Roth, J. S. Russ, and W. Vernon, Phys. Rev. Letters 15, 73 (1965).
27. H. J. Lipkin and A. Abashian, Phys. Letters 14, 151 (1965). J. L. Uretsky, Phys. Letters 14, 154 (1965).

The evidence for CP violation discussed so far comes from the decay $K_L \to \pi^+\pi^-$. The CP violating mode $K_L \to \pi^0\pi^0$ has also been observed.[28] Further evidence comes from decays $K_L \to \pi \ell \nu$, where the lepton ℓ can be either an electron or a muon. CP invariance predicts an equal number of positive and negative leptons. The observed ratio[29]

$$\frac{\Gamma(K_L \to \pi^- \ell^+ \nu)}{\Gamma(K_L \to \pi^+ \ell^- \bar{\nu})} \approx 1.005$$

28. M. Gaillard, F. Krienen, W. Galbraith, A. Hussri, M. R. Jane, N. H. Lipman, G. Manning, T. Fatcliffe, P. Day, A. G. Parham, B. T. Payne, A. C. Sherwood, H. Faissner, and H. Reithler, Phys. Rev. Letters 18, 20 (1967). J. W. Cronin, P. F. Kunz, W. R. Risk, and P. C. Wheeler, Phys. Rev. Letters 18, 25 (1967).
29. D. Dorfan, J. Enstrom, D. Raymond, M. Schwartz, S. Wojcicki, D. H. Millter, and M. Paciotti, Phys. Rev. Letters 19, 987 (1967). S. Bennett, D. Nygren, H. Saal, J. Steinberger, and J. Sunderland, Phys. Rev. Letters 19, 993 (1967).

deviates from unity and thus also indicates CP violation.*

B. Description of Neutral Kaons

The description of neutral kaons given in
Section 5.2 is not sufficient because real kaons
are not eigenstates of the CP operator. Equations
(5.29) must therefore be generalized[30-36] and we
present some of the relevant equations here. We
retain Eqs. (5.16), (5.29), and (5.10): K^o and $\overline{K}^o$

---

*. The charge asymmetry has an application: it
    allows us to tell a world from an antiworld
    without personal contact. Before visiting
    another galaxy, the following exchange will
    take place: "Please measure the charge ratio
    of the particle that is emitted less frequent-
    ly with the charge of the electrons in your
    atom." If the answer comes back: "Charges
    opposite", the visit will be cancelled.
30. T. D. Lee, R. Oehme, and C. N. Yang, Phys.
    Rev. 106, 340 (1957).
31. R. G. Sachs, Ann. Phys. 22, 239 (1963).
32. T. T. Wu and C. N. Yang, Phys. Rev. Letters
    13, 380 (1964). [See also J. Ashkin and P. K.
    Kabir, Phys. Rev. D, 1, 868 (1970).]
33. T. D. Lee and L. Wolfenstein, Phys. Rev. 138B,
    1490 (1965).
34. T. D. Lee and C. S. Wu, Ann Rev. Nucl. Sci.
    16, 511 (1966).
35. L. Wolfenstein, Nuovo Cimento 42A, 17 (1966).
36. L. B. Okun, Proceedings of the Heidelberg Con-
    ference on Elementary Particles, H. Filthuth,
    Ed., North Holland Publishing Company, 1968.

remain eigenstates of Y, K_1^O and K_2^O remain eigenstates of CP. The <u>real</u> particles K_S (for <u>S</u>hort)
and K_L (for <u>L</u>ong), however, are neither eigenstates of Y nor of CP, but they have definite mass
and definite lifetimes. K_S is in a state that is
predominantly CP -even, but has a small CP -odd
admixture; K_L is predominantly CP -odd. They can
therefore be written as

$$|K_S> = \frac{|K_1^O> + \epsilon|K_2^O>}{\left[1 + |\epsilon|^2\right]^{\frac{1}{2}}} \quad ,$$

$$|K_L> = \frac{|K_2^O> + \epsilon'|K_1^O>}{\left[1 + |\epsilon'|^2\right]^{\frac{1}{2}}} \quad , \tag{5.41}$$

where ϵ and ϵ' are two small complex numbers. If
TCP invariance holds, ϵ and ϵ' are equal. To
analyze the observed CP violation, two complex
numbers, η_{+-} and η_{oo}, are introduced through the
definitions

$$\eta_{+-} = |\eta_{+-}|e^{i\phi_{+-}} = \frac{<\pi^+\pi^-|H_w|K_L>}{<\pi^+\pi^-|H_w|K_S>} \quad ,$$

$$\eta_{oo} = |\eta_{oo}|e^{i\phi_{oo}} = \frac{<\pi^O\pi^O|H_w|K_L>}{<\pi^O\pi^O|H_w|K_S>} \quad . \tag{5.42}$$

Note that the experimental result, Eq. (5.40), gives the violation in terms of intensities and refers to K_L alone while η_{+-} and η_{oo} describe amplitude ratios and involve K_L and K_S.

As next step, we perform an isospin analysis. The initial kaon has isospin ½, the two pions in the final state can have isospin values 0, 1, or 2. However, since the two pions have zero angular momentum, the value I=1 is excluded. With Appendix A, we write

$$|\pi^+\pi^-> = \sqrt{\frac{2}{3}}\,|0> + \sqrt{\frac{1}{3}}\,|2> \quad ,$$

$$|\pi^o\pi^o> = -\sqrt{\frac{1}{3}}\,|0> + \sqrt{\frac{2}{3}}\,|2> \quad . \tag{5.43}$$

If isospin were conserved in the weak interactions, the decay $K \to 2\pi$ could not occur. It occurs, isospin is not conserved, and we introduce amplitude ratios

$$\varepsilon_o = \frac{<0|H_w|K_L>}{<0|H_w|K_S>} \quad , \qquad \varepsilon_2 = \frac{<2|H_w|K_L>}{\sqrt{2}\,<0|H_w|K_S>} \quad ,$$

$$\omega = \frac{<2|H_w|K_S>}{<0|H_w|K_S>} \quad . \tag{5.44}$$

With Eq. (5.42) to (5.44), we find

$$\eta_{+-}\left(1 + \frac{\omega}{\sqrt{2}}\right) = \varepsilon_o + \varepsilon_2 \quad ,$$

$$\eta_{oo}\left(1 - 2\omega\right) = \varepsilon_o - 2\varepsilon_2 \quad . \qquad (5.45)$$

The decay of the K_S into two pions satisfies the approximate selection rule $|\Delta I| = \frac{1}{2}$ and ω is therefore very small. With $|\omega| << 1$, Eq. (5.45) becomes

$$\eta_{+-} = \varepsilon_o + \varepsilon_2 \quad , \qquad \eta_{oo} = \varepsilon_o - 2\varepsilon_2 \quad , \qquad (5.46)$$

$$\eta_{+-} - \eta_{oo} = 3\varepsilon_2 \quad , \qquad (5.47)$$

and η_{+-}, η_{oo}, and ε_o satisfy the Wu-Yang triangle condition

$$2\eta_{+-} + \eta_{oo} = 3\varepsilon_o \quad . \qquad (5.48)$$

Once the parameters describing the neutral kaon decay are determined from experiment, they can be plotted in the Wu-Yang diagram,[34] Fig. 5.9; the parameters should satisfy Eq. (5.46). The experimental situation has been reviewed at

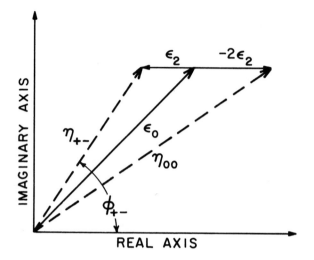

Fig. 5.9.   Wu-Yang diagram for the neutral kaon
decay.[32,25)   The figure does not
represent the latest experimental
situation; it only explains the idea.

several recent conferences.[37,38)   Recent data
appears to indicate that previous parameters, such
as $|\eta_{+-}|$, may have been too small by as much as

37.   C. Rubbia in Proc. XVI Intern. Conf. on High
      Energy Physics, J. D. Jackson, A. Roberts
      and R. Donaldson, eds. (E. Fermi National
      Accelerator Lab., Batavia, Ill., 1972),
      Vol. 4, pp.157-188.
38.   K. Winter in Proc. Amsterdam Intern. Conf. on
      Elementary Particles, A. G. Tenner and M.J.G.
      Veltman, eds. (North-Holland, Amsterdam,
      1972), pp.333-372.

10%. The errors assigned to these parameters has therefore been increased. We list presently accepted values[39)]

$$\eta_{+-} = (2.17 \pm 0.07) \times 10^{-3} \exp[i(46.6 \pm 2.5)^{\circ}]$$

$$\eta_{oo} = (2.25 \pm 0.09) \times 10^{-3} \exp[i(49 \pm 13)^{\circ}]$$

Within the errors, it thus appears that $\eta_{+-} = \eta_{oo}$. Eq. (5.47) then shows that $\varepsilon_2 = 0$, which means that K_L does not decay to a two pion state of isospin I=2; CP is therefore conserved in this decay mode of K_L. The Wu-Yang triangle shown in Fig. 5.9 then collapses to the central line labeled by ε_o.

The CP violation in the neutral kaon system suggests that time-reversal invariance should also be violated.[37,38)] Indeed, it has been argued[40)] that the experimental data are inconsistent with T invariance; these arguments are, however, not fool-proof.[41)]

---

39. Particle Data Group, Phys. Letters 50B, 1 (1974).

40. R. C. Casella, Phys. Rev. Letters 21, 1128 (1968), 22, 554 (1969). K. R. Schubert, B. Wolff, J. C. Chollet, J. M. Gaillard, M. R. Jane, T. J. Ratcliffe, and J. P. Repellin, Physics Letters 31B, 662 (1970). S. V. Dass, Fortschr. Phys. 20, 77 (1972). S. V. Dass and P. K. Kabir, Proc. Roy. Soc. (London) A330, 331 (1972).

41. B. G. Kenny and R. G. Sachs, Phys. Rev. D8, 1605 (1973).

What is the basis of the CP-violation in the
neutral kaon system? The most likely suggestion
at this time is that it is due to a new superweak
or millweak interaction.[33,35] Part of the basis
for this suggestion is that, despite heroic efforts,
no evidence for CP or T violation has been observed
in any other system. Another basis is that a
spontaneous violation could arise in modern weak
interaction theories.[42] Finally, the data
reported above are consistent with such theories.

---

42. T. D. Lee, Phys. Rev. D8, 1226 (1973); Phys.
     Reports 9, 143 (1974).

PROBLEMS

5.1 Show that a weak interaction which induces transitions with $|\Delta Y| = 2$, but which is odd under C, does not lead to a K_1^o-K_2^o mass splitting in first order.

5.2 Justify the entries in Table 5.1.

5.3 Would the K_1^o and the K_2^o masses be different from each other if the pion mass were higher than that of the kaon? How would this situation affect the measurement of the mass difference?

5.4 How could a pure K_1^o state be produced in a reaction?

5.5 Neutron and antineutron are also neutral anti-particles, just as the K^o and the $\overline{K^o}$. Why is it not meaningful to introduce linear combinations n_1 and n_2, similar to the K_1^o and K_2^o?

5.6 Assume that the kaon described by Eq. (5.31) possesses a laboratory energy E. Find the expression for $P(K^o,t)$ in the laboratory system.

5.7 Sketch $P(K_1^o;L)$, given by Eq. (5.38), as a function of L/Λ for $\Delta m \tau_1 = -\frac{1}{2}$.

5.8 Discuss the methods that have been used to determine the sign of the mass difference between the K_1^o and the K_2^o. Sketch the experiments and the corresponding theory.

5.9 Discuss the decay of the K_S into three pions.

5.10 Use the experimentally observed ratio of
      $K_S \rightarrow \pi^+\pi^-$ to $K_S \rightarrow \pi^0\pi^0$ to show that ω,
      Eq. (5.44), is small compared to unity.

5.11 Use the latest experimental data to draw the
      Wu-Yang triangle. Is the triangle over-
      determined? Are transitions with $|\Delta I| = 3/2$
      established?

5.12 Assume that we can neglect weak forces and
      can consider the K and $\bar{K}$ systems alone.
      Write the corresponding state functions as

$$K = \begin{pmatrix} 1 \\ 0 \end{pmatrix}, \qquad \bar{K} = \begin{pmatrix} 0 \\ 1 \end{pmatrix}.$$

(a) Find the operator that corresponds to
     the CP transformation in this represen-
     tation.

(b) Find the eigenstates of CP.

(c) If the time dependence of the states in
     this two-dimensional Hilbert space is
     assumed to be linear,

$$i \dot{\psi} = H \psi, \qquad H = M + i \Gamma,$$

     what are the restrictions imposed on M
     and Γ by Hermiticity. (ψ is a column
     matrix, H is a 2×2 matrix).

(d) Repeat (c) and find the restrictions
     imposed by CP and by TCP invariance.

(e) In terms of the matrices 1 and $\vec{\sigma}$, show
     that the most general representation of
     H is

$$H = A\,1 + \vec{B}\cdot\vec{\sigma} \quad ;$$

     find the conditions imposed on A and $\vec{B}$
     by CP and by TCP invariance.

(f) What is the consequence of time-reversal
     invariance for A and $\vec{B}$, and for M and Γ?

ADDITIONAL REFERENCES

Neutral kaons and the subject of CP violation are treated in the following reviews: T. D. Lee and C. S. Wu, Ann. Rev. Nucl. Sci. 16, 511 (1966); G. Marx, Fortschritte der Physik 14, 695 (1966); L. B. Okun, Soviet Phys. Uspekhi 9, 574 (1967); P. K. Kabir, The CP Puzzle (Academic Press, New York, 1968); R. E. Marshak, Riazuddin and C. P. Ryan, Theory of Weak Interactions in Particle Physics (J. Wiley and Sons, New York, 1969), Ch. 6; The January-February Issue of Soviet Physics Uspekhi 11, 461 - 602, (1969), is devoted to discussions of the theoretical and experimental aspects of CP violation.

Appendix A ADDITION OF ANGULAR MOMENTA

In this appendix we give the rules of angular momentum addition and some tables of Clebsch-Gordan coefficients.[1-5] Numerical tables are available.[6,7]

Let $|j_i m_i>$ be one of an orthonormal set of eigenfunctions of the operators J_i^2 and $(J_i)_z$, with eigenvalues $j_i(j_i+1)$ and m_i, respectively:

$$J_i^2 |j_i m_i> = j_i(j_i+1) |j_i m_i> \quad ,$$

$$(J_i)_z |j_i m_i> = m_i |j_i m_i> \quad . \hspace{3cm} (A1)$$

1. E. U. Condon and G. H. Shortley, The Theory of Atomic Spectra (Cambridge University Press, Cambridge, 1951).
2. A. R. Edmonds, Angular Momentum in Quantum Mechanics (Princeton University Press, Princeton, N. J., 1957).
3. M. E. Rose, Elementary Theory of Angular Momentum (John Wiley and Sons, Inc., New York, 1957).
4. D. M. Brink and G. R. Satchler, Angular Momentum (Oxford University Press, Oxford, 1962).
5. A. de-Shalit and I. Talmi, Nuclear Shell Theory (Academic Press, New York, 1963).
6. M. Rotenberg, R. Bivins, N. Metropolis, and J. K. Wooten, Jr., The 3-j and 6-j Symbols (Technology Press, Cambridge, Massachusetts, 1959).
7. A. F. Nikiforov, V. B. Uvarov, and Yu. L. Levitan, Tables of Racah Coefficients (Macmillan, New York, 1965).

We then combine two systems, with angular momentum operators $\vec{J}_1$ and $\vec{J}_2$, to form a system with angular momentum $\vec{J}_3$, with

$$\vec{J}_1 + \vec{J}_2 = \vec{J}_3$$

$$m_1 + m_2 = m_3 \quad .$$

The Clebsch-Gordan coefficients are the coefficients in the expansion of the eigenfunction $|j_3 m_3> \equiv |j_1 j_2 j_3 m_3>$ in terms of the products $|j_1 m_1> \, |j_2 m_2> \equiv |j_1 m_1 \, j_2 m_2>$:

$$|j_3 m_3> \equiv \sum_{m_1 m_2} |j_1 m_1 j_2 m_2><j_1 m_1 j_2 m_2 | j_3 m_3> \quad . \quad (A2)$$

The Clebsch-Gordan coefficients $<j_1 m_1 j_2 m_2 | j_3 m_3>$ are the elements of a unitary matrix. The phases of the eigenfunctions are so chosen that the coefficients are real numbers and hence the matrix is orthogonal. Various notations are used for these coefficients; these notations are compared on page 133 of ref. 4 and on page 52 of ref. 2. The inversion of Eq. (A2) yields

$$|j_1 m_1 j_2 m_2> = \sum_{j_3' m_3'} |j_3' m_3'><j_1 m_1 j_2 m_2 | j_3' m_3'> \quad . \quad (A3)$$

The Clebsch-Gordan coefficients satisfy the orthogonality relations

$$\sum_{j_3 m_3} \langle j_1 m_1 j_2 m_2 | j_3 m_3 \rangle \langle j_3 m_3 | j_1 m_1' j_2 m_2' \rangle$$

$$= \delta_{m_1 m_1'} \delta_{m_2 m_2'} \quad . \tag{A4}$$

$$\sum_{m_1 m_2} \langle j_3 m_3 | j_1 m_1 j_2 m_2 \rangle \langle j_1 m_1 j_2 m_2 | j_3' m_3' \rangle$$

$$= \delta_{j_3 j_3'} \delta_{m_3 m_3'} \quad . \tag{A5}$$

Calculations often become simpler if the Clebsch-Gordan coefficients are replaced by the more symmetric Wigner 3j-symbols. These symbols are related to the Clebsch-Gordan coefficients by the relation

$$\begin{pmatrix} j_1 & j_2 & j_3 \\ \\ m_1 & m_2 & m_3 \end{pmatrix} = (-1)^{j_1 - j_2 - m_3} (2j_3 + 1)^{-\frac{1}{2}}$$

$$\times \langle j_1 m_1 j_2 m_2 | j_3 -m_3 \rangle \quad . \tag{A6}$$

In Tables A1 and A2, we give expressions for

the most commonly occurring Clebsch-Gordan coefficients.   In Tables A3 we give these coefficients
in numerical form.   Reading <u>down</u> one column of a
numerical table gives the coefficients in the
expansion (A2), reading <u>across</u> one row gives the
coefficients in the expansion (A3).

Table A1. $\langle j_1 m_1 \frac{1}{2} m_2 | jm \rangle$

	$m_2 = \frac{1}{2}$	$m_2 = -\frac{1}{2}$
$j = j_1 + \frac{1}{2}$	$\sqrt{\dfrac{j_1 + m + \frac{1}{2}}{2j_1 + 1}}$	$\sqrt{\dfrac{j_1 - m + \frac{1}{2}}{2j_1 + 1}}$
$j = j_1 - \frac{1}{2}$	$-\sqrt{\dfrac{j_1 - m + \frac{1}{2}}{2j_1 + 1}}$	$\sqrt{\dfrac{j_1 + m + \frac{1}{2}}{2j_1 + 1}}$

Table A2. $<j_1 m_1 1 m_2 | jm>$

$j =$	$m_2 = 1$	$m_2 = 0$	$m_2 = -1$
$j_1 + 1$	$\sqrt{\dfrac{(j_1+m)(j_1+m+1)}{(2j_1+1)(2j_1+2)}}$	$\sqrt{\dfrac{(j_1-m+1)(j_1+m+1)}{(2j_1+1)(j_1+1)}}$	$\sqrt{\dfrac{(j_1-m)(j_1-m+1)}{(2j_1+1)(2j_1+2)}}$
j_1	$-\sqrt{\dfrac{(j_1+m)(j_1-m+1)}{2j_1(j_1+1)}}$	$\dfrac{m}{\sqrt{j_1(j_1+1)}}$	$\sqrt{\dfrac{(j_1-m)(j_1+m+1)}{2j_1(j_1+1)}}$
$j_1 - 1$	$\sqrt{\dfrac{(j_1-m)(j_1-m+1}{2j_1(2j_1+1)}}$	$-\sqrt{\dfrac{(j_1-m)(j_1+m)}{j_1(2j_1+1)}}$	$\sqrt{\dfrac{(j_1+m+1)(j_1+m)}{2j_1(2j_1+1)}}$

Tables A3.   Clebsch-Gordan Coefficients for $j_1 + j_2 = j_3$

$$|j_3 m_3\rangle = \sum_{m_1 m_2} |j_1 m_1 j_2 m_2\rangle \langle j_1 m_1 j_2 m_2 | j_3 m_3\rangle \quad \text{read down.}$$

$$|j_1 m_1 j_2 m_2\rangle = \sum_{j_3 m_3} |j_3 m_3\rangle \langle j_1 m_1 j_2 m_2 | j_3 m_3\rangle \quad \text{read across.}$$

$(\frac{1}{2}) \times (\frac{1}{2})$	$\lvert 1\ 1\rangle$	$\lvert 1\ 0\rangle$	$\lvert 0\ 0\rangle$	$\lvert 1\ -1\rangle$
$\lvert \frac{1}{2}\ \frac{1}{2}\ \ \frac{1}{2}\ \frac{1}{2}\rangle$	1			
$\lvert \frac{1}{2}\ \frac{1}{2}\ \ \frac{1}{2}\ -\frac{1}{2}\rangle$		$\sqrt{\frac{1}{2}}$	$\sqrt{\frac{1}{2}}$	
$\lvert \frac{1}{2}\ -\frac{1}{2}\ \ \frac{1}{2}\ \frac{1}{2}\rangle$		$\sqrt{\frac{1}{2}}$	$-\sqrt{\frac{1}{2}}$	
$\lvert \frac{1}{2}\ -\frac{1}{2}\ \ \frac{1}{2}\ -\frac{1}{2}\rangle$				1

$(1) \times (\frac{1}{2})$	$\left\lvert \frac{3}{2}\ \frac{3}{2}\right\rangle$	$\left\lvert \frac{3}{2}\ \frac{1}{2}\right\rangle$	$\left\lvert \frac{1}{2}\ \frac{1}{2}\right\rangle$	$\left\lvert \frac{3}{2}\ -\frac{1}{2}\right\rangle$	$\left\lvert \frac{1}{2}\ -\frac{1}{2}\right\rangle$	$\left\lvert \frac{3}{2}\ -\frac{3}{2}\right\rangle$
$\lvert 1\ \ 1\ \ \frac{1}{2}\ \frac{1}{2}\rangle$	1					
$\lvert 1\ \ 1\ \ \frac{1}{2}-\frac{1}{2}\rangle$		$\sqrt{1/3}$	$\sqrt{2/3}$			
$\lvert 1\ \ 0\ \ \frac{1}{2}\ \frac{1}{2}\rangle$		$\sqrt{2/3}$	$-\sqrt{1/3}$			
$\lvert 1\ \ 0\ \ \frac{1}{2}-\frac{1}{2}\rangle$				$\sqrt{2/3}$	$\sqrt{1/3}$	
$\lvert 1-1\ \ \frac{1}{2}\ \frac{1}{2}\rangle$				$\sqrt{1/2}$	$-\sqrt{2/3}$	
$\lvert 1-1\ \ \frac{1}{2}-\frac{1}{2}\rangle$						1

$\left(\frac{3}{2}\right) \times \left(\frac{1}{2}\right)$	$\lvert 2\ 2\rangle$	$\lvert 2\ 1\rangle$	$\lvert 1\ 1\rangle$	$\lvert 2\ 0\rangle$	$\lvert 1\ 0\rangle$	$\lvert 2{-}1\rangle$	$\lvert 1{-}1\rangle$	$\lvert 2{-}2\rangle$
$\left\lvert\frac{3}{2}\ \frac{3}{2}\ \frac{1}{2}\ \frac{1}{2}\right\rangle$	1							
$\left\lvert\frac{3}{2}\ \frac{3}{2}\ \frac{1}{2}\ {-}\frac{1}{2}\right\rangle$		$\sqrt{1/4}$	$\sqrt{3/4}$					
$\left\lvert\frac{3}{2}\ \frac{1}{2}\ \frac{1}{2}\ \frac{1}{2}\right\rangle$		$\sqrt{3/4}$	$-\sqrt{1/4}$					
$\left\lvert\frac{3}{2}\ \frac{1}{2}\ \frac{1}{2}\ {-}\frac{1}{2}\right\rangle$				$\sqrt{1/2}$	$\sqrt{1/2}$			
$\left\lvert\frac{3}{2}\ {-}\frac{1}{2}\ \frac{1}{2}\ \frac{1}{2}\right\rangle$				$\sqrt{1/2}$	$-\sqrt{1/2}$			
$\left\lvert\frac{3}{2}\ {-}\frac{1}{2}\ \frac{1}{2}\ {-}\frac{1}{2}\right\rangle$						$\sqrt{3/4}$	$\sqrt{1/4}$	
$\left\lvert\frac{3}{2}\ {-}\frac{3}{2}\ \frac{1}{2}\ \frac{1}{2}\right\rangle$						$\sqrt{1/4}$	$-\sqrt{3/4}$	
$\left\lvert\frac{3}{2}\ {-}\frac{3}{2}\ \frac{1}{2}\ {-}\frac{1}{2}\right\rangle$								1

$(1) \times (1)$	$\lvert 2\ 2\rangle$	$\lvert 2\ 1\rangle$	$\lvert 1\ 1\rangle$	$\lvert 2\ 0\rangle$	$\lvert 1\ 0\rangle$	$\lvert 0\ 0\rangle$	$\lvert 2{-}1\rangle$	$\lvert 1{-}1\rangle$	$\lvert 2{-}2\rangle$
$\lvert 1\ 1\ \ 1\ 1\rangle$	1								
$\lvert 1\ 1\ \ 1\ 0\rangle$		$\sqrt{1/2}$	$\sqrt{1/2}$						
$\lvert 1\ 0\ \ 1\ 1\rangle$		$\sqrt{1/2}$	$-\sqrt{1/2}$						
$\lvert 1\ 1\ \ 1{-}1\rangle$				$\sqrt{1/6}$	$\sqrt{1/2}$	$\sqrt{1/3}$			
$\lvert 1\ 0\ \ 1\ 0\rangle$				$\sqrt{2/3}$	0	$-\sqrt{1/3}$			
$\lvert 1{-}1\ \ 1\ 1\rangle$				$\sqrt{1/6}$	$-\sqrt{1/2}$	$\sqrt{1/3}$			
$\lvert 1\ 0\ \ 1{-}1\rangle$							$\sqrt{1/2}$	$\sqrt{1/2}$	
$\lvert 1{-}1\ \ 1\ 0\rangle$							$\sqrt{1/2}$	$-\sqrt{1/2}$	
$\lvert 1{-}1\ \ 1{-}1\rangle$									1

Appendix B   ROTATION MATRICES

In this appendix, we collect equations describing the properties of the rotation matrices and tabulate some of the most often used rotation matrices. Additional information can be found in the references 1 - 7.

If a physical system is rotated by the Euler angles α, β, and γ, a state $|j,m,a\rangle$ transforms as

$$|j,m,a\rangle^{\mathcal{R}} = \sum_{m'} |j,m',a\rangle \, D^{j}_{m'm}(\alpha,\beta,\gamma) \quad , \qquad (B1)$$

where the elements of the rotation matrix D are

1. E. P. Wigner, Group Theory and its Application to the Quantum Mechanics of Atomic Systems (Academic Press, New York, 1959).
2. A. R. Edmonds, Angular Momentum in Quantum Mechanics (Princeton University Press, Princeton, N. J., 1957).
3. M. E. Rose, Elementary Theory of Angular Momentum (John Wiley, New York, 1957).
4. D. M. Brink and G. R. Satchler, Angular Momentum (Oxford University Press, Oxford, 1962).
5. A. Messiah, Quantum Mechanics, Vol. II (North-Holland, Amsterdam, 1963).
6. K. Gottfried, Quantum Mechanics (W. A. Benjamin, Inc., Reading, Massachusetts, 1966).
7. G. Baym, Lectures on Quantum Mechanics (W. A. Benjamin, Inc., Reading, Massachusetts, 1969).

given by

$$D^j_{m'm}(\alpha,\beta,\gamma) = e^{-i\alpha m'} d^j_{m'm}(\beta) e^{-i\gamma m} , \qquad (B2)$$

$$d^j_{m'm}(\beta) = \langle j\ m'a|e^{-i\beta J_y}|jma\rangle . \qquad (B3)$$

Our definition of the Euler angles is given in Section 4.3; the definition of the D function agrees with Messiah, Gottfried, and Baym.

In the following formulas, we write sometimes $(\alpha,\beta,\gamma) = \mathcal{R}$ and $(-\gamma,\ -\beta,\ -\alpha) = \mathcal{R}^{-1}$. The rotation matrices D satisfy the following equations

$$\left(D^j_{mm'}(\mathcal{R}) \right)^* = D^j_{m'm}(\mathcal{R}^{-1})$$

$$= (-1)^{m-m'} D^j_{-m-m'}(\mathcal{R}) , \qquad (B4)$$

$$\sum_{m'} \left(D^j_{m'm}(\mathcal{R}) \right)^* D^j_{m'n}(\mathcal{R}) = \delta_{mn} ,$$

$$(B5)$$

$$\sum_{m'} D^j_{mm'}(\mathcal{R}) \left(D^j_{nm'}(\mathcal{R}) \right)^* = \delta_{mn} .$$

$$\int_0^{2\pi} \int_0^{2\pi} \int_0^{\pi} \left(D_{mm'}^j (\mathcal{R}) \right)^* D_{nn'}^{j'} (\mathcal{R}) \; \sin\beta d\beta d\alpha d\gamma$$

$$= \frac{8\pi^2}{2j+1} \; \delta_{mn} \; \delta_{m'n'} \; \delta_{jj'} \quad . \tag{B6}$$

$$\sum_n D_{mn}^j (\mathcal{R}_1) D_{nm'}^j (\mathcal{R}_2) = D_{mm'}^j (\mathcal{R}_2 \mathcal{R}_1) \quad . \tag{B7}$$

$$D_{mn}^j (\mathcal{R}) D_{m'n'}^{j'} (\mathcal{R})$$

$$= \sum_k <jmj'm'|k\mu><jnj'n'|k\nu> \; D_{\mu\nu}^k (\mathcal{R}) \; ,$$

$$\mu = m' + m, \quad \nu = n' + n \quad . \tag{B8}$$

For integral values of j, we write ℓ instead of j and have

$$D_{m0}^{\ell}(\alpha,\beta,\gamma) = \sqrt{\frac{4\pi}{2\ell+1}}\, Y_{\ell}^{m*}(\beta,\alpha), \quad D_{0m}^{\ell}(\alpha,\beta,\gamma)$$

$$= (-1)^{m}\sqrt{\frac{4\pi}{2\ell+1}}\, Y_{\ell}^{m*}(\beta,\gamma) \quad , \qquad (B9)$$

$$D_{00}^{\ell}(\alpha,\beta,\gamma) = P_{\ell}(\cos\beta) \quad . \qquad (B10)$$

Some often-used spherical harmonics and Legendre polynomials are given in Tables B1 and B2. The matrices d are real and satisfy the following equations:

$$d_{m'm}^{j}(\beta) = (-1)^{m'-m}\, d_{mm'}^{j}(\beta) = d_{mm'}^{j}(-\beta) \quad , \qquad (B11)$$

$$d_{m'm}^{j}(\beta) = (-1)^{m'-m}\, d_{-m'-m}^{j}(\beta) \quad . \qquad (B12)$$

$$d_{m'm}^{j}(\beta) = (-1)^{j-m'}\, d_{m'-m}^{j}(\pi-\beta)$$

$$= (-1)^{j+m}\, d_{m'-m}^{j}(\pi+\beta) \quad . \qquad (B13)$$

$$\sum_{n} d_{m'n}^{j}(\beta_2)\, d_{nm}^{j}(\beta_1) = d_{m'm}^{j}(\beta_1+\beta_2) \quad . \qquad (B14)$$

$$d_{mn}^{j}(\pi) = (-1)^{j+m} \delta_{m,-n} \quad ,$$

$$d_{mn}^{j}(2\pi) = (-1)^{2j} \delta_{mn} \quad . \tag{B15}$$

Recursion relations for the d matrices can be found in Jacob and Wick[8]; the d matrices for spins ½ to 3 are given explicitly in Table B1.[9]

---

8.   M. Jacob and G. C. Wick, Ann. Phys. 7, 404 (1959).
9.   S. M. Berman and M. Jacob, Phys. Rev. 139, B1023 (1965).

Table B1. The d matrices for spin 1/2 to 3/2.
(From S. M. Berman and M. Jacob, Phys.
Rev. <u>139</u>,B 1023 (1965).) The missing
matrices can be obtained with Eqs. (B11)
and (B12).

Spin $\frac{1}{2}$:

$$d_{\frac{1}{2}\frac{1}{2}}(\beta) = \cos\tfrac{1}{2}\beta, \quad d_{-\frac{1}{2}\frac{1}{2}}(\beta) = \sin\tfrac{1}{2}\beta.$$

Spin 1:

$$d_{11}(\beta) = \tfrac{1}{2}(1+\cos\beta), \quad d_{01}(\beta) = (\sin\beta)/\sqrt{2},$$
$$d_{1-1}(\beta) = \tfrac{1}{2}(1-\cos\beta), \quad d_{00}(\beta) = \cos\beta.$$

Spin $\frac{3}{2}$:

$$d_{\frac{3}{2}\frac{3}{2}}(\beta) = \tfrac{1}{2}(1+\cos\beta)\cos\tfrac{1}{2}\beta, \quad d_{\frac{3}{2}\frac{1}{2}}(\beta) = -\tfrac{1}{2}\sqrt{3}(1+\cos\beta)\sin\tfrac{1}{2}\beta,$$
$$d_{\frac{3}{2}-\frac{1}{2}}(\beta) = \tfrac{1}{2}\sqrt{3}(1-\cos\beta)\cos\tfrac{1}{2}\beta, \quad d_{\frac{3}{2}-\frac{3}{2}}(\beta) = -\tfrac{1}{2}(1-\cos\beta)\sin\tfrac{1}{2}\beta,$$
$$d_{\frac{1}{2}\frac{1}{2}}(\beta) = \tfrac{1}{2}(3\cos\beta-1)\cos\tfrac{1}{2}\beta, \quad d_{\frac{1}{2}-\frac{1}{2}}(\beta) = -\tfrac{1}{2}(1+3\cos\beta)\sin\tfrac{1}{2}\beta.$$

Spin 2:

$$d_{22}(\beta) = \tfrac{1}{4}(1+\cos\beta)^2, \qquad d_{21}(\beta) = -\tfrac{1}{2}(1+\cos\beta)\sin\beta,$$
$$d_{20}(\beta) = (\sqrt{6}/4)\sin^2\beta, \qquad d_{2-1}(\beta) = -\tfrac{1}{2}(1-\cos\beta)\sin\beta,$$
$$d_{2-2}(\beta) = \tfrac{1}{4}(1-\cos\beta)^2, \qquad d_{11}(\beta) = \tfrac{1}{2}(1+\cos\beta)(2\cos\beta-1),$$
$$d_{10}(\beta) = -(\tfrac{3}{2})^{1/2}\sin\beta\cos\beta, \quad d_{1-1}(\beta) = \tfrac{1}{2}(1-\cos\beta)(2\cos\beta+1),$$
$$d_{00}(\beta) = \tfrac{1}{2}(3\cos^2\beta-1).$$

Spin $\frac{5}{2}$:

$$d_{\frac{5}{2}\frac{5}{2}}(\beta) = \tfrac{1}{4}(1+\cos\beta)^2\cos\tfrac{1}{2}\beta, \qquad d_{\frac{5}{2}\frac{3}{2}}(\beta) = -(5)^{1/2}\tfrac{1}{4}(1+\cos\beta)^2\sin\tfrac{1}{2}\beta,$$
$$d_{\frac{5}{2}\frac{1}{2}}(\beta) = (10^{1/2}/4)\sin^2\beta\cos\tfrac{1}{2}\beta, \qquad d_{\frac{5}{2}-\frac{1}{2}}(\beta) = -(10^{1/2}/4)\sin^2\beta\sin\tfrac{1}{2}\beta,$$
$$d_{\frac{5}{2}-\frac{3}{2}}(\beta) = 5^{1/2}\tfrac{1}{4}(1-\cos\beta)^2\cos\tfrac{1}{2}\beta, \qquad d_{\frac{5}{2}-\frac{5}{2}}(\beta) = -\tfrac{1}{4}(1+\cos\beta)^2\sin\tfrac{1}{2}\beta,$$
$$d_{\frac{3}{2}\frac{3}{2}}(\beta) = \tfrac{1}{2}(5\cos\beta-3)\cos^3\tfrac{1}{2}\beta, \qquad d_{\frac{3}{2}\frac{1}{2}}(\beta) = (1\sqrt{2})(-(5\cos\beta-1))\cos^2\tfrac{1}{2}\beta\sin\tfrac{1}{2}\beta,$$
$$d_{\frac{3}{2}-\frac{1}{2}}(\beta) = (1/\sqrt{2})(1+5\cos\beta)\sin^2\tfrac{1}{2}\beta\cos\tfrac{1}{2}\beta, \quad d_{\frac{3}{2}-\frac{3}{2}}(\beta) = -\tfrac{1}{2}(5\cos\beta+3)\sin^3\tfrac{1}{2}\beta,$$
$$d_{\frac{1}{2}\frac{1}{2}}(\beta) = \tfrac{1}{2}(5\cos^2\beta-2\cos\beta-1)\cos\tfrac{1}{2}\beta, \quad d_{\frac{1}{2}-\frac{1}{2}}(\beta) = -\tfrac{1}{2}(5\cos^2\beta+2\cos\beta-1)\sin\tfrac{1}{2}\beta.$$

Spin 3:

$$d_{33}(\beta) = \tfrac{1}{8}(1+\cos\beta)^3, \qquad d_{32}(\beta) = -(6^{1/2}/8)\sin\beta(1+\cos\beta)^2,$$
$$d_{31}(\beta) = (15^{1/2}/8)\sin^2\beta(1+\cos\beta), \qquad d_{30}(\beta) = -(5^{1/2}/4)\sin^3\beta,$$
$$d_{3-1}(\beta) = (15^{1/2}/8)\sin^2\beta(1-\cos\beta), \qquad d_{3-2}(\beta) = -(6^{1/2}/8)\sin\beta(1-\cos\beta)^2,$$
$$d_{3-3}(\beta) = \tfrac{1}{8}(1-\cos\beta)^3, \qquad d_{22}(\beta) = \tfrac{1}{4}(1+\cos\beta)^2(3\cos\beta-2),$$
$$d_{21}(\beta) = -(5^{1/2}/4\sqrt{2})\sin\beta(3\cos^2\beta+2\cos\beta-1), \quad d_{20}(\beta) = (15^{1/2}/2\sqrt{2})\cos\beta\sin^2\beta,$$
$$d_{2-1}(\beta) = (5^{1/2}/4\sqrt{2})\sin\beta(3\cos^2\beta-2\cos\beta-1), \quad d_{2-2}(\beta) = \tfrac{1}{4}(1-\cos\beta)^2(3\cos\beta+2),$$
$$d_{11}(\beta) = \tfrac{1}{8}(1+\cos\beta)(15\cos^2\beta-10\cos\beta-1), \quad d_{10}(\beta) = -(\sqrt{3}/4)\sin\beta(5\cos^2\beta-1),$$
$$d_{1-1}(\beta) = \tfrac{1}{8}(1-\cos\beta)(15\cos^2\beta+10\cos\beta-1), \quad d_{00}(\beta) = (5\cos^3\beta-3\cos\beta)/2.$$

Table B2.   Legendre Polynomials and Spherical Harmonics

Ordinary Legendre Polynomials

$P_0(\cos\theta) = 1$	$P_3(\cos\theta) = \frac{1}{2}(5\cos^3\theta - 3\cos\theta)$
$P_1(\cos\theta) = \cos\theta$	$P_4(\cos\theta) = \frac{1}{8}(35\cos^4\theta - 30\cos^2\theta + 3)$
$P_2(\cos\theta) = \frac{1}{2}(3\cos^2\theta - 1)$	

Normalized Spherical Harmonics

$Y_0^{\,0}(\theta,\phi) = \sqrt{1/4\pi}$	$Y_3^{\pm2}(\theta,\phi) = \frac{1}{4}\sqrt{105/2\pi}\cos\theta(1-\cos^2\theta)e^{\pm2i\phi}$
$Y_1^{\,0}(\theta,\phi) = \frac{1}{2}\sqrt{3/\pi}\cos\theta$	$Y_3^{\pm3}(\theta,\phi) = \mp\frac{1}{8}\sqrt{35/\pi}\sin^3\theta\,e^{\pm3i\phi}$
$Y_1^{\pm1}(\theta,\phi) = \mp\frac{1}{2}\sqrt{3/2\pi}\sin\theta\,e^{\pm i\phi}$	$Y_4^{\,0}(\theta,\phi) = \frac{3}{16}\sqrt{1/\pi}(35\cos^4\theta - 30\cos^2\theta + 3)$
$Y_2^{\,0}(\theta,\phi) = \frac{1}{4}\sqrt{5/\pi}(3\cos^2\theta - 1)$	$Y_4^{\pm1}(\theta,\phi) = \mp\frac{3}{8}\sqrt{5/\pi}\sin\theta(7\cos^3\theta - 3\cos\theta)e^{\pm i\phi}$
$Y_2^{\pm1}(\theta,\phi) = \mp\frac{1}{2}\sqrt{15/2\pi}\cos\theta\sin\theta\,e^{\pm i\phi}$	$Y_4^{\pm2}(\theta,\phi) = -\frac{3}{8}\sqrt{5/2\pi}(7\cos^4\theta - 8\cos^2\theta+1)e^{\pm2i\phi}$
$Y_2^{\pm2}(\theta,\phi) = \frac{1}{4}\sqrt{15/2\pi}(1 - \cos^2\theta)e^{\pm2i\phi}$	$Y_4^{\pm3}(\theta,\phi) = \mp\frac{3}{8}\sqrt{35/\pi}\sin\theta(\cos\theta - \cos^3\theta)e^{\pm3i\phi}$
$Y_3^{\,0}(\theta,\phi) = \frac{1}{4}\sqrt{7/\pi}(5\cos^3\theta - 3\cos\theta)$	$Y_4^{\pm4}(\theta,\phi) = \frac{3}{16}\sqrt{35/2\pi}\sin^4\theta\,e^{\pm4i\phi}$
$Y_3^{\pm1}(\theta,\phi) = \mp\frac{1}{8}\sqrt{21/\pi}(5\cos^2\theta-1)\sin\theta\,e^{\pm i\phi}$	

INDEX

Page numbers in parentheses refer to problems.